To friends
in RESTORE
program
with wishes of
success and joy
sincerely
Dr. J

Fluorine Magnetic Resonance Imaging

Fluorine Magnetic Resonance Imaging

edited by

Ulrich Flögel
Eric Ahrens

Pan Stanford Publishing

Published by

Pan Stanford Publishing Pte. Ltd.
Penthouse Level, Suntec Tower 3
8 Temasek Boulevard
Singapore 038988

Email: editorial@panstanford.com
Web: www.panstanford.com

British Library Cataloguing-in-Publication Data
A catalogue record for this book is available from the British Library.

Fluorine Magnetic Resonance Imaging

Cover image: "Universe of Imaging," courtesy of Prof. Juerg Schwitter, Division of Cardiology and Cardiac MR Center of the University Hospital Lausanne, CHUV, Switzerland.

ISBN 978-981-4745-31-4 (Hardcover)
ISBN 978-981-4745-32-1 (eBook)

Printed in the USA

Contents

Part 2: ^{19}F Imaging Agents

Part 5: PHARMACOLOGY

Preface

This book provides an overview of the state of the art in engineering and biomedical applications of fluorine-19 (^{19}F) magnetic resonance imaging (MRI). The ^{19}F nucleus is intrinsically an excellent MRI probe because it has 100% natural abundance, nuclear spin-1/2, comparable sensitivity to proton, and there is no background ^{19}F signal in the body. MRI using ^{19}F was first demonstrated in the 1970s during MRI's early development. A variety of ^{19}F-laden materials were investigated as potential "tracer" agents. Further interest in ^{19}F MRI evolved in the 1980s and 1990s, when there was significant commercial interest in developing perfluorocarbon (PFC) emulsion formulations for use as artificial blood substitutes. This effort helped to devise strategies for the production of stable, non-toxic PFC emulsion formulations, in addition to elucidating the safety profiles of these materials, including pharmacokinetics. During the same period, ^{19}F MRI was used to study PFCs, for example, as potential angiographic and oximetry agents; however, ^{19}F imaging during this epoch could still be characterized as esoteric.

The use of ^{19}F MRI has been reinvigorated in recent years. A key motivator of this renewed interest surrounds imaging probe development for the bourgeoning field of in vivo cellular and molecular imaging. The ^{19}F MRI resurgence also coincides with the increased use of fluorous chemistry in biomedicine; drug candidates are increasingly incorporating the fluorine atom. Fluorine can impart unique properties to compounds, including metabolic stability, enhanced binding interactions, and selective reactivities. A further motivator for ^{19}F probe development stems from increased interest in alternatives to metal-ion based MRI contrast agents due to the recent awareness of potential toxicities associated with clinically approved agents, as well as technological challenges associated with image specificity and quantification with traditional contrast agents.

This book presents a snapshot of the current activities surrounding the use of ^{19}F MRI in biomedical research. It covers a breadth of topics, including MRI pulse sequence development, radio frequency hardware considerations, probe chemistry and formulation, nuclear hyperpolarization, targeted and "smart" probes, imaging fluorinated drug biodistribution, in vivo biomedical device imaging, tracking stem and immune cells in the body, and imaging inflammatory events in the body.

Our view is that the future for ^{19}F is bright. There has been a groundswell of research interest in ^{19}F MRI, which has evolved into totally new uses and markets for the MRI technique. As described in this volume, there are significant avenues for future improvements in the sensitivity of ^{19}F MRI detection. However, in the near term, a limiting factor in the growth of ^{19}F MRI is economics; large-scale MRI research and clinical macro trends are driven, to a significant degree, by the hardware and software features offered by MRI manufactures. A bottleneck in the clinical adoption is the availability of suitable scanner hardware to image ^{19}F in a hospital setting. However, as technical developments, compelling biomedical applications, and clinical trial data continue to proliferate, this bottleneck will subside.

Finally, we offer our sincerest appreciation to all of the contributors to this book. We feel honored to have had the opportunity to have shepherded these excellent manuscripts. We hope that readers will find this volume enlightening and that it will help stimulate new efforts in ^{19}F MRI.

Ulrich Flögel
Eric Ahrens

Part 1
Technical Issues

Chapter 1

Pulse Sequence Considerations and Schemes

Cornelius Faber and Florian Schmid

Department of Clinical Radiology, University Hospital Münster, 48149 Münster, Germany

faberc@uni-muenster.de

1.1 Introduction

Since publication of the first ^{19}F magnetic resonance images, a wide range of potential applications of fluorine MRI in biomedical imaging have been explored. In particular, since Ahrens and co-workers have shown that distinct cellular populations can be labeled with fluorine-containing compounds and subsequently detected by ^{19}F MRI, this technology has been an accepted method in the molecular imaging toolbox (Ahrens et al., 2005; Ruiz-Cabello et al., 2011; Yu et al., 2013). As detailed in other chapters in this book, ^{19}F possesses a high gyromagnetic ratio (94% of ^{1}H), which provides an intrinsically high MR signal. A 100% isotopic abundance and the almost complete lack of any detectable fluorine signal in the mammalian body makes ^{19}F the ideal nucleus for cell tracking and molecular imaging approaches by MRI. Most importantly, ^{19}F provides unambiguous identification of

Fluorine Magnetic Resonance Imaging
Edited by Ulrich Flögel and Eric Ahrens

ISBN 978-981-4745-31-4 (Hardcover), 978-981-4745-32-1 (eBook)
www.panstanford.com

the cellular or molecular label, and can therefore be considered a gold standard for MR detection specificity of a marker, comparable to the specificity of positron emission tomography (PET) or fluorescence imaging. This excellent detection specificity solves one of the major problems in molecular and cellular MRI approaches that use detection of ^{1}H in combination with iron oxide or lanthanide-based labels (Ahrens and Bulte, 2013). Only recently have novel ^{1}H MRI detection methods, such as PARA-CEST or highly shifted proton MRI, shown that for ^{1}H the specificity problem can be solved (Aime et al., 2005; Schmidt et al., 2014). The practical usefulness of these methods, however, remains to be shown in further applications.

Due to these favorable properties, high MR signal and excellent detection specificity, fluorine MRI has been used in animal models to follow the fate of grafted cells such as neural stem cells, dendritic cells, and lymphocytes, as described in more detail in other chapters of this volume. Furthermore, fluorine MRI was applied to characterize immune response by observing inflammatory cells migrating into lesions after diverse challenges. ^{19}F MRI was used, for example, to detect graft rejection (Flögel et al., 2011), pulmonary inflammation (Ebner et al, 2010), and inflammation following cardiac or cerebral ischemia (Flögel et al., 2008). Also, the immune response upon bacterial infection has been characterized using injections of perfluorocarbons to label migrating immune cells (Hertlein et al., 2011).

However, despite the intrinsically high ^{19}F MR signal, cell tracking and molecular imaging always face the problem of detection sensitivity, especially when small cell numbers are to be detected. Therefore, it is of particular importance that optimized detection schemes are employed to minimize (i) signal losses due to relaxation and (ii) recording of unnecessary noise, for example, by using excessive acquisition bandwidths. Irrespective of the need for sequence optimization, many previous studies have used turbo spin echo-like sequences, in some cases with non-optimal parameters. Other studies have applied different pulse sequences with varying parameters to account for the use of different molecular compounds. Besides turbo spin echo sequences such as rapid acquisition using relaxation enhancement (RARE), for example, standard balanced steady-state free precession (bSSFP) sequences have been applied to detect labeled mesenchymal stem cells (Ribot et al., 2014). However, the question of finding the optimal pulse sequence and optimizing detection sensitivity has only recently been addressed in a few publications. The general considerations to optimize detection sensitivity are identical to those that apply for ^{1}H MRI. For RARE sequences, echo train lengths and repetition time (TR) have been analyzed (Mastropietro et al., 2014). The possibility to exploit paramagnetic relaxation enhancement (PRE) has been explored using spoiled gradient echo (Chalmers et al., 2011), as well as radial zero echo time (ZTE) sequences (Schmid et al., 2013). Further, a bSSFP ultra-short echo time (UTE) sequence has been proposed and compared to detection sensitivity of other gradient echo or free induction decay (FID)-based detection schemes (Goette et al., 2014).

1.2 General Considerations

The most important parameters for detection sensitivity are, first, relaxation times, determining the time frame for signal losses, and, second, the resonance frequencies over which potentially detectable ^{19}F signals are distributed. This second issue, which encompasses multiple resonance lines and multiplicity patterns of individual lines, will not be discussed here. A number of methods to account for multiple lines or scalar coupling have been proposed previously (Giraudeau et al., 2010; Goette et al., 2014), and are addressed in other chapters of this book. We will begin with defining the most important pulse sequences for fluorine MRI. After some general considerations on MR detection efficiency, we will analyze and compare the efficiency of the different sequences. Finally, we will briefly treat the nuclear Overhauser effect (NOE) as a possible method to boost the fluorine signal.

1.2.1 The Pulse Sequences

Generally in MRI a number of different read out schemes are used, each providing specific advantages with respect to robustness, ease of use, and minimum achievable scan time. On the other hand, every sequence has specific shortcomings, making a careful selection of the best pulse sequence for the given experimental conditions an important issue. First, we will consider a spoiled

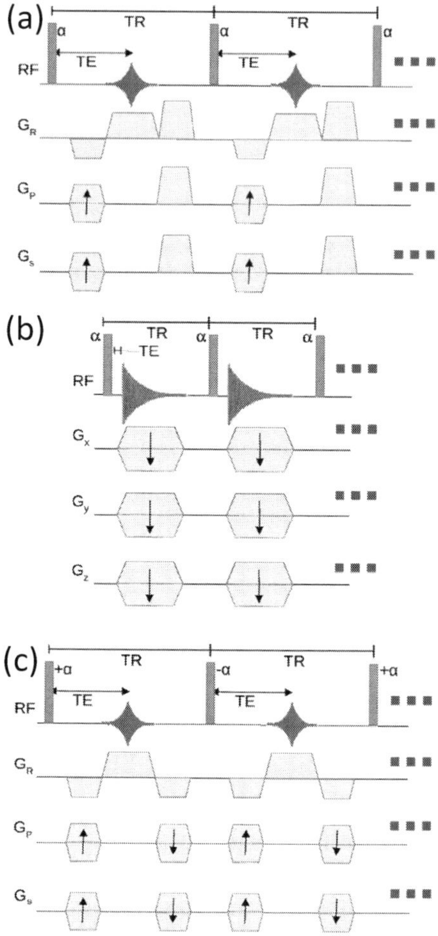

Figure 1.1 Pulse sequence diagrams for (a) FLASH, (b) UTE (FISP), and (c) bSSFP. Upper line shows radio frequency (RF) pulses as grey bars and the acquired signal as echo or FID. Here, α indicates flip angle, TE echo time and TR repetition time. Gradient pulses are indicated by trapezoid shapes either along Cartesian axes (G_x, G_y, G_z) or as read (G_R) and varying phase encoding gradients (G_P, G_S).

gradient echo or fast low angle shot (FLASH) sequence (Fig. 1.1a). Transverse magnetization is created by a low flip angle excitation pulse α (usually <20°, depending on the repetition time and T_1). A gradient recalled echo is created at echo time TE. After the end of the read out, all residual transverse magnetization is dephased by a spoiler gradient. To minimize the repetition time TR, the next excitation is applied immediately after the delay required by the scanner for radio frequency and gradient switching.

Both minimum achievable TE and TR are limited by these delays, defined by the scanner hardware.

Second, we will consider the case of fast imaging with steady-state precession (FISP), which is achieved if a pattern of α-TR-α-TR-... is applied continuously (Fig. 1.1b). Our treatment will consider the particular case where the FISP signal is read out as an FID signal immediately after the excitation pulse, as applicable for UTE MRI. TR is minimized and the next excitation pulse is applied immediately after the end of the read out and the required switching delays. The minimum TR and the delay before read out of signal starts are again limited by scanner hardware parameters.

Third, we will consider bSSFP sequences, which are a special case of FISP with completely balanced gradients and the echo generated at TE = TR/2 (Fig. 1.1c). Note that the excitation pulse is applied with alternating sign in subsequent excitations to maximize steady-state magnetization. Again scanner specifications limit minimum TR and TE.

Further, we will consider turbo spin echo sequences, such as RARE (Fig. 1.2). Here, spin echoes are created at echo time TE and multiples thereof (n*TE). The number of echoes after one excitation pulse is given by the RARE factor. RARE can be used with an additional flip-back module (indicated by gray shaded boxes in Fig. 1.2). There, an additional refocusing pulse is applied after the last acquired echo, and transverse magnetization is flipped back along the longitudinal direction with a 90° pulse at the time (RARE factor +1) times TE. In the following, all pulse sequences will be treated as 3D sequences, as shown in Fig. 1.1a–c and Fig. 1.2a, as 3D acquisitions are generally more efficient than 2D acquisitions. However, for RARE the multi-slice case will also be considered. Since RARE uses a 90° excitation, TR cannot simply be minimized, as is the case for the other three sequences described above. Therefore, under certain conditions a dead time Δ during which no signal is acquired may result, and multi-slice excitation (Fig. 1.2b) may be advantageous because signal can be acquired continuously.

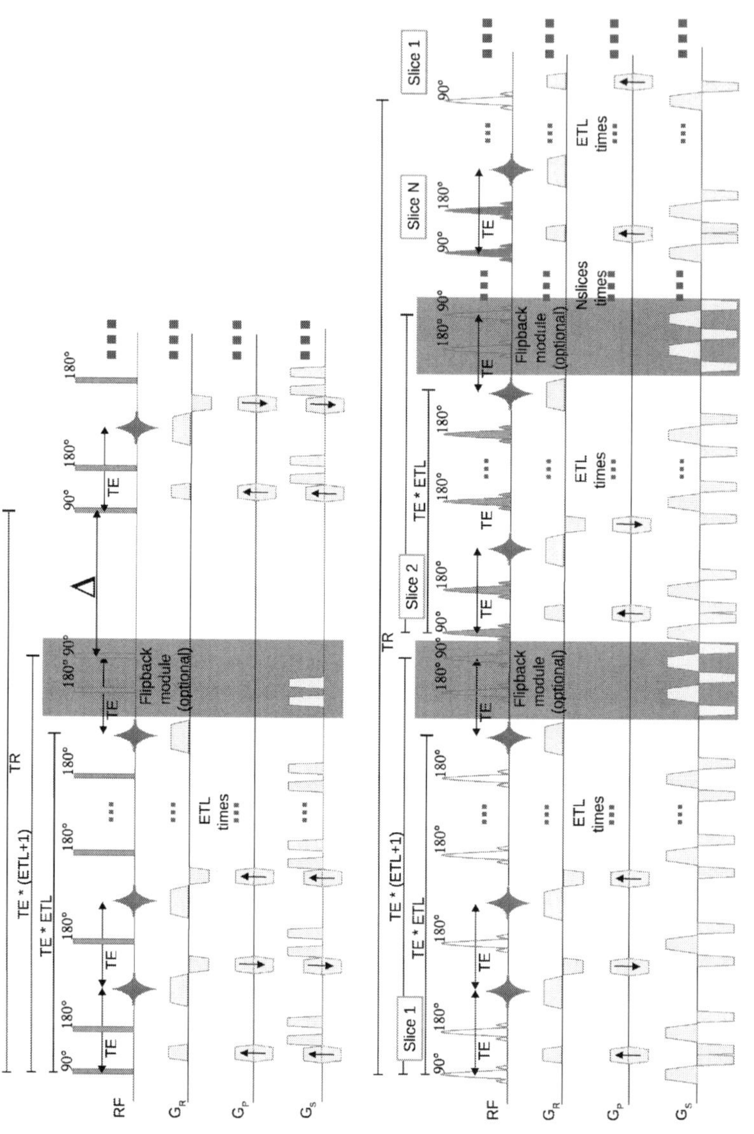

Figure 1.2 Pulse sequence diagrams for (a) 3D RARE and (b) 2D multi-slice RARE. Upper line shows radio frequency (RF) pulses as grey bars and the acquired signal as echo. Gradient pulses are indicated by trapezoid shapes along G_R, G_P, and G_S. Gray box highlights optional flip-back module and Δ denotes a relaxation delay. The echo train length (ETL) indicates the RARE factor.

1.2.2 More General Pulse Sequence Considerations

To achieve the most efficient read out of signal in an MR measurement, dead time, during which no signal is acquired, has to be minimized. Furthermore, the average signal intensity that is recorded has to be maximized, and finally, the acquired signal has to be distributed in k space most effectively. For the moment, we will ignore this third point, which will be discussed later, and focus on the issues of how to use the available scan time for sampling most efficiently, and how to maximize the sampled signal. Schematic illustrations of the sampled signal for the different pulse sequences are shown in Fig. 1.3. With all sequences a steady-state magnetization can be sampled, which is lower than M_0 and defined by T_1 and the sequence parameters. In FLASH, repeated excitations with short TR and spoiling of all transverse magnetization prior to the next excitation lead to a low steady-state magnetization. After each excitation, the envelope of the detectable signal (black line) decays with T_2^*. Since time is needed for gradient encoding prior to acquisition and spoiling after acquisition, only a part of the potentially available signal can be sampled (gray area) (Fig. 1.3a). In FISP, here shown using an FID read out (UTE with radial read out), acquisition starts almost immediately after excitation, allowing for a larger part of the signal to be sampled. This avoids losses due to transverse relaxation and is especially useful for sampling quickly decaying signals in fast relaxing agents (Fig. 1.3b). Furthermore, transverse magnetization is not spoiled and may contribute to steady-state magnetization. In bSSFP, a delay prior to acquisition is needed for encoding, and the position of the acquisition window between two excitations is similar to the FLASH sequence. However, the overall signal amplitude is generally higher due to balanced refocusing and the reuse of the residual transverse magnetization instead of spoiling it (Fig. 1.3c). In RARE, a 90° excitation pulse is applied and several spin echoes are sampled, decaying with T_2. Due to the 90° excitation, a long TR is necessary to allow for T_1 relaxation to generate substantial longitudinal magnetization, even with use of flip-back pulses. However, 3D RARE benefits from higher signal amplitudes compared to the sequences described above (Fig. 1.3d).

In multi-slice 2D RARE, signal from the individual slices can be sampled while magnetization in the other slices experiences

T_1 relaxation. This leads to higher longitudinal magnetization prior to excitation. However, as single slices are excited, only a fraction of the volume magnetization can be sampled, leading to a substantial reduction of available signal compared to full volume excitation in 3D RARE. The advantage of multi-slice RARE is that no delay without signal acquisition is required to allow for relaxation. TR can therefore be sampled more efficiently than with 3D RARE (Fig. 1.3e).

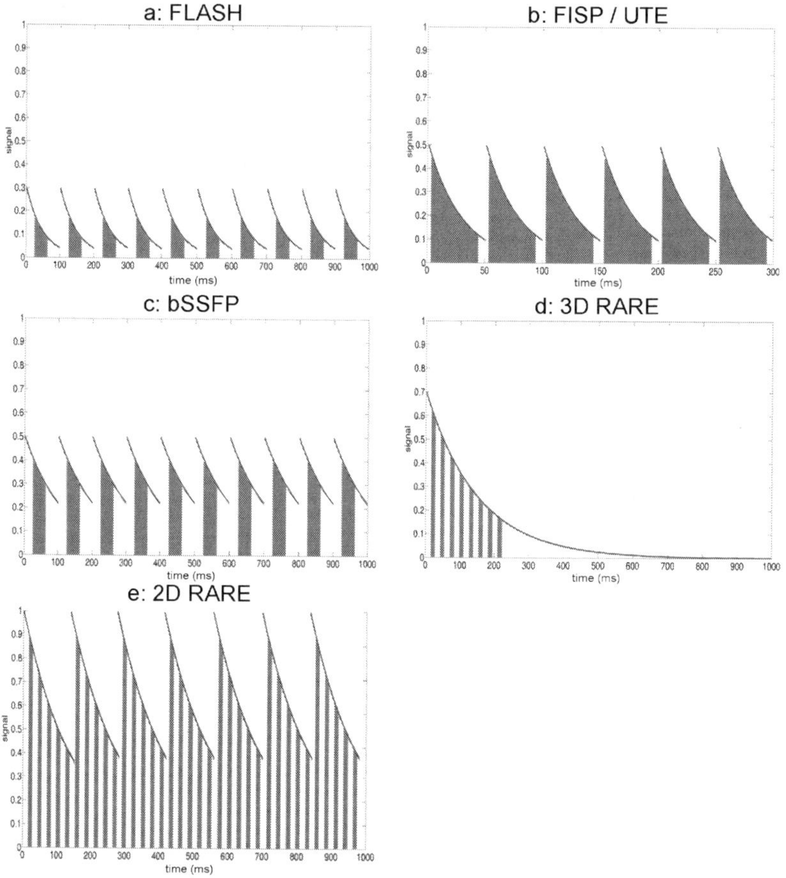

Figure 1.3 Schematic sampling for different pulse sequences used for efficient MR detection. Shown are (a) FLASH, (b) UTE, (c) bSSFP, (d) 3D RARE, and (e) 2D RARE. Signal evolution is in black, and the sampled signal is indicated by gray areas.

1.3 Sensitivity of Particular Sequences in Parameter Space

In the following section we will compare SNR efficiencies, defined as SNR per square root of scan time, for the four different read out principles, FLASH, bSSFP, UTE, and RARE. The equations and parameter estimations for these four basic schemes apply to most commonly used pulse sequences and therefore provide an excellent estimate of the most efficient pulse sequence for ^{19}F detection. As detection sensitivity crucially depends on the actual relaxation times, we will calculate acquisition parameters for a wide range of T_1 and T_2 (or T_2^*) combinations. In an actual measurement, relaxation times depend on magnetic field strength, temperature, pH, and potentially on additional experimental conditions. It is therefore essential to measure relaxation times before the optimum pulse sequence can be identified.

Generally, SNR efficiency $\mathrm{SNR_{eff}}$ is given by

$$\mathrm{SNR_{eff}} = \frac{\mathrm{SNR}}{\sqrt{\mathrm{NA^*TR}}} \propto V\,S^{\mathrm{avg}}\sqrt{\frac{\mathrm{NA^*N_{enc}}}{\mathrm{BW^*NA^*TR}}} = V\,S^{\mathrm{avg}}\sqrt{\frac{N_{\mathrm{enc}}}{\mathrm{BW^*TR}}}, \tag{1.1}$$

where V is the voxel volume, S^{avg} the average signal amplitude, NA the number of averages, BW the acquisition bandwidth, and N_{enc} the number of encoding samples given by $N_{\mathrm{enc}} = N_{\mathrm{read}} * N_{\mathrm{phase}}$. For 3D acquisition, N_{phase} is the product of phase encoding steps in two dimensions and may be expressed in Cartesian space as $N_{\mathrm{phase}} = N_y * N_{\mathrm{slice}}$. To keep the sequence comparison tractable, we will hold V and N_{enc} constant for all calculations. While TR and BW can be set in the measurement, S^{avg} is subject to a complex interplay of many parameters and will be different for each sequence and experimental condition. To compare different sequences, we will use the simplification that S^{avg} can be obtained by calculating the maximum signal amplitude that is acquired during the read out. This value corresponds to the signal at echo maximum or the beginning of the FID, and is given by the steady-state magnetization at a given echo time (or acquisition delay for UTE). S^{avg} is determined by the experimental conditions and shows different dependencies for each experiment. It must therefore be calculated separately for the different pulse sequences.

For FLASH, the maximum signal in the steady state can be calculated as

$$S_{\text{FLASH}}^{\text{avg}} = M_0 \frac{\sin\alpha(1-E_1)}{(1-E_1\cos\alpha)} e^{-(\text{TE}/T_2^*)}, \tag{1.2}$$

where $E_1 = e^{-(\text{TR}/T_1)}$ and is maximized if the flip angle is chosen to be the well-known Ernst angle

$$\alpha_{\text{E}} = \arccos E_1.$$

Note that the signal decays according to T_2^* the relaxation time. For FISP (UTE in our case), maximum signal can be calculated as

$$S_{\text{FID,FISP}} = M_0 \tan\frac{\alpha}{2}\left(1 - \frac{(1-\sin\alpha)(1-E_2^2)}{\sqrt{(p^2-q^2)}}\right), \tag{1.3}$$

where $E_2 = e^{-(\text{TR}/T_2)}$, $P = 1 - E_1 \cos\alpha - E_2^2(E_1 - \cos\alpha)$, and $q = E_2(1 - E_1)(1 + \cos\alpha)$.

Note, as transverse magnetization is not spoiled, steady-state magnetization depends on both T_1 and T_2.

Also for bSSFP, with sign-alternating excitation pulse, maximum signal depends on both T_1 and T_2 and can be calculated as

$$S_{\text{SSFP,alt}} = M_0 \frac{\sin\alpha(1-E_1)}{1-(E_1 - E_2)\cos\alpha - E_1E_2} e^{-(\text{TE}/T_2)} \tag{1.4}$$

In all practically relevant cases TR is chosen such that TR << T_2, and the optimum flip angle can be calculated as

$$\alpha_{\text{optimum}} = \arccos\left(\frac{T_1 - T_2}{T_1 + T_2}\right).$$

For RARE, the calculation of the average signal is a little more complicated. For the first echo in each TR, if no flip-back module is used, the steady state is simply given by

$$M^{\text{SS}} = M_0\left(1 - e^{-\frac{\text{TR-ETL*TE}}{T_1}}\right), \tag{1.5}$$

where ETL (echo train length) represents the RARE factor. With flip-back pulse (see Fig. 1.2a,b), steady-state magnetization additionally depends on T_2 and is described by

$$M^{SS} = M_0 \frac{\left(1 - e^{-\frac{TR - (ETL+1)*TE}{T_1}}\right)}{\left(1 - e^{-\frac{(ETL+1)*TE}{T_2}} e^{-\frac{TR-(ETL+1)*TE}{T_1}}\right)}, \tag{1.6}$$

However, during one echo train several echoes are read out, which are subject to T_2 relaxation. To account for this signal reduction, we use the simplification that the average signal can be calculated as the arithmetic mean of the echo train given by

$$S_{RARE}^{avg} = \frac{M^{SS}}{ETL} \sum_{n=1}^{ETL} \left(e^{-\frac{(n*TE)}{T_2}} \right) \tag{1.7}$$

Using these equations to calculate SNR efficiency, we have to account for additional averaging by acquisition of ETL echoes, which adds another factor of square root of ETL. Consequently, for a 3D RARE sequence

$$SNR_{eff,3DRARE} \propto V\, S^{avg} \sqrt{\frac{ETL * N_{read} * N_{phase} * N_{slice}}{TR * BW}} \tag{1.8}$$

Note that N_{slice} represents the number of phase encoding steps in slice direction. If multislice RARE is used, additional averaging by acquisition of multiple slices during one TR has to be accounted for by an additional factor of $\sqrt{\frac{TR}{ETL * TE}}$, while the factor $\sqrt{N_{slice}}$ drops out, resulting in

$$SNR_{eff,MSRARE} \propto V\, S^{avg} \sqrt{\frac{N_{read} * N_{phase}}{TE * BW}} \tag{1.9}$$

This is a remarkable result, since SNR efficiency does not depend on TR but only on TE. Obviously this is only valid, if TR is an exact multiple of TE and ETL, and if sufficient slices are acquired to fill TR completely. For our considerations below, we take these two requirements for granted, to keep the treatment tractable. However, in the actual measurement this issue needs to be addressed.

Using the above formulas, it becomes evident that SNR efficiency is not a function of all parameters T_1, T_2, TR, TE, BW, α, and ETL (for RARE), but only depends on a subset. TE is determined by the actual acquisition time $t_{\mathrm{acq}} = N_{\mathrm{read}}/\mathrm{BW}$, and delays given by the MR scanner hardware setup (see Fig. 1.1). TE, therefore, becomes a function of the bandwidth and cannot be adjusted independently. Considering now that all sequences (except UTE, which does not use echo detection) have an exponential damping factor that scales with TE, it becomes evident that TE must be minimized to maximize SNR efficiency. The same holds for TR, for FLASH, UTE, and bSSFP. As SNR efficiency for these sequences scales with $1/\sqrt{\mathrm{TR}}$, TR must be minimized. However, since minimum TR is limited by scanner parameters and t_{acq}, TR cannot be adjusted independently, but must be optimized by choosing the best acquisition bandwidth. We can therefore conclude that SNR efficiency depends on different parameters for the different sequences.

For FLASH and bSSFP:

$$\mathrm{SNR}_{\mathrm{eff}}(T_1, T_2, \mathrm{BW})$$

For UTE:

$$\mathrm{SNR}_{\mathrm{eff}}(T_1, T_2, \mathrm{BW}, \alpha)$$

For 3D RARE:

$$\mathrm{SNR}_{\mathrm{eff}}(T_1, T_2, \mathrm{BW}, ETL, TR)$$

For multi-slice RARE:

$$\mathrm{SNR}_{\mathrm{eff}}(T_1, T_2, \mathrm{BW}, \mathrm{ETL})$$

1.3.1 SNR Efficiencies and Optimum Parameters for UTE, FLASH and bSSFP

In the following, we have plotted the optimum parameters as a function of T_1 and T_2 as parametric maps (Figs. 1.4–1.8). For the calculations we have used delay times typical for modern high field small animal MRI systems, including: pre-acquisition delays of 0.5 ms for FLASH and bSSFP, 5 ms for RARE and of 20 μs for UTE. For the delay between end of acquisition and subsequent excitation 1 ms was assumed. For bandwidths, a range of 5 to 200 kHz was used.

For FLASH, according to Eq. (1.2) optimum bandwidth is independent of T_1 and mostly independent of T_2^*. For most relaxation times the lowest possible bandwidth (5 kHz in the calculations) yields highest SNR efficiency. Only for very short T_2* values in the millisecond range, higher bandwidths are required to achieve a sufficiently short TE for a fast read out. Note that we have not considered the possibility of shifting the echo position; an earlier echo formation during the read out window may further reduce the impact of T_2^* on the choice of optimum bandwidth.

For bSSFP, according to Eq. (1.4) the lowest possible bandwidth was found to optimize SNR efficiency almost exclusively. Only for extremely short T_2 values, in the millisecond range, slightly larger bandwidths than the minimum of 5 kHz yielded higher efficiency. In practical measurements, however, minimum bandwidth will be limited (see below), and therefore we can conclude that for bSSFP, the lowest possible bandwidth should always be chosen where artifact-free images can be obtained.

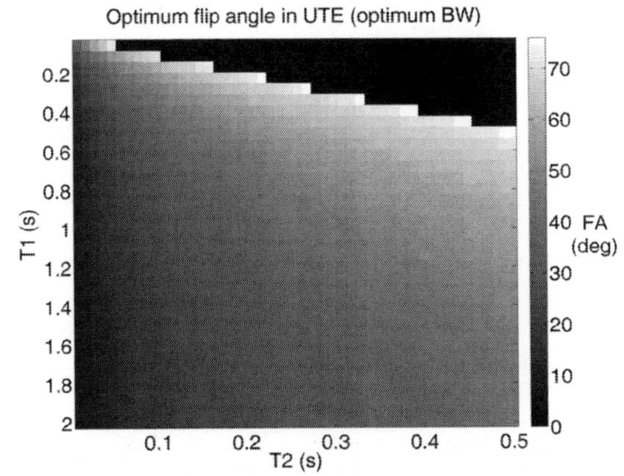

Figure 1.4 Parametric map of optimum flip angle for UTE (FISP) while using optimum bandwidth. Note, forbidden region, $T_1 < T_2$, colored black.

For UTE, according to Eq. (1.3) the lowest possible bandwidth was also found to yield the highest SNR efficiency. Equation (1.3) also instructs that the optimum flip angle ranges from very small values for short T_2 and long T_1 to values up to 75° if the condition is met that $T_1 \approx T_2$ (Fig. 1.4).

Calculating SNR efficiencies for the optimum parameters for all three sequences from Eqs. (1.2), (1.3), and (1.4) allows for a direct comparison. The SNR efficiencies for all three sequences are shown in Fig. 1.5 as function of T_1 and T_2.

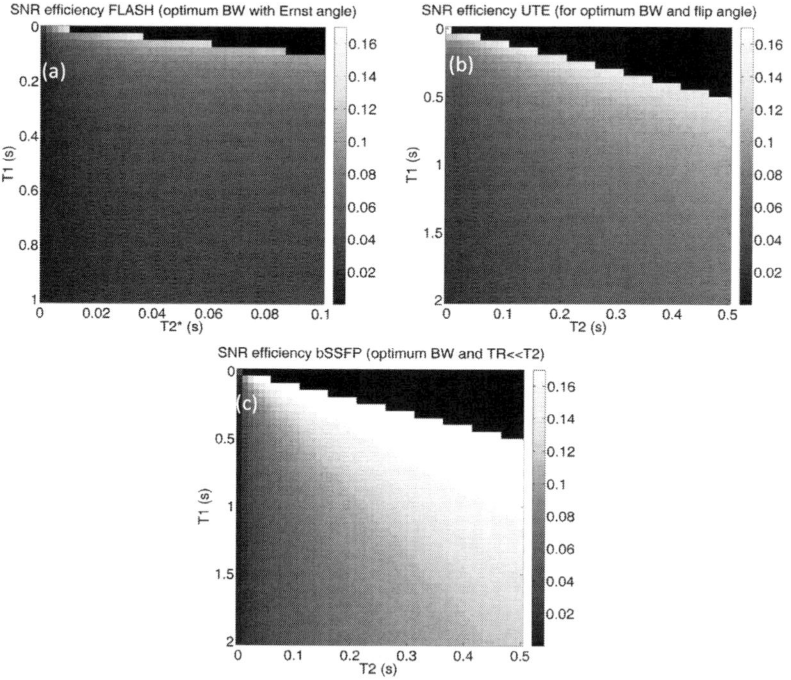

Figure 1.5 Parametric maps of SNR efficiencies for (a) FLASH, (b) UTE, and (c) bSSFP. SNR efficiencies are shown in units of M_0 per square root of scan time. Note, for FLASH, T_2^* was used with smaller values on horizontal axis. Note, forbidden region, $T_1 < T_2$, colored black.

Values up to 0.23 $\frac{M_0}{\sqrt{t}}$ were obtained with different relaxation time dependencies for the three sequences. From these plots it becomes obvious that FLASH, while robust and often most easy to implement, provides the lowest SNR efficiency of all sequences. This result originates from the fact that SNR in FLASH depends on T_2^*, and not T_2, as for all other sequences. Note that Fig. 1.5a shows the parametric map only for T_2^* up to 100 ms, while Fig. 1.5b,c show T_2 up to 500 ms.

Comparison of the other two sequences is slightly more complex. UTE (FISP) and bSSFP are compared in Fig. 1.6, illustrating that bSSFP is the superior sequence for most relaxation times. Only for the extreme case of ultra-short T_2 in the millisecond range, such as for paramagnetic molecules, UTE performs superior. This result is not surprising, as for very fast T_2 relaxation, the shorter pre-acquisition delay in UTE becomes the dominating parameter.

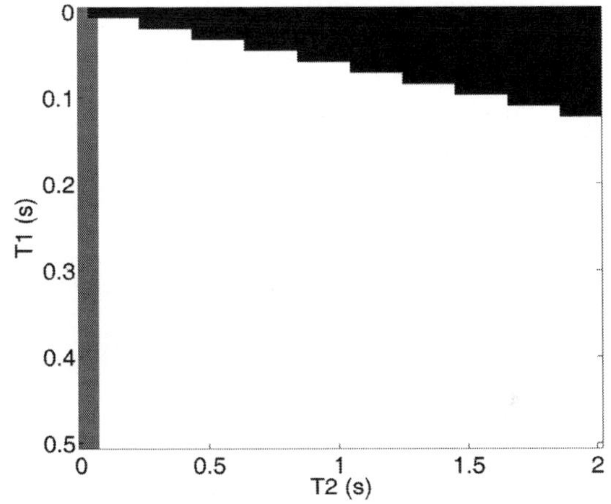

Figure 1.6 Comparison of bSSFP and UTE in terms of SNR efficiency. Gray indicates higher SNR for UTE and white for bSSFP. Note, forbidden region, $T_1 < T_2$, colored black.

1.3.2 SNR Efficiencies and Optimum Parameters for RARE

Performing the above calculations for RARE using Eqs. (1.5), (1.6), (1.8), and (1.9), it is found that the use of a flip-back module always provides better SNR efficiencies. Only for short T_2 the gain becomes small and eventually negligible. We therefore only compare 3D and multi-slice 2D RARE with flip-back pulse in the following. The best bandwidths for the (arbitrarily) chosen parameters ETL = 16 and TR = 4 sec are shown in Fig. 1.7. For 3D RARE only for extremely short T_2 a bandwidth above 10 kHz is found favorable.

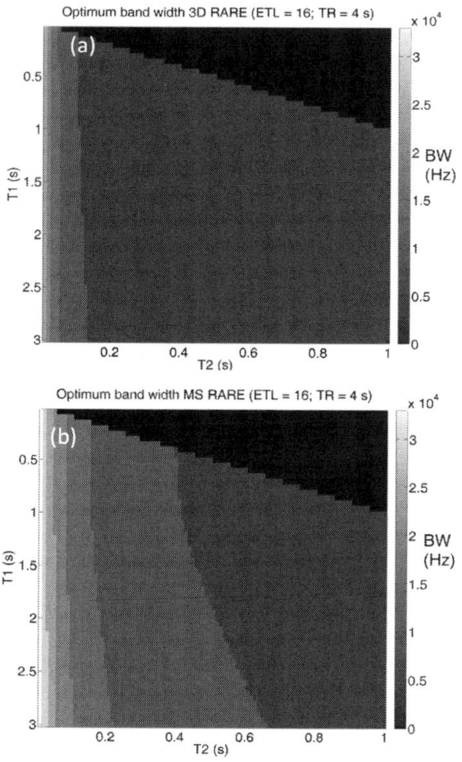

Figure 1.7 Optimum acquisition bandwidth (in kHz) for (a) 3D RARE and (b) multi-slice 2D RARE, both using flip-back pulses. Note, forbidden region, $T_1 < T_2$, colored black.

For multi-slice RARE, however, bandwidths between 10 and 20 kHz were found to be in the best range for typical T_2 values encountered for fluorine containing compounds and standard magnetic field strengths. As a compromise, for the calculation of the optimum RARE factor and TR we used a bandwidth of BW = 10 kHz. Lower bandwidths will most likely result in artifacts in the images under realistic experimental conditions. The calculation of optimum bandwidth does vary only slightly for other ETL and TR values. The results presented here are therefore representative for the range of parameters usually applied in most practical measurements.

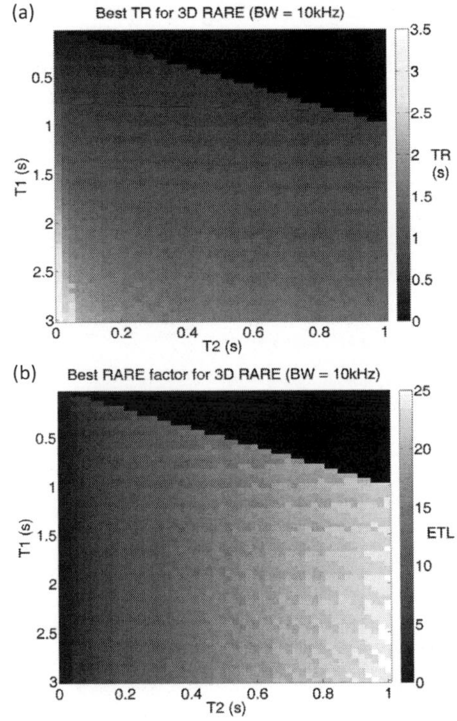

Figure 1.8 Parametric map of (a) optimum TR with optimized RARE factor for each value and (b) optimum RARE factor with optimized TR for each value, for 3D RARE using flip-back pulses. Note, forbidden region, $T_1 < T_2$, colored black.

For 3D RARE, optimum TR and RARE factor are shown in Fig. 1.8a,b, respectively. From these parametric maps it becomes evident that TR should be set to values between 0.5 to 1 times T_1, except for very short T_1 or extremely short T_2 values, where TR of up to 1.2 T_1 will provide highest SNR efficiency (Fig. 1.8a). Further, it can be seen that the optimum RARE factor depends mainly on T_2 and is only slightly influenced by T_1 (Fig. 1.8b). For typical T_2 values of most fluorine containing compounds, RARE factors between 8 and 20 are favorable, increasing with T_2. Note that for RARE without flip-back pulse, longer TR values, up to almost twice T_1, and substantially larger RARE factors, will provide the highest SNR efficiency; however, SNR is always lower than for measurements using a flip-back pulse. Figure 1.9 shows a map of best SNR efficiency values that can be achieved with 3D and multi-slice 2D RARE, both using flip-back pulses. In these maps the optimum values for both RARE factor and TR were used for each T_1 - T_2 combination. It becomes evident that 3D RARE is superior in the long T_2 regime, while multi-slice RARE is better for short T_2 values.

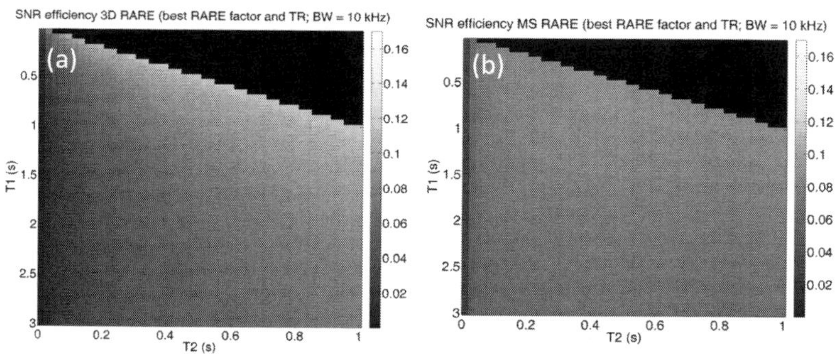

Figure 1.9 SNR efficiencies for (a) 3D RARE and (b) multi-slice 2D RARE, both using flip-back pulses. Scale bar gives SNR efficiency in units of M_0 per square root of scan time. Note, forbidden region, $T_1 < T_2$, colored black.

1.4 The Best Pulse Sequence for ^{19}F MRI

Figure 1.10 summarizes the optimum SNR efficiencies for the sequences analyzed herein for pairs of T_1 and T_2 values. Comparing RARE with UTE and bSSFP it becomes evident that each sequence is superior in a particular relaxation time regime. bSSFP is the superior sequence for most relaxation times, particularly for long T_2 values. In the short T_2 regime ($T_2 < 0.14\ T_1$), multi-slice 2D RARE yields highest SNR efficiency. Only for ultra-short T_2, UTE is found to be the superior sequence. Depending on T_1, the UTE regime is limited to T_2 values below 15 ms. Figure 1.10 also shows that 3D RARE is never the best choice. However, if a spin echo type sequence must be used, 3D RARE is more efficient than multi-slice 2D RARE for long T_2 (see Fig. 1.9 and Table 1.1).

Table 1.1 SNR efficiency values (in units $M_0/\sqrt{t}$) for five sequences and selected relaxation times

T_1/T_2 (ms)	3D RARE	MS 2D RARE	FLASH[1]	UTE	bSSFP
1.5/1.5	<0.001	<0.001	0.136	0.477	0.144
350/20	0.027	0.053	0.048	0.052	0.055
600/300	0.105	0.090	0.051	0.105	0.170
840/700	0.130	0.092	0.043	0.134	0.220
1000/600	0.115	0.092	0.040	0.114	0.187

[1]$T_2^* = T_2$ was assumed.

Figure 1.10 may serve as a general starting point to identify the most efficient sequence for fluorine measurements. Using Figs. 1.5 and 1.9 the expected differences in efficiency for the different MR sequences can be estimated for a given compound.

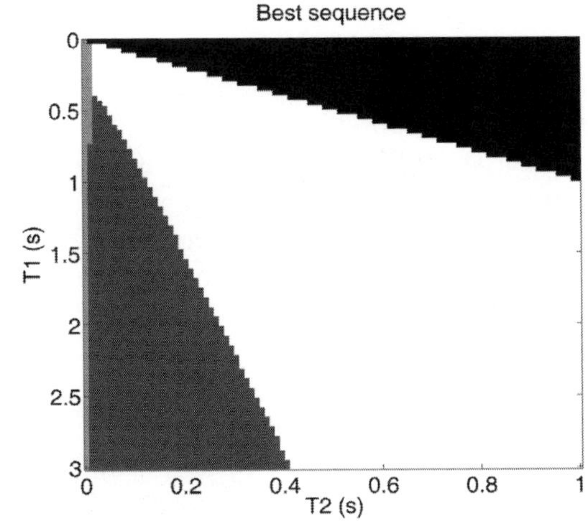

Figure 1.10 Best sequence for each $T_1 - T_2$ combination. Color code is: UTE, light gray; bSSFP, white; MS RARE, dark gray; forbidden region, $T_1 < T_2$, black. Note, there is no $T_1 - T_2$ combination favoring 3D RARE.

Table 1.1 summarizes the expected SNR efficiencies for different relaxation time combinations. These roughly correspond to emulsions of a perfluorocrown ether (PFCE) at 3 T (T_1 = 1000 ms,

T_2 = 600 ms) (de Vries et al., 2014), or at 9.4 T (T_1 = 600 ms, T_2 = 300 ms) (Jacoby et al., 2014), perfluorooctyl bromide (PFOB) at 3 T (T_1 = 840 ms, T_2 = 700 ms) (Giraudeau et al., 2010), gadolinium-functionalized PFCE nanoparticles at 9.4 T (T_1 = 350 ms, T_2 = 20 ms) (de Vries et al., 2014), and small paramagnetic molecules with ^{19}F relaxation (T_1 = 1.5 ms, T_2 = 1.5 ms) (Schmid et al., 2013). Table 1.1 shows that for ultra-short T_2, UTE is by far more efficient than any other sequence. For short T_2, all sequences except 3D RARE perform similarly. For long T_2, bSSFP is clearly the best choice, followed by UTE and 3D RARE which perform similarly.

1.5 Implications for Actual ^{19}F MRI Measurements

In the practical measurement, it must be considered that our calculations made a number of assumptions that may be invalid for a particular experimental setup. Also, very low bandwidths of 10 kHz and below may lead to artifacts in the images. It will therefore often not be possible to use theoretically optimum parameters as derived here. Furthermore, if certain probes are used, for example, PFOB with multiple resonance lines and active J couplings, additional constraints on sequence design may apply. As shown recently, UTE read out may dramatically improve SNR due to reduced effects of J coupling, and combination of UTE with bSSFP may be the optimum imaging sequence choice (Goette et al., 2014).

Strictly, all the above considerations are valid for radial *k* space sampling only. In that case, the signal maximum is sampled at the center of *k* space and therefore fully contributes to signal intensity in the image. Our above treatment, therefore, correctly reflects the situation for UTE. Also for radial FLASH or radial bSSFP the above estimates are correct. For RARE, however, radial read out is difficult, as varying signal intensities between the subsequent echoes result in amplitude modulation in *k* space, which in turn results in image artifacts. However, such artifacts can be minimized, when appropriate filtering is applied (Neumann et al., 2014). The signal gain of radial versus Cartesian sampling has been shown recently for a number of different sequences and was estimated to be approximately 50% (Goette et al., 2014).

Yet, often radial sampling is not desired or feasible. In such cases, the effect of k space sampling patterns has to be considered in addition to the calculations made above. In most cases, this will most likely lead to reduced SNR efficiency, and may have pronounced impact on the sequence chosen for highest efficiency.

1.6 Further Methods to Increase SNR: Heteronuclear Overhauser Enhancement

Beyond optimizing the signal acquisition to achieve maximum SNR efficiency, the absolute ^{19}F signal amplitude can be manipulated. As mentioned above, paramagnetic relaxation enhancement (Chalmers et al., 2011; Schmid et al., 2013) or minimization of the effects of J coupling (Giraudeau et al., 2010) may be useful techniques. One further method to increase the fluorine MR signal may be the use of the NOE. For compounds where the fluorine atoms are in close proximity to hydrogen nuclei, dipolar coupling between these nuclei may lead to heteronuclear cross relaxation. Saturation of one nucleus may increase or decrease the magnetization of the other nucleus, depending on molecular structure and experimental parameters. The maximum possible enhancement of the ^{19}F signal is 53% if ^{1}H is saturated; Krems et al. (1995) have measured the transfer rates for two fluorinated compounds relevant for cancer treatment, 5-fluorouracil and α-fluoro-β-alanine, in model solutions and used proton saturation to increase the fluorine signal in a patient measurement at 1.5 T. They observed a gain in fluorine signal close to the theoretically achievable maximum.

This enhancement strategy may possibly be combined with any read out scheme described above. Here, we show a signal enhancement by the NOE for small, water-soluble fluorinated compounds. A gain in fluorine signal upon saturation of the ^{1}H frequency can be observed. For saturation of ^{1}H magnetization with series of increasing numbers of 180° pulses with 1 s delay, ^{19}F magnetization reaches a plateau (i.e., Truncated Driven NOE). Magnetization transfer that enhances the ^{19}F signal, and the homonuclear T_1 relaxation that drives ^{19}F signal towards equilibrium reach a balanced state. For measurements at 3 T, signal increases of 20% for fluorodeoxyglucose (FDG), 9% for trifluoroacetic acid and 12% for trifluoroethanol (TFE) were observed (shown for FDG in Fig. 1.11a). A single ^{1}H saturation pulse caused smaller enhancement of ^{19}F signal, peaking after approximately 2 s. The signal progression is shown for FDG in Fig. 1.11b. Although these signal enhancements are substantial when long preparation times (^{1}H saturation) are used, this approach is not appropriate to increase SNR efficiency. Extra time is required for the preparation period, strongly reducing overall SNR efficiency and making the calculations in the previous sections obsolete.

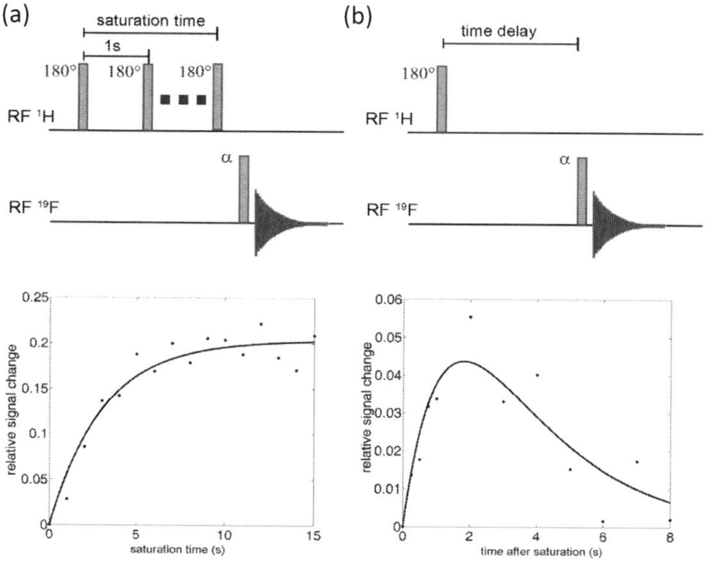

Figure 1.11 Signal increase in FDG ^{19}F signal by NOE. Panel (a) shows increase of ^{19}F magnetization with increasing saturation time. Panel (b) shows evolution of ^{19}F magnetization after single ^{1}H saturation pulse.

However, NOE enhancement can easily be used to increase ^{19}F signal if a steady-state saturation of ^{1}H magnetization is employed. For this purpose continuous or pulsed saturation of the ^{1}H frequency is applied, while any selected MRI sequence is performed on the ^{19}F frequency (steady-state NOE). This approach provides an increased signal for the ^{19}F measurement, as both T_1 relaxation, as well as the NOE generate longitudinal magnetization.

For measurements with steady-state NOE at 3 T, a signal gain of 29% for TFE and of 31% for FDG was detected. Note that the steady-state NOE can only be used if the MR scanner is capable of pulsing simultaneously on two frequencies (^{1}H and ^{19}F). Although technically easy to solve, clinical MR scanners may be incapable of this function (Goette et al., 2014; Keupp et al., 2011). Furthermore, heteronuclear NOE will not provide signal increases for all compounds that may be used for fluorine MRI. If an appropriate substance is used, the NOE enhancement basically comes for free, since ^{1}H saturation can be performed while optimized imaging at the ^{19}F frequency is carried out.

References

Ahrens ET, Bulte JW. (2013). Tracking immune cells in vivo using magnetic resonance imaging. *Nat Rev Immunol,* **13,** 755–763.

Ahrens ET, Flores R, Xu HY, Morel PA. (2005). In vivo imaging platform for tracking immunotherapeutic cells. *Nat Biotech,* **23,** 983–987.

Aime S, Carrera C, Delli Castelli D, Crich SG, Terreno E. (2005). Tunable imaging of cells labeled with MRI-PARACEST agents. *Angew Chem Int Edit,* **44,** 1813–1815.

Chalmers KH, Kenwright AM, Parker D, Blamire AM. (2011). ^{19}F-lanthanide complexes with increased sensitivity for ^{19}F-MRI: Optimization of the MR acquisition. *Magn Reson Med,* **66,** 931–936.

de Vries A, Moonen R, Yildirim M, Langereis S, Lamerichs R, Pikkemaat JA, Baroni S, Terreno E, Nicolay K, Strijkers GJ, Grull H. (2014). Relaxometric studies of gadolinium-functionalized perfluorocarbon nanoparticles for MR imaging. *Contrast Media Mol Imag,* **9,** 83–91.

Ebner B, Behm P, Jacoby C, Burghoff S, French BA, Schrader J, Flögel U. (2010). Early assessment of pulmonary inflammation by ^{19}F MRI in vivo. *Circ Cardiovasc Imag,* **3,** 202–210.

Flögel U, Ding Z, Hardung H, Jander S, Reichmann G, Jacoby C, Schubert R, Schrader J. (2008). In vivo monitoring of inflammation after cardiac and cerebral ischemia by fluorine magnetic resonance imaging. *Circulation,* **118,** 140–148.

Flögel U, Su S, Kreideweiss I, Ding Z, Galbarz L, Fu J, Jacoby C, Witzke O, Schrader J. (2011). Noninvasive detection of graft rejection by in vivo ^{19}F MRI in the early stage. *Am J Transplant,* **11,** 235–244.

Giraudeau C, Flament J, Marty B, Boumezbeur F, Meriaux S, Robic C, Port M, Tsapis N, Fattal E, Giacomini E, Lethimonnier F, Le Bihan D, Valette J. (2010). A new paradigm for high-sensitivity ^{19}F magnetic resonance imaging of perfluorooctylbromide. *Magn Reson Med,* **63**, 1119–1124.

Goette MJ, Keupp J, Rahmer J, Lanza GM, Wickline SA, Caruthers SD. (2015). Balanced UTE-SSFP for F MR imaging of complex spectra. *Magn Reson Med,* **74**, 537–543.

Hertlein T, Sturm V, Kircher S, Basse-Lusebrink T, Haddad D, Ohlsen K, Jakob P. (2011). Visualization of abscess formation in a murine thigh infection model of *Staphylococcus* aureus by ^{19}F-magnetic resonance imaging (MRI). *Plos One,* **6**, e18246.

Jacoby C, Temme S, Mayenfels F, Benoit N, Krafft MP, Schubert R, Schrader J, Flögel U. (2014). Probing different perfluorocarbons for in vivo inflammation imaging by ^{19}F MRI: Image reconstruction, biological half-lives and sensitivity. *NMR Biomed,* **27**, 261–271.

Keupp J, Rahmer J, Grasslin I, Mazurkewitz PC, Schaeffter T, Lanza GM, Wickline SA, Caruthers SD. (2011). Simultaneous dual-nuclei imaging for motion corrected detection and quantification of ^{19}F imaging agents. *Magn Reson Med,* **66,** 1116–1122.

Krems B, Bachert P, Zabel HJ, Lorenz WJ. (1995). ^{19}F-[1H] nuclear Overhauser effect and proton decoupling of 5-fluorouracil and alpha-fluoro-beta-alanine. *J Magn Reson B,* **108**, 155–164.

Mastropietro A, De Bernardi E, Breschi GL, Zucca I, Cametti M, Soffientini CD, de Curtis M, Terraneo G, Metrangolo P, Spreafico R, Resnati G, Baselli G. (2014). Optimization of rapid acquisition with relaxation enhancement (RARE) pulse sequence parameters for ^{19}F-MRI studies. *J Magn Reson Imaging,* **40**, 162–170.

Neumann D, Breuer FA, Volker M, Brandt T, Griswold MA, Jakob PM, Blaimer M. (2014). Reducing contrast contamination in radial turbo-spin-echo acquisitions by combining a narrow-band KWIC filter with parallel imaging. *Magn Reson Med,* **72**,1680–1686.

Ribot EJ, Gaudet JM, Chen Y, Gilbert KM, Foster PJ. (2014). In vivo MR detection of fluorine-labeled human MSC using the bSSFP sequence. *Int J Nanomed,* **9**, 1731–1739.

Ruiz-Cabello J, Barnett BP, Bottomley PA, Bulte JWM. (2011). Fluorine ^{19}F MRS and MRI in biomedicine. *NMR Biomed,* **24**, 114–129.

Schmid F, Holtke C, Parker D, Faber C. (2013). Boosting ^{19}F MRI-SNR efficient detection of paramagnetic contrast agents using ultrafast sequences. *Magn Reson Med,* **69**, 1056–1062.

Schmidt R, Nippe N, Strobel K, Masthoff M, Reifschneider O, Castelli DD, Holtke C, Aime S, Karst U, Sunderkotter C, Bremer C, Faber C. (2014). Highly shifted proton MR imaging: Cell tracking by using direct detection of paramagnetic compounds. *Radiology,* **272**, 785–795.

Yu JX, Hallac RR, Chiguru S, Mason RP. (2013). New frontiers and developing applications in ^{19}F NMR. *Prog Nucl Magn Reson Spectrosc,* **70**, 25–49.

Chapter 2

Advanced Detection Techniques and Hardware: Simultaneous $^{19}F/^{1}H$ MRI

Lingzhi Hu,[a] Jochen Keupp,[b] Shelton D. Caruthers,[a] Matthew J. Goette,[c] Gregory M. Lanza,[c] and Samuel A. Wickline[c]

[a]*Philips Healthcare, Cleveland, USA*
[b]*Philips Research, Hamburg, Germany*
[c]*Department of Medicine, Washington University, St. Louis, USA*

jochen.keupp@philips.com

While magnetic resonance spectroscopy and imaging of fluorine (^{19}F) nuclei have been pursued for medical research for decades, the simultaneous combination of ^{1}H and ^{19}F MRI has opened up new opportunities for the application of multi-nuclear techniques in the context of molecular and cellular imaging for diagnosis or therapy delivery. The applications of $^{19}F/^{1}H$ MRI rely heavily on the development of novel fluorinated contrast agents and imaging methodologies. Currently, perfluorocarbon (PFC) nanoparticle (NP) based contrast agents have been applied extensively in several key areas of diagnostic imaging because of their biological safety, versatile chemical properties, and high ^{19}F content [1]. The massive quantities of ^{19}F nuclei per NP may allow much higher detection sensitivity as compared to imaging agents based on fluorinated chemical compounds, which typically contain fewer number of ^{19}F nuclei per molecule. The enthusiasm

Fluorine Magnetic Resonance Imaging
Edited by Ulrich Flögel and Eric Ahrens

ISBN 978-981-4745-31-4 (Hardcover), 978-981-4745-32-1 (eBook)
www.panstanford.com

for ^{19}F imaging of PFC NP is focused mainly in the fields of cell labeling [2–4, 12–14], targeted molecular imaging [5, 6, 11], and blood pool functional imaging [7, 8]; all of these techniques have been validated extensively in animal models, with some entering human clinical trials.

2.1 Imaging Applications of Perfluorocarbon Nanoparticles and Introduction of Simultaneous ^{19}F/^{1}H MRI

A particularly promising use of ^{19}F/^{1}H imaging is targeted molecular imaging, first demonstrated by Wickline and Lanza et al. [5, 15]. After incorporating additional targeting ligands and payloads onto the surface of PFC NPs, these agents can actively bind to a large variety of biomarkers expressed throughout the course of cancer and cardiovascular disease. ^{19}F/^{1}H enables not only a noninvasive observation of the binding, but, more importantly, the direct quantification of bound targeted PFC NPs in the lesion. The initial proof of concept work on ^{19}F/^{1}H molecular imaging was conducted by Morawski et al. in unstable atherosclerotic plaque using fibrin-targeted PFC NPs in 2004 [16]. In the early stages of ^{19}F molecular imaging, limitations to imaging technologies presented challenges for in vivo detection of such targeted nanoparticles, given the low local concentration of ^{19}F nuclei. However, this pioneering ex vivo study successfully demonstrated the nexus between the underlying biology and the feasibility of targeted molecular imaging with fluorine. With improved ^{19}F imaging technology [1], targeted molecular imaging of atherosclerotic plaque was later demonstrated in vivo at 3 T on clinical MRI scanners [17]. Targeted NPs also have been used to image other biomarkers, such as VCAM [18], $\alpha_v\beta_3$ integrin [19–21] and ICAM [22], and have enabled promising potential diagnostic applications as well as efficient drug delivery to lesions in inflammatory disease and cancer [23] with image-based confirmation of delivery.

Along with cellular and molecular imaging applications, the use of PFC as a blood pool contrast agent is a natural extension from its FDA approved usage as a blood substitute [24]. Because there is no natural ^{19}F background in soft tissue, intravenously injected PFC NPs can be effectively imaged in the first pass without subtraction. ^{19}F coronary angiography at clinical field strength was demonstrated with minimal background contamination by Neubauer et al. [7]. PFC NPs have also been used as a novel probe for observing vascular leakage in various pathology processes, such as vulnerable atherosclerotic plaque [25] and ischemia-reperfusion kidney injury [26]. Although typically constrained by size within the healthy intact vasculature, PFC NPs can penetrate into inflamed tissues through damaged endothelial barriers, and ^{19}F/^{1}H MRI can be used to quantify vascular leakage by measuring the retained concentration of PFC NPs [27].

In addition to offering a platform for noninvasive and quantitative assessment of molecular target and agent concentrations, PFC NPs also provide sensing opportunities for measuring oxygenation because of their high affinity to oxygen molecules in the blood stream [24]. Sensing oxygenation via fluorinated molecules (e.g., hexafluorobenzene) has been useful in oncological research for many years [28]. With the use of dedicated ^{19}F/^{1}H pulse sequences and PFC NPs, the absolute blood oxygenation in major blood vessels and microcirculation in specific organs can be obtained through quantifying the ^{19}F longitudinal relaxation time (T_1) [8].

Advancement of these various ^{19}F/^{1}H imaging applications for basic research and translational investigations necessitates continued improvement in MRI technology. Research in ^{19}F MRI of PFC NPs has undergone rapid growth over the past 10 years and suggests a promising future for eventual deployment in clinical settings. In conjunction with significant improvements in contrast agent design and proof-of-concept demonstration for disease targeting, many of the technical challenges have been addressed to bridge the gap between laboratory investigations and clinical utility.

Compared to ^{1}H MRI, ^{19}F manifests no appreciable physiological background signal, so the signal strength for ^{19}F MRI is solely determined by the modest doses of fluorinated compounds administered. To illustrate, because of the inherently low ^{19}F concentration, the seminal work on MRI of ^{19}F labeled immunotherapeutic cells required 3 h of imaging at 11.7 T [9]. In other early exploration of molecular and cellular ^{19}F MRI [10, 29, 30], the scan time was prohibitively long for clinical translation. With the emerging interest in fluorine contrast agents, on-going clinical trials of ^{19}F MRI for cell tracking, and molecular and functional imaging, significant attention has been paid to developing imaging technology to improve the detection limits for ^{19}F nuclei. Because of the complex chemical structure of fluorinated compounds, advanced pulse sequences must be designed to accommodate the unique features of ^{19}F signals such as variable T_1 corresponding to different oxygenation level, J-coupling, and fast apparent T_2 decay. Advanced ^{19}F imaging techniques aimed at the most efficient use of available signal from ^{19}F nuclei and combining species with different chemical shifts have been explored in recent years [31] leading to more realistic imaging time requirements of 10–20 min at clinical field strengths like 3 T.

Further acquisition time and workflow constraints in ^{19}F MRI were previously related to the fact that ^{19}F imaging had to be performed sequentially with conventional ^{1}H MRI, in particular for anatomical co-localization of the ^{19}F imaging data. In addition to the prolonged imaging time, subsequent ^{1}H and ^{19}F image acquisitions bear the risk of misalignment between the ^{1}H and ^{19}F information due to off-resonance conditions, physiological motion and/or radio-frequency (RF) coil exchange. The latter two challenges can potentially be overcome by combined and simultaneous ^{19}F and ^{1}H acquisition techniques [32]. Simultaneous dual-nuclei acquisition techniques can be performed on suitable MR systems with individual RF transmission architecture for proton and non-proton nuclei, which is available in many state-of-the-art clinical MRI scanners. In addition, specialized RF coils are required, capable of sending and receiving ^{19}F and ^{1}H signals at the same time via dual-tuned RF circuits. Furthermore, the acquisition software must allow for simultaneous ^{19}F/^{1}H imaging methodology. A prototype for simultaneous ^{19}F/^{1}H imaging based on a clinical MRI scanner has been described in [32] and was successfully utilized in subsequent pre-clinical studies [19].

In addition to the direct advantages for ^{19}F image quality and precision, dual-tuned ^{19}F/^{1}H scanner architectures may also be essential for optimizing workflow and reliability for future clinical ^{19}F MRI applications. As the ^{19}F MR signal is weak and not available prior to the agent injection, MR system and RF coil parameter calibration procedures, as well as data acquisition for signal homogeneity correction (e.g., RF field at the ^{19}F frequency), need to be performed via the ^{1}H signal, which is straightforward in a dual-tuned architecture. Furthermore, physiological motion effects on the ^{19}F signal may be corrected using simultaneous ^{1}H based motion tracking. The benefits for simultaneous signal acquisition enhance detectability and quantification of the ^{19}F signal at low concentrations (i.e., [pM] level for PFC NP [16]).

In this chapter, we will explain the basic concept for a simultaneous $^{19}F/^{1}H$ MRI system. Specifically, we will focus on the architecture, RF coil design and calibration, as well as on advanced and sensitive simultaneous $^{19}F/^{1}H$ MRI techniques designed for 3 T clinical MRI systems.

2.2 MRI Hardware and Reconstruction for Simultaneous $^{19}F/^{1}H$ Imaging

Simultaneous $^{19}F/^{1}H$ dual-nuclei MRI capability has been implemented based on a commercial 3 T MRI platform (Achieva, Philips Healthcare, The Netherlands) that enables simultaneous acquisition of ^{1}H- and ^{19}F-images with high sensitivity [32]. Limited hardware extensions and essential software enhancements allow the scanner to perform simultaneous transmission and detection of RF signals at two distinct resonant frequencies, in particular at 128 MHz (^{1}H) and 120 MHz (^{19}F). In this scanner architecture, acquisition of the ^{19}F signal can be complemented with real-time assessment of the underlying anatomy as well as physiological motion via the ^{1}H channel without interrupting continuous averaging at the ^{19}F frequency.

2.2.1 Scanner Hardware Design

Hardware modifications [32] for a simultaneous $^{19}F/^{1}H$ MRI architecture are limited to the RF sub-system, while other components of the scanner (e.g., the gradient system) remained unchanged. Separate transmitter and receiver devices for the ^{19}F nucleus are added as shown schematically in Fig. 2.1 (green). To achieve simultaneous excitation of ^{1}H and ^{19}F nuclei, waveforms for the prescribed RF pulses are generated independently for each nucleus and simultaneously fed into a single RF amplifier. Employing a broadband solid-state RF amplifier, as available in many commercial MR systems, the dual-nuclei detection system can be implemented through the incorporation of a standard power combiner on the low-power input side of the amplifier (Fig. 2.1, "C"), permitting simultaneous RF transmission on both resonant frequencies, 128 MHz and 120.1 MHz. Suitable dual-resonant RF coil designs will be discussed in Section 2.3. The collected composite MR signal is separated into two channels (Fig. 2.1, using power splitter "S", Mini-Circuits ZFSC-2-1) with different gain settings (gain 1, gain 2) to accommodate for the different orders of signal magnitude at the ^{19}F and ^{1}H frequency, respectively.

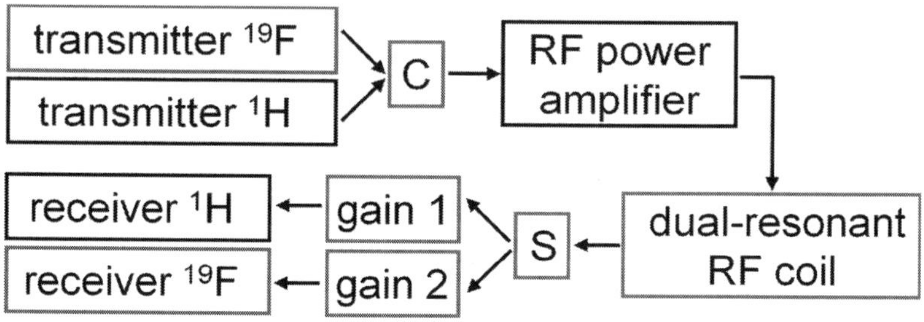

Figure 2.1 Architecture of a spectrometer system capable of simultaneous dual-nuclei MR. RF-waveforms generated in two separate transmitters (^{19}F and ^{1}H) are combined by a power combiner "C" before the RF power amplifier. After transmission and reception by a dual-resonant RF coil, the signal is divided into two paths with individual gain settings using the splitter "S" and recorded by two independent receivers for the two nuclei. The highlighted components (green and red) are the additional RF-components that transform a standard MR system into a simultaneous dual-nuclei system.

The major differences for simultaneous dual-nuclei MR sequences, as compared with conventional MRI, entail concurrent RF pulses and data acquisition time windows on both resonance frequencies. As an example, the ^{19}F excitation pulse can be complemented with a ^{1}H RF pulse at an identical starting time but with different input power, RF waveform, and center frequency (see, e.g., Fig. 2.10).

2.2.2 MR Reconstruction Methods

In a standard MR sequences, timing and gradient strength are adjusted for the gyromagnetic ratio γ of the respective nucleus. For a dual-nuclei MRI sequence, the timing and gradient strength can be only be preset for one specific nucleus, the other will be acquired with non-standard settings. Thus, corrections need to be applied for the second nucleus. For example, the actual overall size of the field of view and spatial resolution is slightly different (because of the different γ for ^{1}H and ^{19}F). For example, if the pulse sequence is calculated based on ^{19}F, the resulting concurrent ^{1}H-image needs to be shrunken by about 6% ($\gamma[^{19}\mathrm{F}]/\gamma[^{1}\mathrm{H}]$), because the gradient strength is actually too low for ^{1}H. As this error for the ^{1}H acquisition is equivalent to a linear gradient mismatch, the image and voxel scaling may be corrected using the standard 3D gradient-linearity correction performed in the reconstruction software of the MR system [32].

Because of the inherent low concentration of ^{19}F nuclei in pre-clinical and clinical setting, the ^{19}F MRI signal is usually much lower compared to ^{1}H. Therefore, image reconstruction techniques are highly demanding for dual ^{19}F/^{1}H MRI to enhance SNR and detectability of ^{19}F nuclei present in low concentrations. Thus, ^{19}F MRI usually requires a lower spatial resolution than typically used for ^{1}H MRI to obtain a sufficient signal-to-noise ratio (SNR). However, in simultaneous dual-nuclei acquisitions, the ^{1}H spatial resolution should not be compromised, because precise anatomical co-localization and motion correction require a higher spatial resolution. To address this challenge, a dual-resolution reconstruction technique was demonstrated that allows efficient *a posterior* reconstruction of ^{1}H and ^{19}F images at different spatial resolution (but identical FOV) from the same dual-nuclei dataset acquired using a 3D isotropic radial *k*-space trajectory [33]. Different reconstructed spatial resolution is achieved by applying different spherically symmetric weighting for the two nuclei in *k*-space. For ^{1}H data, full resolution reconstruction can be performed using standard parabolic weighting [34] for *k*-values within the Nyquist-radius (radius of the sphere, where sampling density fulfills the Nyquist criterion), and uniform weighting is applied outside. For ^{19}F data, the Nyquist radius may be reduced by a factor η to reconstruct an image with lower spatial resolution but increased SNR.

An important application of dual-nuclei ^{19}F/^{1}H image acquisition is ^{1}H-signal-based correction of patient motion and physiological motion during ^{19}F signal averaging. One example is shown in Fig. 2.2 [32]. Here, the ^{19}F signal that was blurred to an invisible level (SNR < 1) could be recaptured by using concurrent ^{1}H-based motion tracking and retrospective motion correcting reconstruction. Briefly, the ^{1}H data can be first divided into sub-sampled volumes at subsequent time points. After individual gridding reconstruction of the sub-volumes, motion detection can be performed in the spatial domain by mutual registration [35]. Subsequently, the measured motion vectors can be used to correct the fully sampled ^{19}F and ^{1}H data through transformation in *k*-space. To correct translational motion, a linear phase transformation can be applied to the *k*-space lines according to the time point of acquisition and orientation to the motion direction. Final gridding reconstruction of the ^{19}F and ^{1}H data yield motion corrected images for both nuclei.

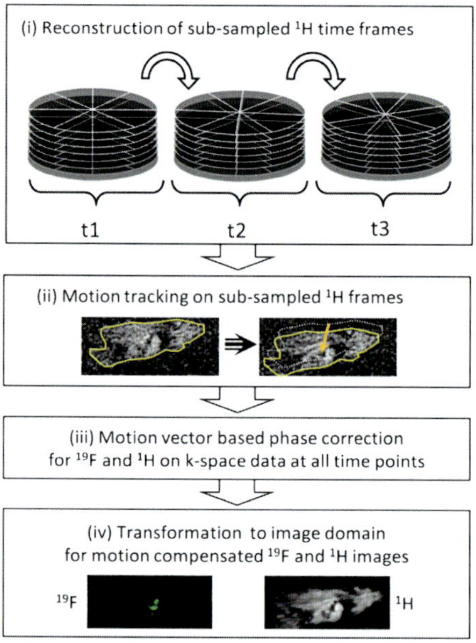

Figure 2.2 Schematic overview of the dual-nuclei MRI based motion correction technique. The proton data is (i) divided in to sub-sampled volumes at subsequent time points (t_1, t_2, ...), on which motion detection is performed in the image domain (ii). The derived motion vectors are then used for correction of the fully sampled ^{19}F data as well as the ^{1}H data (iii) using phase transformations in *k*-space according to the time point of acquisition. Final gridding reconstruction (iv) of the ^{19}F and ^{1}H data yields motion corrected dual-nuclei images.

In vivo simultaneous $^{19}F/^{1}H$ imaging has been demonstrated using targeted PFC NP and the modified clinical 3 T as mentioned above. Figure 2.3 illustrates combined ^{19}F and ^{1}H imaging with the use of $\alpha_v\beta_3$-integrin targeted nanoparticles for quantitative assessment of angiogenesis in a rabbit tumor model on the modified clinical 3 T scanner [31]. For a measurement time of 35 min, the sensitivity of the ^{19}F image acquisition is sufficient to register angiogenesis at the tumor surface with a spatial resolution of 2.2 × 2.2 × 4 mm^3 as visualized in the color overlay in Fig. 2.3b,d,e. A 3-D gradient echo sequence was used for ^{19}F, and a 3-D T_1-weighted gradient echo sequence for the ^{1}H image acquisition was used to depict morphology (specifically, the popliteal fossa of the hind leg, Fig. 2.3a,c,e). The ^{1}H image was acquired without changing the examination setup and was merged with the ^{19}F image for anatomical co-localization (Fig. 2.3b, d, e). The color scale corresponds quantitatively to the millimolar (mM) concentration of fluorine nuclei at the molecular target site and can be correlated with the extent of angiogenesis.

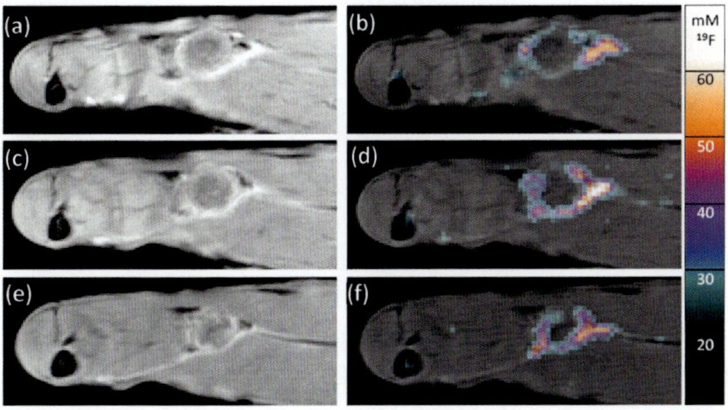

Figure 2.3 Quantification of $\alpha_v\beta_3$-targeted nanoparticle (NP) concentrations via ^{19}F MRI on a Vx-2 tumor in the hind leg of a New Zealand White rabbit, measured on a clinical 3 T MR scanner. The millimolar fluorine concentration from the NP at the target site (color scale 10–70 mM, right b, d, f) is superimposed with T_1-weighted gradient echo images (left a, c, e) for anatomical co-localization in three selected image slices (out of 15). For an in-plane resolution of 2.2 mm and a slice thickness of 4 mm, the data acquisition was completed in 35 min. The NP density can be derived from the fluorine concentrations and correlated with the degree of angiogenesis, which shows a heterogeneous pattern at the tumor rim as expected from tumor pathophysiology.

2.3 ^{19}F/^{1}H Dual-Frequency RF Coil Design and System Calibration for Simultaneous ^{19}F/^{1}H Imaging

For simultaneous ^{19}F/^{1}H imaging, the RF coil used for signal transmission and reception needs to be resonant effectively at both ^{19}F and ^{1}H frequency without requiring manual retuning or switching. Furthermore, being able to transmit with identical spatial RF field distributions (B_1 field) at both ^{1}H and ^{19}F frequencies is also essential for a dual-frequency ^{19}F/^{1}H RF coil to ensure accurate power delivery and quantitative signal calibration, which will be discussed in detail later in this section. Therefore, an ideal ^{19}F/^{1}H coil should be designed to meet four major requirements: (1) high quality factor (Q-value) for both frequencies; (2) identical B_1 field for both frequencies; (3) versatility in the geometrical structure, fitting closely to the anatomy to be imaged; and (4) good accessibility and minimized degrees of freedom (number of tuning elements) for tuning and matching at both frequencies.

2.3.1 ^{19}F/^{1}H Dual-Frequency RF Coil Design

As widely employed in analog electronics, the coupled resonator model offers a suitable general strategy for designing an ^{19}F/^{1}H coil that satisfies the requirements for simultaneous ^{19}F/^{1}H imaging. Specifically, a conventional ^{1}H RF probe can be connected to a secondary inductor and capacitor (LC) resonator with a coupling capacitor in series (Fig. 2.4). We note that even though capacitive coupling is used as an example, connecting two resonators through inductive coupling is particularly effective [36]. For nuclei with close resonance frequencies, such as ^{1}H and ^{19}F, the inductance of a MR sample probe can be treated as a constant value at both frequencies. Following the theoretical analysis proposed by Haase et al. [37], when the secondary resonator is built with identical impedance to the sample coil, two possible oscillating modes are formed within the circuit, namely in-phase and anti-phase modes. When resonating at the in-phase mode, the two resonators of the circuit have equal electrical potentials, and thus the voltage and current across the coupling capacitor always remains zero. In this case, the coupling capacitor can be virtually removed and then the equivalent circuit of the coupled resonator can be drawn as Fig. 2.4b with the resonant frequency

$$\omega_{\text{hi}} = \gamma_{\text{H}} B = 1/\sqrt{LC_{t1}}. \tag{2.1}$$

In contrast, for the anti-phase mode, the electric potential at both sides of the coupling capacitor is opposite, so the voltage at the middle point of coupling capacitor remains zero. For circuit analysis, a virtual short wire can be added between the ground and the middle point of the coupling capacitor (Fig. 2.4c), and thus it gives rise to a different resonant frequency

$$\omega_{\text{lo}} = \gamma_{\text{F}} B = 1/\sqrt{L(C_{t1} + 2C_{\text{c}})}. \tag{2.2}$$

Given the gyromagnetic ratios of ^{1}H and ^{19}F, when

$$C_{\text{c}}/C_{t1} = 0.064 \tag{2.3}$$

the two frequencies of the coupled resonator correspond to the ^{1}H and ^{19}F frequencies [38].

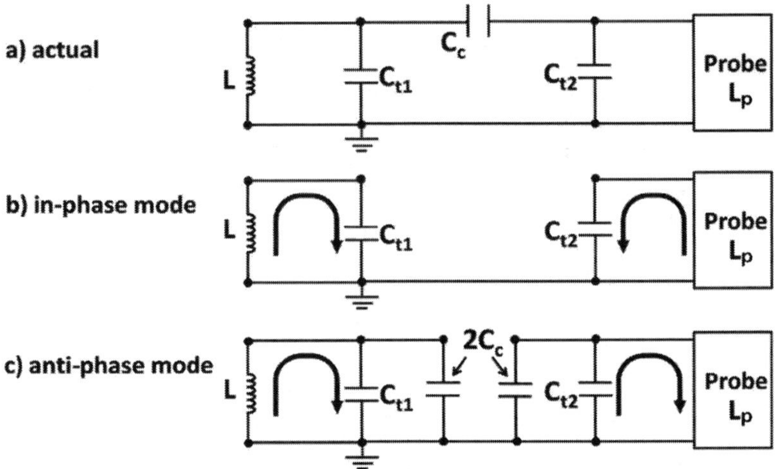

Figure 2.4 Schematic diagram of a coupled resonator circuit. (a) The resonator formed by capacitor C_{t2} and probe (with inductance L_p) is capacitively coupled, by capacitor C_c, to the secondary resonator formed by capacitor C_{t1} and inductor L. (b and c) The equivalent circuit of in-phase and anti-phase mode.

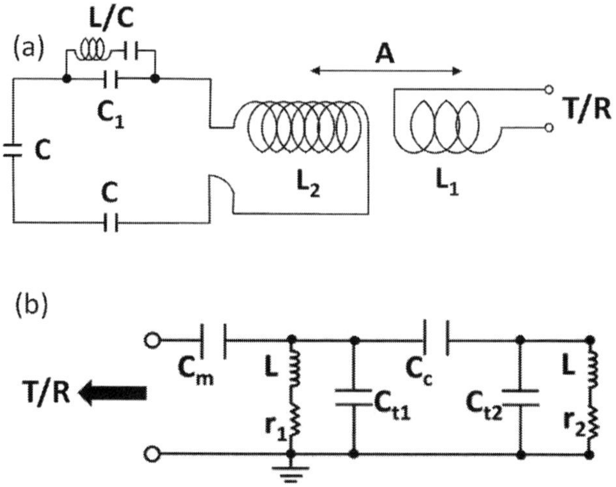

Figure 2.5 (a) Inductive coupling for matching a coupled-resonator to the RF T/R interface of an MR scanner. (b) Capacitive coupling for matching a coupled-resonator to the RF T/R interface of an MR scanner.

The coupled resonator needs to be connected to the MR scanner at the end of the secondary resonator in series with a matching circuit for impedance matching to 50 ohms at both frequencies to minimize RF reflection. To allow the minimum number of degrees of freedom for tuning and matching, capacitive and mutual inductive matching are two feasible approaches (Fig. 2.5). Figure 2.5a shows a dual resonant RF coil which is inductively coupled by means of an inductive coupling device L1, L2 instead of capacitive coupling [39]. Both coils L1, L2 are arranged such that they can be mechanically shifted in relation to each other as indicated by the arrow A in order to adjust the magnetic flux; consequently the inductive coupling between both is used for tuning and matching of the RF coil. In this way, the tuning of the RF coil to the two resonant frequencies and the matching to the transmit/receive units at these frequencies can be realized more easily than in the case of capacitive coupling. For inductive coupling, the symmetry of the RF coil is not substantially influenced, and the current distribution and homogeneity remains mostly unchanged. Only four degrees of freedom are required for the tuning and matching at both frequencies (three tunable capacitors and the variable coupling via A). This is especially relevant for surface coils, which are exposed to a large range of coil loading conditions due to their close proximity to the examination object and the potential variability in coil positioning. Furthermore, the spatial sensitivity and transmission B_1 field are practically the same at the two Larmor frequencies due to the substantially same electrical current patterns in the coil at both frequencies. Figure 2.5b shows a capacitive matching scheme to achieve matching of a coupled-resonator to the MR scanner RF interface. In this case, a small detuning of two resonators, which renders them slightly off-resonance, is able to introduce an extra degree of freedom and compensate for the impedance difference [37].

The capacitive and inductive coupling resonator schemes are applicable to different RF coil geometries, usage and configuration and can be employed to design $^{19}F/^{1}H$ dual-frequency coils at all field strengths. Figure 2.6 shows the detailed electronic circuit design (with four degrees of freedom for tuning and matching at both frequencies) of a simultaneous dual-frequency ^{1}H and ^{19}F open coil designed for imaging the rabbit knee at 3 T [40]. It is worth mentioning that the matched quality factor (Q) measured from S11 is a necessary but not sufficient condition to provide a useful measurement for a dual-frequency RF coil based on the coupled resonator. The Q value of the coil is affected by losses associated with the sample probe as well as the secondary LC circuit [41]. In this case, alternative metrics can be exploited to assess the performance of the dual-frequency coil, e.g., imaging sensitivity test and measurement of the pulse width with given power and flip angle. The same coupled resonator model can be employed to design transmit- or receive-only coils with active decoupling through additional DC control inputs. In this case, a switchable TTL bias voltage generated from the scanner spectrometer can be utilized to keep the volume/surface coil on resonance during transmit, and off resonance during receive, to avoid unwanted coupling between transmit and receive circuits that might compromise the quality factor. Being able to design efficient $^{19}F/^{1}H$ homogeneous transmit- and receive-only coils is a crucial hardware prerequisite for sensitive detections of ^{19}F nuclei using spin echo based sequences [42] and other functional imaging sequences [8] that depend on homogenous B_1 excitation pulses. Figure 2.7a shows an example of a specific coil electronic circuit design and its performance for a transmit-only $^{19}F/^{1}H$ dual-frequency saddle coil for mouse imaging at 4.7 T [43]. Figure 2.7b–d depicts the performance measurement for the dual-frequency saddle coil at its working frequencies.

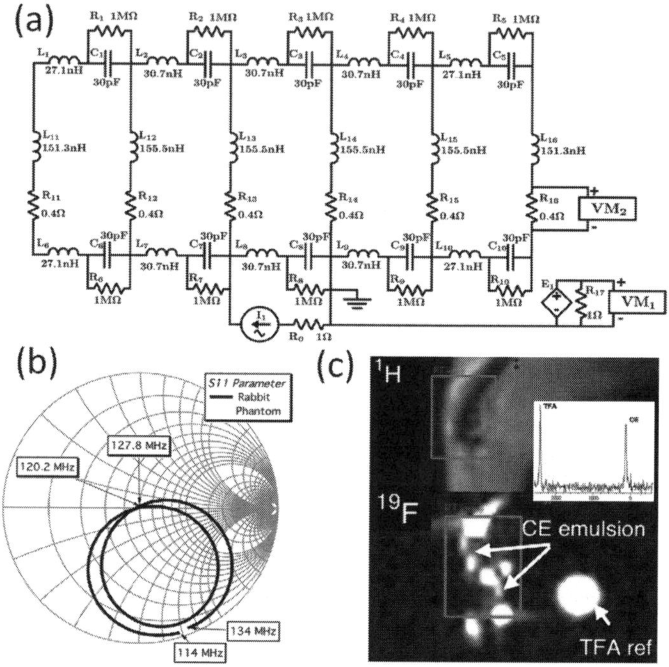

Figure 2.6 Circuit design of $^{19}F/^{1}H$ RF coil. (a) Electronic circuit design of a 140° arc length $^{19}F/^{1}H$ dual-frequency open coil at 3 T. SPICE simulation can be used to evaluate the electrical performance of the coil [38]. (b) Smith-chart of S11 measurement of the coil demonstrates good impedance matching at both ^{19}F and ^{1}H frequency. (c) Spectral lines corresponding to the measured ^{19}F resonance from perfluoro-15-crown-5-ether NP injected into the rabbit knee and a trifluoroacetic acid (TFA) reference standard next to the knee.

The coil achieves good impedance matching at both ^{1}H and ^{19}F frequencies through careful adjustment of four tunable capacitors (there are only four degrees of freedom for solving a dual frequency tuning and matching problem in general [43]). When the PIN diode is switched on, the saddle coil can be detuned from both ^{1}H and ^{19}F frequencies.

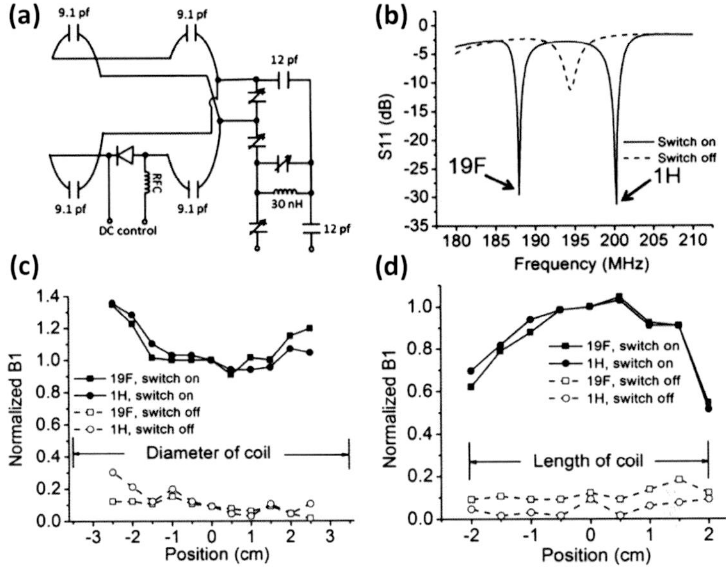

Figure 2.7 Coil circuit design and performance for a transmit-only $^{19}F/^{1}H$ dual-frequency saddle coil for mouse imaging. (a) Circuit design of a $^{19}F/^{1}H$ dual-frequency actively decoupled transmit-only saddle coil for mouse image at 4.7 T. (b) S11 of actively decoupled dual-frequency saddle coil when switched on and off. (c and d) Normalized B_1 field along the diameter and axis of the actively decoupled dual-frequency saddle coil when switched on and off.

2.3.2 MR System and RF Coil Calibration for Simultaneous $^{19}F/^{1}H$ Imaging

For ^{19}F MRI to mature as a robust detection approach for noninvasively quantifying molecular binding events in various locations in the body, it is important to establish a reliable RF power calibration strategy for dual-frequency $^{19}F/^{1}H$ RF coils. Although ^{19}F MRI is potentially quantitative in nature because ^{19}F spin density (and signal) is linearly correlated to the deposition and resultant concentration of ^{19}F nuclei at a particular site in the body [44], the imaging chain of a clinical MR system comprises many steps that can confound absolute signal calibration, even when specialized transmit/receive (T/R) coils are used [43]. For example, RF drive power (i.e., flip angle) settings must be calibrated for ^{19}F tracers that are present in miniscule concentrations and sparsely distributed in the body which yield weak signals. In addition, B_1 inhomogeneity that is inherent to many T/R RF coils must be accounted for [45, 46, 55]. Because the concentration of ^{19}F nuclei within a voxel is usually determined through a comparison to an external standard of known ^{19}F concentration, local differences in the B_1 field between the targeted region of interest and the location of the standard can lead to errors in the quantified ^{19}F signal.

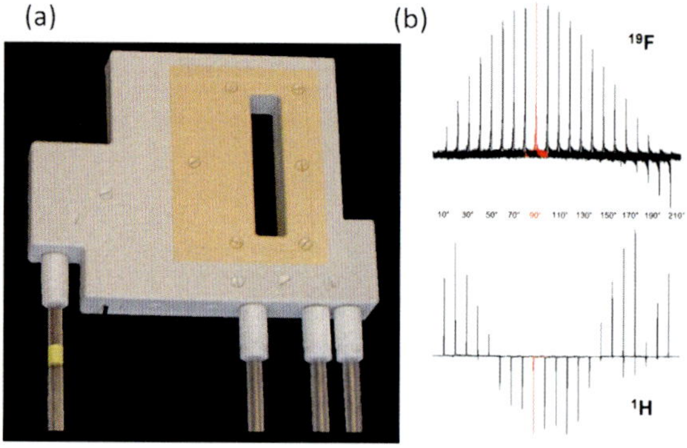

Figure 2.8 Flip angle (FA) sweep using a ^{19}F/^{1}H dual-frequency surface coil at 3 T. (a) Dual-frequency ^{19}F/^{1}H surface coil for 3 T imaging. (b) Representative FA sweep (10°–210°) on ^{19}F nucleus (top) and ^{1}H nucleus (bottom) using the same power setting (peak power = 122.9 W) with a ^{19}F/^{1}H dual-tuned surface coil.

Proper scale factors can be determined at ^{19}F and ^{1}H frequencies to achieve optimized RF power so that a requested 90° flip angle (FA) yields maximum magnetization in FA sweep measurements of a point source PFC phantom. Figure 2.8 shows a representative FA sweep using a ^{19}F/^{1}H dual-frequency surface coil at 3 T that was optimized for ^{19}F (top) and the resultant ^{1}H FA sweep (bottom); the same power setting (peak power = 122.9 W) was too high for ^{1}H, yielding a maximum ^{1}H signal intensity at what was expected to be 20°. These plots exemplify a possible difference in ptimal power settings between ^{19}F and ^{1}H nuclei. In this case the ratio of optimum power levels for ^{19}F and ^{1}H was (1.48 ± 0.06). However, due to the dual-tuned coil design, the coil provides identical spatial B_1 field distributions at both frequencies if driven at the respective optimal power levels. Once determined for a specific coil, the power level ratio can be used to indirectly calibrate the ^{19}F power level via standard ^{1}H calibration procedures [55].

Dual-frequency $^{19}F/^{1}H$ coils enable simultaneous transmit and receive of RF energy at both frequencies. However, low concentrations of fluorinated agents in the body, even in the absence of any physiological background signal, and in conjunction with varying RF coil sensitivity profiles (i.e., B_1-field inhomogeneity), creates obstacles for optimized imaging sensitivity and accurate quantification. As mentioned previously, one major advantage of the coupled resonator model is that it generates identical B_1 profiles at both resonant frequencies, even though there may be differences in power deposition at both frequencies. Therefore, it is feasible to quantify the sparse ^{19}F signal more accurately with a ^{1}H image-based actual flip angle (AFI) [47] with the same B_1 compensation applied to both ^{19}F and ^{1}H images.

Using the flip angle map where AFI = $\alpha_{\text{requested}}/\alpha_{\text{nominal}}$, and a model of the balanced gradient echo signal as per Eq. 4.24 in Vlaardingerbroek et al. [48], a spatially dependent calibration mask (ρ, Eq. 2.4) can be calculated to compensate for the ^{1}H and ^{19}F signal intensities of a balanced steady-state (bSSFP) molecular imaging sequence [49] by dividing each image by ρ, on a voxel-by-voxel basis.

$$\rho = \text{AFI} * \sin(\text{AFI} * \alpha_{\text{nom}}) \frac{1 - E_1}{1 - E_1 E_2 - (E_1 - E_2)\cos(\text{AFI} * \alpha_{\text{nom}})} \quad (2.4)$$

where $E_1 = e^{-\text{TR}/T1}$; $E_2 = e^{-\text{TE}/T2}$.

Figure 2.9 illustrates a potential application of B_1-mapping correction in targeted molecular $^{19}F/^{1}H$ imaging. $\alpha_v\beta_3$-integrin targeted PFCE nanoparticle emulsion was injected into New Zealand White rabbits implanted with VX2 tumors. A simultaneous $^{19}F/^{1}H$ 3D balanced ultrashort echo time (UTE) steady-state free precession (SSFP) sequence (see Section 2.4.1) with Wong-type 3D radial readout [33] was used in conjunction with a B1-mapping correction to perform quantitative ^{19}F imaging in vivo. After correction of the ^{19}F signal, the measured concentration of nanoparticles when compared to a standard was 25.5 ± 0.10 mM ^{19}F, versus 20.0 ± 0.12 mM ^{19}F before correction, a 25% change.

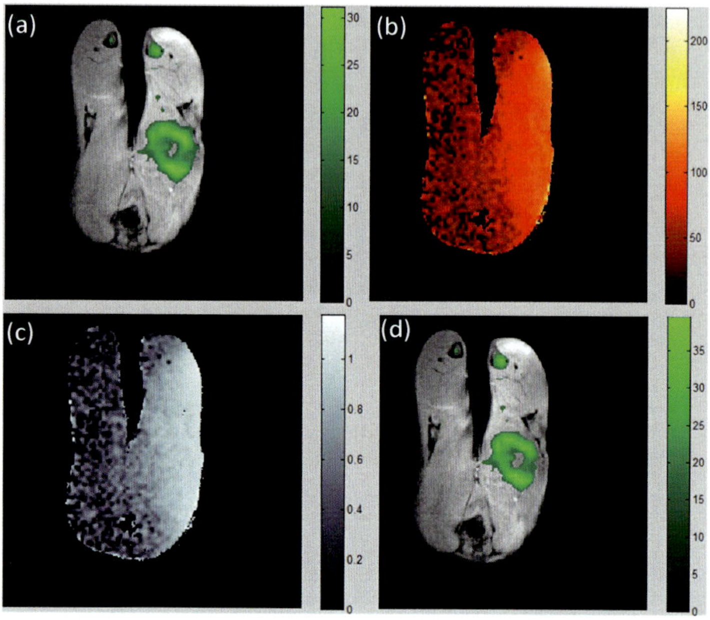

Figure 2.9 B_1-mapping compensation of in vivo cancer model in rabbit with the use of $\alpha_v\beta_3$-integrin-targeted PFC NP nanoparticles captured by ^{19}F MRI with a $^{19}F/^1H$ dual-tuned surface coil (located at image right). In the uncorrected ^{19}F image overlaid on the high-resolution ^1H image (a), concentration was quantified as 20.0 ± 0.12 mM ^{19}F, compared to an external standard of known concentration (150 mM ^{19}F). Illustrated are: B_1-field mapping with AFI (% actual/requested FA) (b) and calibration mask calculated from a balanced UTE-SSFP signal model (c). In the corrected ^{19}F image overlaid on the high-resolution ^1H image (d), PFC concentration was quantified as 25.5 ± 0.10 mM ^{19}F.

2.4 Advanced MR Sequences for Simultaneous $^{19}F/^1H$ Imaging

Imaging fluorinated compounds poses unique challenges for designing MRI acquisition methodologies. First, in the context of functional and molecular imaging in which most fluorinated agents are used, the ^{19}F signal is several orders of magnitude weaker than that of ^1H. As such, the biggest challenge for ^{19}F MRI is achieving sufficient SNR for detection, as opposed to improving imaging contrast as in conventional ^{1}H imaging. The ability to visualize targeted PFC NP at a minimal dose level is a pre-requisite for the translation of these novel contrast agents into clinical use. Secondly, fluorinated compounds typically exhibit complex chemical structures with large chemical shifts. Therefore, different fluorine nuclei within a particular fluorinated agent may resonate at slightly different frequencies (chemical shift) and the resultant weak J-coupling among them may complicate signal patterns in conventional spin echo based imaging sequences.

In this section, we review ^{19}F imaging sequences that have demonstrated high sensitivity in molecular and functional $^{19}F/^{1}H$ imaging, including balanced ultrashort TE steady-state free precession (UTE-SSFP), fluorine ultra-fast turbo spectroscopic imaging (F-uTSI), and blood-flow enhanced saturation recovery (BESR). These sequences were designed to achieve high SNR efficiency and offer dedicated compensation mechanisms for the complex chemical shift spectra of fluorine imaging agents.

2.4.1 Balanced Ultrashort TE Steady State-Free Precession Sequence

As previously mentioned, PFC presents a rich source of ^{19}F signal but unfortunately also may exhibit complex fluorine spectra that complicate imaging studies. An important example is PFOB (perfluoro-octyl bromide, C_8BrF_{17}). Historically, PFOB has been imaged on the CF_3 group, which can be isolated by narrow bandwidth excitation of the single peak. However this approach captures only three of the total 17 ^{19}F nuclei comprising the signal. A simultaneous 3D $^{19}F/^{1}H$ balanced UTE-SSFP pulse sequence [49] was designed to capture the CF_2 resonances from PFOB (Fig. 2.10), which contain 12 of the 17 fluorine nuclei, resulting in five spectral components (at 3 T: 0, ± 100 Hz, ± 500 Hz). As shown in Fig. 2.10c, the five proximate chemical shift (CS) components of the CF_2 group lead to destructive signal overlay at larger echo times (e.g., 2.8 ms). However, all CS components remain within a phase range of ±90° for 0.5 ms and are not yet significantly affected by apparent T_2 relaxation. The UTE-SSFP sequence consists of simultaneous $^{19}F/^{1}H$ RF excitation [32] and subsequent FID acquisition at an ultrashort echo time (~90 μs), using balanced gradients with a Wong-type [33] radial readout trajectory. UTE excitation and FID acquisition are used to acquire the ^{19}F signal before dephasing develops, while balanced SSFP gradients are utilized to exploit the achievable high steady-state signal. The simultaneous ^{1}H excitation and acquisition is not necessarily part of the sensitive ^{19}F detection sequence but is beneficial for an efficient scan time and precise co-localization of fluorine signals with the underlying anatomy.

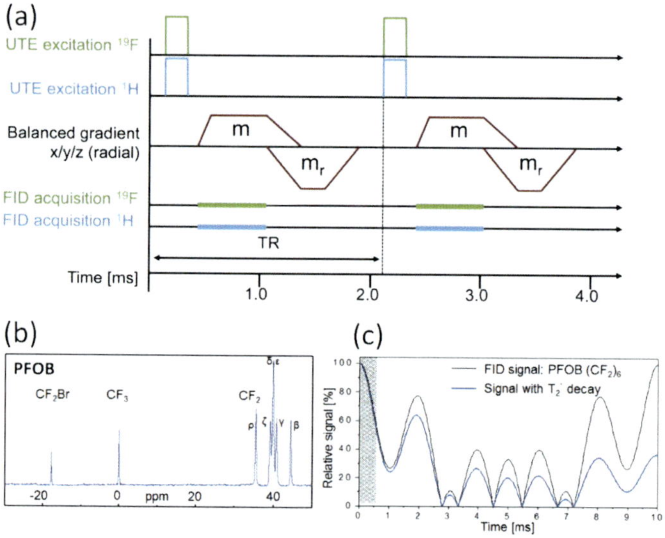

Figure 2.10 Pulse sequence for imaging complex ^{19}F spectra. (a) A simultaneous 3D $^{19}F/^{1}H$ balanced UTE-SSFP pulse sequence, consisting of simultaneous $^{19}F/^{1}H$ RF excitation and subsequent FID acquisition at an ultrashort echo time, using balanced gradients with a Wong-type [33] radial readout trajectory. Logical gradient lobes (m, m_r) are superimposed into a single continuous gradient waveform when executed on the physical gradient coils. (b) Perfluorooctyl bromide (PFOB: CF_3-$(CF_2)_6$-CF_2Br) ^{19}F spectrum. (c) ^{19}F signal evolution of the $(CF_2)_6$ line group with and without apparent T_2 relaxation. During a fast FID readout as in the balanced UTE-SSFP technique (shaded region), the relative signal remains above 60%, which cannot be recovered for later echo times.

Table 2.1 summarizes the experimentally measured detection sensitivity for several commonly used pulse sequence in conventional ^{1}H imaging and simultaneous $^{19}F/^{1}H$, including the balanced UTE-SSFP technique. With $S = 51$ μmol $PFOB^{-1}min^{-1/2}$, the balanced UTE-SSFP technique demonstrates a sensitivity of at least twice that of other sequence types. 3D radial UTE gradient echo sequences without balanced gradients at $\alpha = 30°$ and $\alpha = 5°$ (Ernst angle) exhibit substantially lower sensitivities (20 and 8 μmol $PFOB^{-1}min^{-1/2}$, respectively).

Table 2.1 Sensitivity of ^{19}F MR acquisition techniques

PFOB Line(s)	^{19}F Sequence	Sensitivity ($\mu mol_{PFOB}{}^{-1}min^{-1/2}$)[a]
CF_3	Cartesian fast spin-echo	7
	Cartesian gradient-echo	12
$(CF_2)_6$	Cartesian fast spin-echo	16
	Cartesian balanced SSFP	23
	Radial UTE gradient-echo ($\alpha = 5°$)[b]	8
	Radial UTE gradient-echo ($\alpha = 30°$)	20
	Radial balanced UTE-SSFP	51

[a]Sensitivity measured as $S = SNR \times (mol/voxel)^{-1} \times T_{exp}^{-1/2}$, where SNR is the achieved signal-to-noise ratio, T_{exp} the sequence duration, and (mol/voxel) the amount of PFOB agent within an imaging voxel.

[b]Ernst angle calculated as $\alpha = \cos^{-1}(\exp(-TR/T1))$, where TR = 3.6 ms, and T_1 was measured as 840 ms.

2.4.2 Fluorine Ultrafast Turbo Spectroscopic Imaging Sequence

Another sequence that can be used effectively to image all fluorine nuclei from PFOB molecules at once is the F-uTSI sequence [50], as developed by Lamerichs et al. It has been demonstrated on the simultaneous $^{19}F/^{1}H$ 3 T MRI scanner prototype. Specifically, the F-uTSI sequence was designed to handle the wide NMR spectrum of PFOB (about 80 ppm, which corresponds to a 10 kHz bandwidth at 3 T). Because of the relatively large gaps between the resonance lines, it is possible to acquire spectra with reasonable resolution using short sampling durations. This short acquisition time allows high numbers of echoes per excitation to be generated (i.e., turbo factor or multi-echo factor) (Fig. 2.11a). The typical acquisition time is as short as 4 ms with a turbo factor of 16. To ensure maximum speed, no slice selective gradients are used during RF excitation and refocusing. Spatial encoding is realized solely by employing phase encoding in 2D (projection imaging) or 3D.

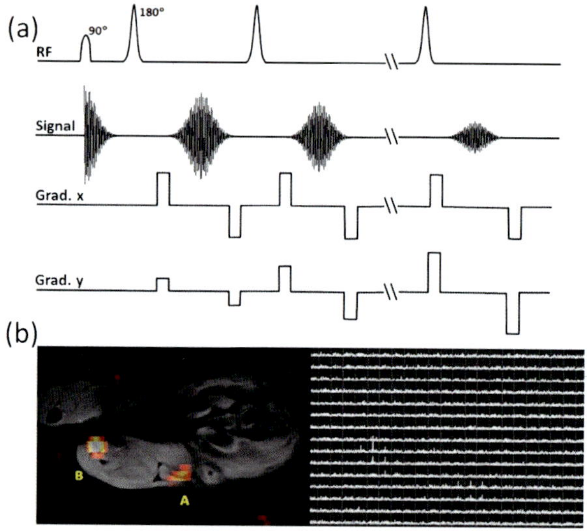

Figure 2.11 (a) F-uTSI pulse sequence diagram. (b) 3D F-uTSI of angiogenesis. Shown is a single slice from a 31 slice 3D data set. The ^{19}F data are shown as a color overlay on the anatomical image (left panel) and full spectra within each voxel (right panel).

The F-uTSI sequence has sufficient sensitivity to detect angiogenesis in vivo. An example (overlaid on a T_1w anatomic scan) is shown in Fig. 2.11b. Active binding of $\alpha_v\beta_3$-integrin targeted PFOB nanoparticle to VX2 tumor implanted in rabbit leg can be detected in vivo at a 3 T field strength. The main advantage of the F-uTSI technique is that, based on integrating defined spectral peaks, the resulting images are free of chemical shift artifacts, therefore, eliminating the need for an accurate determination of the resonance frequencies of the ^{19}F signal.

2.4.3 Blood-Flow Enhanced Saturation Recovery Sequence

The blood flow-enhanced-saturation-recovery (BESR) (Fig. 2.12) sequence was designed by Hu et al. for measuring ^{19}F T_1 of flowing PFC NPs in blood pool and was optimized for rapid measurement of pO_2 in pulsatile blood flow [8]. Previous reports have demonstrated that the slope of ^{19}F, $R_1 = 1/T_1$, of PFC as a function of pO_2 could be 10-fold greater than the dependence of ^{1}H R_1 on pO_2 [51–53]. Therefore, measuring ^{19}F T_1 effectively and efficiently is crucial for further exploring the functional applications of PFC NP. The BESR sequence is composed of two parts that includes homogeneous spin preparation and time-of-flight image acquisition. The spin preparation is achieved with a non-selective whole body saturation pulse train, and the image acquisition is realized with gradient echoes at high repetition rate. Both the pre-pulses and imaging excitation pulses are gated to the ECG with an appropriate trigger delay to minimize the motion artifact associating with the beating heart and vascular contraction and dilation. Therefore, the BESR sequence is able to preserve the time efficiency of the classical Look-Locker sequence and the high SNR of a saturation recovery sequence.

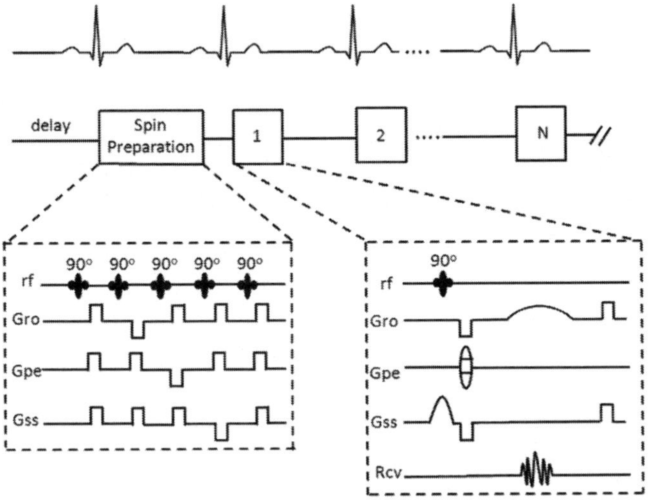

Figure 2.12 Illustration of the blood flow-enhanced-saturation-recovery sequence. Spin preparation is achieved with multiple non-selective saturation pulses; image acquisition uses a gradient echo approach with cardiac triggering. Saturation pre-pulse train and gradient echo imaging are both triggered by ECG. Each image acquisition is acquired at the same trigger delay time t but corresponds to different saturation delay time TS.

Figure 2.13 shows representative BESR images of a mouse heart under conditions of hyperoxia and normoxia. After spin saturation, blood in the left ventricle (LV) exhibits a faster recovery of signal intensity than blood in the right ventricle (RV), which was further accelerated when the mouse was breathing pure O_2 (i.e., hyperoxia). ^{1}H gradient echo cine images were acquired initially to position the slices to cover both LV and RV. Blood ^{19}F T_1 measurements were carried out under separate conditions of ventilation with room air and then pure oxygen. An actively decoupled ^{19}F/^{1}H saddle coil and a ^{19}F/^{1}H surface coil were tuned to the proper frequency of 201.5 and 189.6 MHz for ^{1}H and ^{19}F imaging at 4.7 T, respectively.

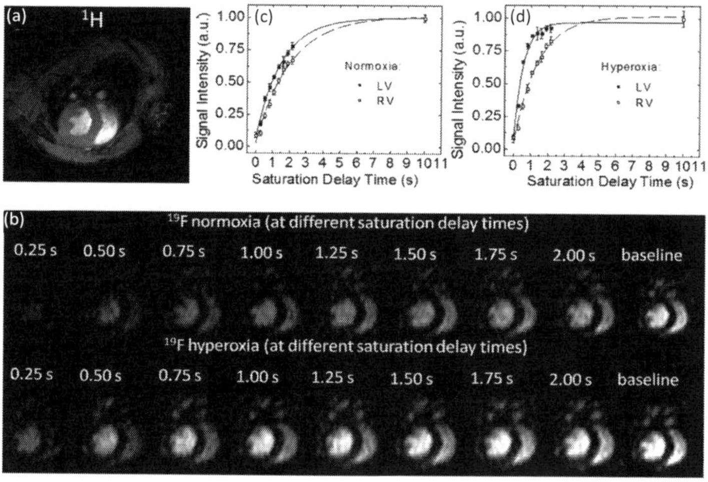

Figure 2.13 In vivo ^{19}F BESR MRI of heart ventricles. (a) A short-axis ^{1}H image shows the LV and RV ventricles. (b) ^{19}F BESR images of the same mouse heart under hyperoxia and normoxia. (c–d) Fitted ^{19}F saturation recovery curve for LV and RV blood under normoxia and hyperoxia. Error bars represent standard deviation.

2.5 Conclusion

Although still at an early stage of development, simultaneous ^{19}F/^{1}H technology can yield clear advantages over traditional sequential ^{1}H and ^{19}F imaging. First, one of the major challenges for utilizing MRI as a molecular imaging modality is posed by the inherent low signal intensity from the targeted material. Therefore, long scan times and extensive signal averaging are typical in the literature of MRI molecular imaging. A systematic anatomical and functional imaging protocol for a clinical patient usually exceeds 30 min, and thus adding an additional 30–60 min to the scan time for molecular imaging would render such technology problematic from a clinical workflow and economic perspective. To this end, simultaneous ^{1}H and ^{19}F signal acquisition and image reconstruction enables more rapid subject scans. Furthermore, utilizing dual tuned coils removes the need for coil exchange, further streamlining the imaging workflow.

Second, simultaneous $^{19}F/^{1}H$ improves the accuracy of molecular imaging and expands the potential translational opportunities for treatment monitoring by enabling quantification of tissue biomarkers. Scanner calibration can be a challenging task when performed on the weak ^{19}F signal. With simultaneous $^{19}F/^{1}H$ however, the opportunity exists to conduct power/flip angle calibration, gain calibration and local sensitivity calibration on the strong and abundant ^{1}H signal and immediately reuse this data for the ^{19}F imaging channel. Effective system calibration is expected to simplify the imaging workflow and to produce more accurate measurements of targeted contrast agents, which in turn generate more precise quantification of potential imaging biomarkers.

Last, similar to other simultaneous imaging technologies (such as simultaneous PET/CT and PET/MR) simultaneous $^{19}F/^{1}H$ MRI offers precise registration between anatomical and molecular information. Acquiring both images at the same time minimizes the potential for misalignment between different scans by patient or physiological motion, especially when each scan requires prolonged data acquisitions in clinical settings. Moreover, the high resolution anatomical information derived from ^{1}H can also be utilized to correct motion artifacts. This feature might be especially relevant for imaging biomarkers in the abdominal and chest regions.

MR scanner architectures, sequences and coils developed thus far for simultaneous $^{19}F/^{1}H$ imaging offer novel opportunities for translation at common clinical field strengths. Targeted NPs may produce high local ^{19}F payloads (10–100 mM of ^{19}F nuclei) which can be detected at 1.5–3 T as reviewed in this chapter. The remaining major problem is the clinical evaluation, approval and production of the ^{19}F imaging agents that could realize clinical ^{19}F molecular imaging.

References

1. Chen J, GM Lanza, and SA Wickline, Quantitative magnetic resonance fluorine imaging: Today and tomorrow. *Wiley Interdiscip Rev Nanomed Nanobiotechnol*, 2010. 2(4), 431–440.
2. Ahrens ET and JW Bulte, Tracking immune cells in vivo using magnetic resonance imaging. *Nat Rev Immunol*, 2013. 13(10), 755–763.
3. Ahrens ET and J Zhong, In vivo MRI cell tracking using perfluorocarbon probes and fluorine-19 detection. *NMR Biomed*, 2013. 26(7), 860–871.
4. Srinivas M, et al., In vivo 19F MRI for cell tracking. *J Vis Exp*, 2013. 81, e50802.
5. Lanza GM, et al., ^{1}H/^{19}F magnetic resonance molecular imaging with perfluorocarbon nanoparticles. *Curr Top Dev Biol*, 2005. 70, 58–76.
6. Caruthers SD, et al., In vitro demonstration using ^{19}F magnetic resonance to augment molecular imaging with paramagnetic perfluorocarbon nanoparticles at 1.5 Tesla. *Invest Radiol*, 2006. 41(3), 305–312.
7. Neubauer AM, et al., Fluorine cardiovascular magnetic resonance angiography in vivo at 1.5 T with perfluorocarbon nanoparticle contrast agents. *J Cardiovasc Magn Reson*, 2007. 9(3), 565–573.
8. Hu L, et al., Rapid quantification of oxygen tension in blood flow with a fluorine nanoparticle reporter and a novel blood flow-enhanced-saturation-recovery sequence. *Magn Reson Med*, 2013. 70(1), 176–183.
9. Ahrens ET, et al., In vivo imaging platform for tracking immunotherapeutic cells. *Nat Biotechnol*, 2005. 23(8), 983–987.
10. Partlow KC, et al., ^{19}F magnetic resonance imaging for stem/progenitor cell tracking with multiple unique perfluorocarbon nanobeacons. *FASEB J*, 2007. 21(8), 1647–1654.
11. Ebner B, et al., Early assessment of pulmonary inflammation by ^{19}F MRI in vivo. *Circ Cardiovasc Imaging*, 2010. 3(2), 202–210.
12. Helfer BM, et al., Functional assessment of human dendritic cells labeled for in vivo ^{19}F magnetic resonance imaging cell tracking. *Cytotherapy*, 2010. 12(2), 238–250.

13. Bonetto F, et al., A novel ^{19}F agent for detection and quantification of human dendritic cells using magnetic resonance imaging. *Int J Cancer*, 2011. 129(2), 365–373.

14. Hitchens TK, et al., ^{19}F MRI detection of acute allograft rejection with in vivo perfluorocarbon labeling of immune cells. *Magn Reson Med*, 2011. 65(4), 1144–1153.

15. Lanza GM, et al., Targeted antiproliferative drug delivery to vascular smooth muscle cells with a magnetic resonance imaging nanoparticle contrast agent: Implications for rational therapy of restenosis. *Circulation*, 2002. 106(22), 2842–2847.

16. Morawski AM, et al., Quantitative "magnetic resonance immunohistochemistry" with ligand-targeted ^{19}F nanoparticles. *Magn Reson Med*, 2004. 52(6), 1255–1262.

17. Goette M, et al., In vivo quantitative imaging of angiogenesis-targeted PFOB nanoparticles in a hypercholesterol rabbit model using ^{19}F-MRI with ultra-short echo time balanced SSFP. *J Cardiovasc Magn Reson*, 2012. 14(Suppl 1), M8.

18. Southworth R, et al., Renal vascular inflammation induced by Western diet in ApoE-null mice quantified by ^{19}F NMR of VCAM-1 targeted nanobeacons. *Nanomedicine*, 2009. 5(3), 359–367.

19. Schmieder AH, et al., Characterization of early neovascular response to acute lung ischemia using simultaneous $^{19}F/^{1}H$ MR molecular imaging. *Angiogenesis*, 2014. 17(1), 51–60.

20. Waters EA, et al., Detection of targeted perfluorocarbon nanoparticle binding using ^{19}F diffusion weighted MR spectroscopy. *Magn Reson Med*, 2008. 60(5), 1232–1236.

21. Wagner EM, et al., Angiogenesis and airway reactivity in asthmatic Brown Norway rats. *Angiogenesis*, 2015. 18(1), 1–11.

22. Leung K, VHPKQHRGGSKGC-liquid perfluorocarbon nanoparticles, in *Molecular Imaging and Contrast Agent Database (MICAD)*. 2004, Bethesda (MD).

23. Lanza GM, et al., Theragnostics for tumor and plaque angiogenesis with perfluorocarbon nanoemulsions. *Angiogenesis*, 2010. 13(2), 189–202.

24. Spahn DR, Blood substitutes. Artificial oxygen carriers: Perfluorocarbon emulsions. *Crit Care*, 1999. 3(5), R93–R97.

25. Zhang H, et al., Quantifying the evolution of vascular barrier disruption in advanced atherosclerosis with semipermeant nanoparticle contrast agents. *PLoS One*, 2011. 6(10), e26385.

26. Hu L, et al., Assessing intrarenal nonperfusion and vascular leakage in acute kidney injury with multinuclear $^1H/^{19}F$ MRI and perfluorocarbon nanoparticles. *Magn Reson Med*, 2014. 71(6), 2186–2196.

27. Hu LZ, et al., Thrombin-inhibiting perfluorocarbon nanoparticles reduce intrarenal coagulation in acute kidney injury and provide image-based readouts (^{19}F MRI) of renal endothelial disruption. *Circulation*, 2012. 126(21), A16488.

28. Mason RP, et al., Regional tumor oxygen tension: Fluorine echo planar imaging of hexafluorobenzene reveals heterogeneity of dynamics. *Int J Radiat Oncol Biol Phys*, 1998. 42(4), 747–750.

29. Bulte JW, Hot spot MRI emerges from the background. *Nat Biotechnol*, 2005. 23(8), 945–946.

30. Higuchi M, et al., ^{19}F and ^1H MRI detection of amyloid beta plaques in vivo. *Nat Neurosci*, 2005. 8(4), 527–533.

31. Caruthers S, et al., MR molecular imaging of angiogenesis using targeted perfluorocarbon nanoparticles. *Medicamundi*, 2010. 54(2), 7.

32. Keupp J, et al., Simultaneous dual-nuclei imaging for motion corrected detection and quantification of ^{19}F imaging agents. *Magn Reson Med*, 2011. 66(4), 1116–1122.

33. Wong ST and MS Roos, A strategy for sampling on a sphere applied to 3D selective RF pulse design. *Magn Reson Med*, 1994. 32(6), 778–784.

34. Glover GH, JM Pauly, and KM Bradshaw, Boron-11 imaging with a three-dimensional reconstruction method. *J Magn Reson Imaging*, 1992. 2(1), 47–52.

35. Thevenaz P, UE Ruttimann, and M Unser, A pyramid approach to subpixel registration based on intensity. *IEEE Trans Image Process*, 1998. 7(1), 27–41.

36. Mazurkewitz PC, et al., A Double-Resonant $^{19}F/^1H$ Transmit/Receive Solenoid Coil for MRI. *Proc. Intl. Soc. Mag. Reson. Med.*, 2006. 14, 2596.

37. Haase J, NJ Curro, and CP Slichter, Double resonance probes for close frequencies. *J Magn Reson*, 1998. 135(2), 273–279.

38. Hockett FD, et al., Simultaneous dual frequency ^1H and ^{19}F open coil imaging of arthritic rabbit knee at 3T. *IEEE Trans Med Imaging*, 2011. 30(1), 22–27.

39. Leussler C, Mazurkewitz PC and Keupp J, Multi-resonant t/r antenna for MR image generation. 2013. Patent WO2013108142 A1.

40. Hockett F, et al., Simultaneous dual frequency and open coil imaging of arthritic rabbit knee at 3T. *IEEE Trans Med Imaging*, 2010.

41. Mispelter J, M Lupu, and A Briguet, *NMR Probeheads for Biophysical and Biomedical Experiments: Theoretical Principles & Practical Guidelines*. 2006, London Hackensack, NJ: Imperial College Press; distributed by World Scientific. xiv, 596 p.
42. Giraudeau C, et al., A new paradigm for high-sensitivity ^{19}F magnetic resonance imaging of perfluorooctylbromide. *Magn Reson Med*, 2010. 63(4), 1119–1124.
43. Hu L, et al., A generalized strategy for designing $^{19}F/^{1}H$ dual-frequency MRI coil for small animal imaging at 4.7 Tesla. *J Magn Reson Imaging*, 2011. 34(1), 245–252.
44. Morawski AM, et al., Quantitative "magnetic resonance immunohistochemistry" with ligand-targeted ^{19}F nanoparticles. *Magn Reson Med*, 2004. 52(6), 1255–1262.
45. Keupp J and Mazurkewitz PC. Simultaneous $^{19}F/^{1}H$ imaging for quantification: Calibration and sensitivity assessment. *Proc. Intl. Soc. Mag. Reson. Med.*, 2007. 15, 1334.
46. Sung K, BL Daniel, and BA Hargreaves, Transmit B1+ field inhomogeneity and T_1 estimation errors in breast DCE-MRI at 3 tesla. *J Magn Reson Imaging*, 2013. 38(2), 454–459.
47. Yarnykh VL, Actual flip-angle imaging in the pulsed steady state: A method for rapid three-dimensional mapping of the transmitted radiofrequency field. *Magn Reson Med*, 2007. 57(1), 192–200.
48. Vlaardingerbroek MT and JAD Boer, *Magnetic Resonance Imaging: Theory and Practice*. 3rd ed. 2003, Berlin; New York, Springer. xxi, p. 499.
49. Goette MJ, et al., Balanced UTE-SSFP for F MR imaging of complex spectra. *Magn Reson Med*, 2015. 74(2), 537–543.
50. Yildirim M, et al., Chemical Shift Independent Imaging of ^{19}F Contrast Agents Using Ultrafast MRSI (F-uTSI). *Proc. Intl. Soc. Mag. Reson. Med.*, 2007. 15, 1249.
51. Duong TQ and SG Kim, In vivo MR measurements of regional arterial and venous blood volume fractions in intact rat brain. *Magn Reson Med*, 2000. 43(3), 393–402.
52. Eidelberg D, et al., ^{19}F NMR imaging of blood oxygenation in the brain. *Magn Reson Med*, 1988. 6(3), 344–352.
53. Liu S, et al., Quantitative tissue oxygen measurement in multiple organs using ^{19}F MRI in a rat model. *Magn Reson Med*, 2011. 66(6), 1722–1730.

54. Keupp J, et al., Hadamard-type pulse-phase encoding for imaging of multi-resonant Fluorine-19 nanoparticles in targeted molecular MRI. *Proc. Intl. Soc. Mag. Reson. Med.*, 2010. 18, 982.

55. Goette MJ, et al., Improved quantitative ^{19}F MR molecular imaging with flip angle calibration and B1-mapping compensation. *J Magn Reson Imaging*, 2015. 42(2), 488–494.

Chapter 3

Hyperpolarization for Signal Enhancement in Fluorine MR Applications

Ute Bommerich,[a,b] Johannes Bernarding,[a] Denise Lego,[b] Thomas Trantzschel,[a] and Markus Plaumann[a]

[a]*Department for Biometrics and Medical Informatics, Otto von Guericke University Magdeburg, Germany*
[b]*Leibniz Institute for Neurobiology, Magdeburg, Germany*

ute.bommerich@med.ovgu.de

3.1 Introduction

^{19}F MR techniques very often have to contend with small signal intensities resulting from low spin densities and the general low sensitivity of magnetic resonance. In recent decades, different approaches have been developed to overcome this sensitivity issue by increasing the MR signal significantly beyond the limit of Boltzmann polarization (cf. Fig. 3.1). These hyperpolarization (HP) techniques, such as optical pumping, dynamic nuclear polarization (DNP), chemically induced dynamic nuclear polarization (CIDNP), and parahydrogen-induced polarization (PHIP), utilize various physical concepts to achieve HP not only

Fluorine Magnetic Resonance Imaging
Edited by Ulrich Flögel and Eric Ahrens

ISBN 978-981-4745-31-4 (Hardcover), 978-981-4745-32-1 (eBook)
www.panstanford.com

for the commonly used hydrogen nucleus, but also for heteronuclei [1]. This chapter provides a brief overview of the history of chemical and biochemical applications documented in ^{19}F HP studies, as well as the mechanisms behind these different HP techniques.

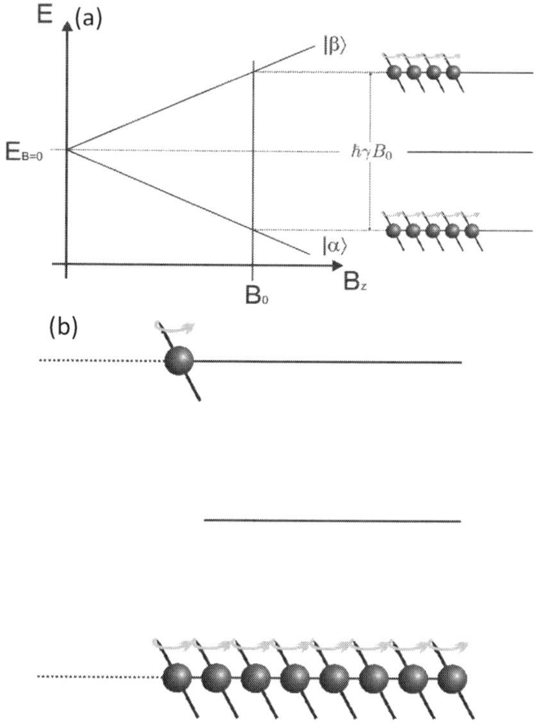

Figure 3.1 Concept of HP. (a) Population in case of thermal Boltzmann-polarization and (b) HP with high population differences.

3.2 Hyperpolarization Techniques: History and Physical Principles

To reiterate, MR applications are often limited by very low sensitivity compared to other methods like optical spectroscopy or positron emission tomography (PET). This is caused by the source of the MR signals—the population difference of the nuclear energy levels—which is usually very small. Considering the two nuclear Zeeman levels α and β of a proton in an external magnetic field (Fig. 3.1a), the population excess of the lower level at common field strengths is only a few ppm. Here, the polarization[1] in thermal equilibrium follows from the Boltzmann statistic and, assuming high temperature approximation, is expressed by

$$P_\mathrm{B} = \frac{n_\alpha - n_\beta}{n} \approx \frac{\hbar\gamma B_0}{2k_\mathrm{B}T}. \tag{3.1}$$

Therefore, this value is even smaller for low-γ nuclei. Instead of increasing the external magnetic field B_0 or decreasing temperature T to increase MR signals, the approach taken in these various HP methods is to exploit the potential of a higher polarization. This is achieved by generating spin ensembles with strong non-equilibrium population of the energy levels (Fig. 3.1b). The established methods for signal enhancement in ^{19}F MR are DNP, CIDNP, and PHIP, which are discussed in the following sections.

3.2.1 Dynamic Nuclear Polarization

DNP utilizes the fact that the polarization of an unpaired electron in a certain magnetic field and at a certain temperature is notably larger than that of any nucleus, due to its much higher magnetic moment (e.g., 8.6 T, 3 K: P_electron = 95%, P_proton = 0.06%). The transfer of this polarization to neighboring nuclei can be induced by microwave-driven resonant electron spin transitions. In 1953, Overhauser predicted the possibility of transferring the strong polarization of free unpaired electrons to nuclei in metals [2], and Carver and Slichter confirmed the Overhauser effect experimentally in metallic lithium [3]. Two years later, Abragam extended the concept to nonmetals, using electron spins in solution [4, 5]. In liquids, the Overhauser effect is the only operative DNP mechanism, while in solid-state experiments, other mechanisms also occur: the solid effect, the cross effect, and thermal mixing, all of which were discovered in the subsequent two decades [6–10]. Thereafter, DNP became an important tool for solid-state research in chemistry and for the investigation of biological structures such as large proteins and membranes

[1]The polarization of a sample is usually stated as the percentage: $P\% = P \cdot 100\%$.

[10–13]. A newer approach for generating high polarizations even in solution is dissolution DNP. In this approach, the cooled sample (<4 K) is irradiated in a strong magnetic field, a process which is followed by a rapid dissolution or melting step [14]. Through this procedure, biologically relevant substrates in aqueous solution can be hyperpolarized, allowing them to be used as contrast agents in biomedical research. The advent of commercial setups for dissolution DNP (e.g., Oxford Instruments HyperSense) recently propelled this method to a first clinical trial, where ^{13}C-hyperpolarized pyruvate was monitored in prostate cancer patients [15]. The largest achieved polarizations were on the order of 40% for ^{13}C and 9% for ^{19}F [16].

As mentioned earlier, the strong polarization of unpaired, or free electrons can be transferred to nuclei due to four DNP mechanisms (Overhauser, solid, and cross effect, and thermal mixing), which depend on a variety of experimental conditions, in particular, the state of matter, in which the polarization is carried out (liquid or solid) and the choice of radicals. These radicals serve as the polarizing agents, providing stable unpaired electrons and notably, determining the proceeding DNP mechanism. These agents can be either endogenous, appearing in samples with intrinsic free electrons, or exogenous, e.g., added into the sample as organic radicals yielding a homogeneous mixture. In terms of the transfer mechanism, the ESR (electron spin resonance) characteristics (such as spectral width $\Delta\omega_e$ but also concentration, solubility, and relaxivity) are what particularly affect the experiment. As a detailed discussion is far beyond the scope of this chapter, we list only a few basic aspects here [17–19].

- $\omega_n > \Delta\omega_e$. Regime of solid effect, where irradiation induces forbidden transitions (ω_n is the nuclear resonance frequency). Examples are triarylmethyl (trityl) and bis-α, γ-diphenylene-β-phenyl-allyl (BDPA), which exhibit narrow ESR lines.
- $\omega_n < \Delta\omega_e$. Regime of cross effect, which does not involve forbidden transitions and is therefore more effective than the solid effect. Here, biradicals (e.g., 1-(TEMPO-4-oxy)-3(TEMPO-4-amino)propan-2-ol (TOTAPOL)) can be applied, where the electron spin lattice relaxation time (T_{1e}) is shorter than the interaction time of different paramagnetic centers (T_{2e}).

- $\omega_n << \Delta\omega_e$. Regime of thermal mixing, where strong electron spin–spin interactions are ensured by agents such as 2,2,6,6-tetramethylpiperidin-1-oxyl (TEMPO) being added in high concentrations (>10 mM).

The transfer itself is driven by an alternating electromagnetic field (at or close to the electron Larmor frequency in the microwave range) and nuclear and electron spin transitions, which occur because of irradiation and relaxation [20]. Here, technical factors such as microwave frequency and power, as well as the properties of the particular spin system, determine the transfer efficiency [17].

The first discovered Overhauser effect relies on the relaxation of a saturated electron spin system that is driven by scalar and dipolar electron-nuclear interactions and results in the polarization of the nuclear spins. As shown in Fig. 3.2, this relaxation is a product of zero, single, and double quantum transitions (ZQ, SQ, and DQ) with corresponding probabilities W_0, $W_{n,e}$, and W_2. After saturation of the electron SQ transitions by microwave irradiation with ω_e, the nuclear spin populations are affected by ZQ and DQ transitions, which leads to a non-equilibrium population of the nuclear spin states that itself results in NMR signals enhanced by a factor ε [17, 20, 21].

$$\varepsilon = 1 - \xi f s \frac{\gamma_e}{\gamma_n} \tag{3.2}$$

The coupling parameter ξ, the leakage factor f and the saturation factor s are defined as

$$\xi = \frac{\sigma}{\rho} = \frac{W_2 - W_0}{W_0 + 2W_n + W_2}; \; f = \frac{\rho}{\rho + W_n^0}; \; s = \frac{S_0 - \langle S_z \rangle}{S_0}, \tag{3.3}$$

where σ and ρ are the cross-relaxation and auto-relaxation rates, and W_n^0 is the longitudinal nuclear relaxation rate in absence of a radical. $\langle S_z \rangle$ is the longitudinal nuclear magnetization and S_0 is the corresponding value at thermal equilibrium [20]. The coupling factor ξ can take values between −1 (pure scalar relaxation) and +0.5 (pure dipolar relaxation) [17]. The saturation factor s reflects the saturation of the electronic levels by irradiation.

Equations (3.3) demonstrate that for the Overhauser DNP effect, a difference in the transition probabilities of ZQs and DQs is required. As the effect of molecular motion (quantified by the correlation time τ) on the transition probabilities W depends on the particular Larmor frequencies (ω_e and ω_n), the coupling factor, like the enhancement factor, shows a field dependence. Hence, an important condition for the efficiency of the Overhauser effect is $\tau\omega_e < 1$ [17]. Therefore, Overhauser DNP enhancements are higher at lower magnetic fields. The prerequisite of coupling due to molecular motion-induced relaxation can be fulfilled by mobile electrons in solids (e.g., electrons in the conduction bands of metals). In liquids, where the Overhauser effect is the only observable DNP mechanism, the described requirement is also met by rotational and translational motion (dipolar interaction) or chemical exchange/fast relaxation of one spin (scalar interaction) [17].

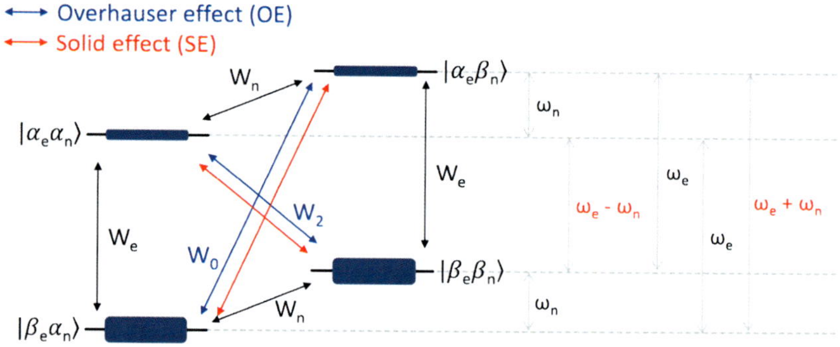

Figure 3.2 Energy levels of an electron-nuclear two-spin system (e and n spin 1/2) with scalar (hyperfine) and dipolar couplings. W denotes transition probabilities, and ω the energy differences in units of frequencies. The blue boxes qualitatively represent the Boltzmann-distributed population in thermal equilibrium.

In the two-spin system depicted in Fig. 3.2, the solid effect can be explained by the energy differences given in units of frequency. In contrast to the Overhauser effect, the transfer here is achieved by specifically exciting forbidden transitions, either ZQ or DQ. This is done by irradiating with a frequency of $\omega_e + \omega_n$ or $\omega_e - \omega_n$ leading to a negative or a positive nuclear enhancement, respectively. Consequently, the (homogeneous and inhomogeneous) bandwidth of the EPR resonance has to be much smaller than ω_n to excite only one of the two transitions [20]. The enhancement of the Overhauser effect scales with ω_n^{-2} decreasing the efficiency at high fields [22].

In order to understand the cross effect, a three-spin system (two dipolar-coupled electrons and one nuclear spin 1/2) has to be considered. In addition, the prerequisite $|\omega_{e1} - \omega_{e2}| = \omega_n$, which corresponds to the degeneracy of either the states $|\alpha_e\beta_e\alpha_n\rangle$ and $|\beta_e\alpha_e\beta_n\rangle$ (as shown in Fig. 3.3) or $|\beta_e\alpha_e\alpha_n\rangle$ and $|\alpha_e\beta_e\beta_n\rangle$, has to be fulfilled. This can be achieved experimentally by working with high radical concentrations (biradicals) [20]. The cross effect is induced by irradiation with ω_{e1} or ω_{e2}, whereby the corresponding SQ transitions of one electron are saturated, as well as the ZQ or DQ transitions of the other electron and the nuclear spin. For the sake of clarity, only the first situation is illustrated in Fig. 3.3. Together with the mixing of the degenerated states, this saturation leads to the equal population of the four ZQ-transition-involved states (polarization of the nuclear spin states). As a consequence, the MR signals show a negative enhancement for the illustrated example. An irradiation with ω_{e2} would, in contrast, result in a positive enhancement [17]. The cross effect enhancement scales with B_0^{-1}.

In samples with higher concentrations of radicals, non-negligible electron spin–spin interactions force another transfer mechanism—thermal mixing. In essence, it can be seen as an extension of the cross effect to an electron spin ensemble that results from state manifolds [20]. Thermal mixing is often explained using the concept of spin temperature, which allows for the treatment of systems with many electrons [23]. Usually, the existence of three thermodynamic ensembles, called reservoirs, is assumed: an electron Zeeman, an electron non-Zeeman, and a nuclear Zeeman reservoir. In the first step, microwave irradiation close to the resonance frequency of the electrons changes the spin temperature of the non-Zeeman reservoir from the equilibrium value (cooling/heating). The large number of electron spin–spin interactions leads to a quasi-continuous broadening of energy levels, which in turn causes a thermal contact with the nuclear Zeeman reservoir (the condition $|\omega_{e1} - \omega_{e2}| = \omega_n$ is fulfilled for any pair of electrons) [24]. This thermal mixing produces a cooling of the nuclear reservoir and in turn enhancing the nuclear polarization. To allow for thermal mixing, the condition $1/T_{2e} >> 1/T_{1e}$ has to be fulfilled, which means that the electron spin–spin interactions take on a relevant role in the process [18].

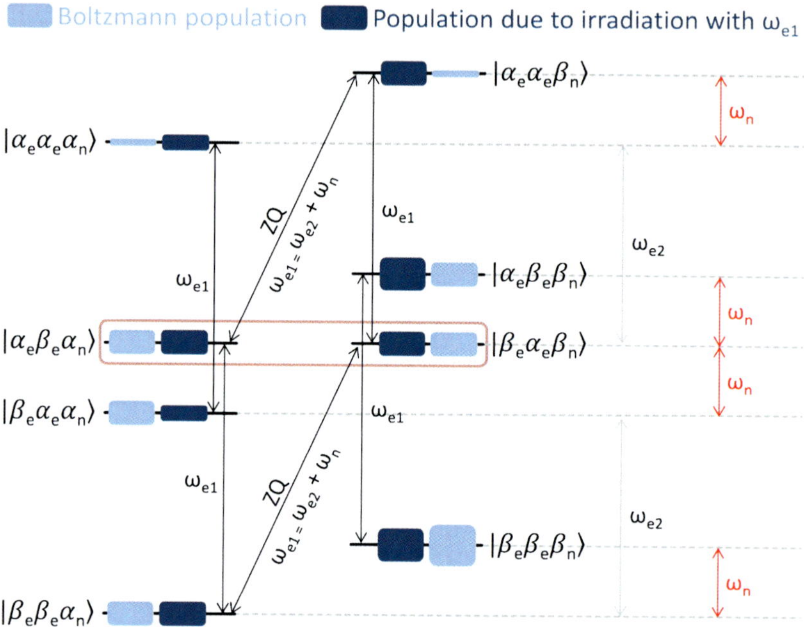

Figure 3.3 Schematic illustration of cross effect. Energy levels of a three-spin system consisting of two dipolar coupled electrons (e) and one nuclear spin 1/2 (n). ZQ denotes the zero quantum transitions of one electron and the nuclear spin induced by irradiation of the other electron. ω is the energy difference in units of frequencies.

It is important to note that the transfer of electron polarization can also be achieved using pulsed DNP approaches, which promise notable advantages. A detailed description of these techniques can be found in [17].

3.2.2 Chemically Induced Dynamic Nuclear Polarization

In CIDNP experiments the polarization forms out of the reactions between radical pairs. The first CIDNP-related findings were independently documented in 1967 by Bargon and

Fischer, as well as Ward and Lawler. Bargon and Fischer investigated the thermal decomposition of azo compounds and peroxides and observed NMR signals in emission [25]. Ward and Lawler investigated reactions between *n*-butyl bromide and *n*-butyllithium, as well as *t*-butyllithium, which yielded 1-butene due to a halogen-metal exchange. The authors reported increased ^{1}H olefin signals [26]. These observations were at first erroneously associated with the Overhauser-effect and dynamic nuclear polarization [27–29]. In 1969, both Kaptein and Closs identified radical pair mechanisms (RPM) as the correct origin of CIDNP signals [30, 31]. From then on, CIDNP was utilized to investigate kinetics, to identify mechanisms or unstable intermediates, and to detect tiny amounts of reaction products in various radical pair reactions (e.g., photo reactions). As later demonstrated, the CIDNP signal formation is determined by radical and radical pair parameters such as hyperfine couplings and g-factors [29, 32]. Beside radical pair generation by thermal dissociation, in 1978 Kaptein established ^{1}H photo-CIDNP to investigate reversible photochemical reactions of excited photosensitizers (e.g., flavine) and aromatic systems (e.g., tyrosine) [29, 33]. After that breakthrough, even biological systems became the focus of CIDNP investigations like: photosystem II, the magnetic field effect in RPMs (discussed as an explanation for magnetoreception), or electron transfer (as source of non-enzymatic DNA repair) [29, 34, 35]. The largest observed signal enhancements in photo-CIDNP experiments are about 0.07% for ^{19}F [36].

CIDNP is strongly related to the singlet and triplet states, which are formed from the interaction of two electrons with spin 1/2 in a radical pair. These radical pairs occur because of their limited motion in the solvent cage. Radicals can be formed either by thermal dissociation or by photolytic excitation of a precursor molecule. With the first, spin-correlated radical pairs with singlet character are produced, while pairs with triplet character are produced by the second [37]. During the lifetime of the solvent cage the correlated radical pair can recombine (recombination/geminate products) or diffuse apart and pass to side reactions (escape/transfer products), which is illustrated for a common RPM in a photo-CIDNP experiment (see Fig. 3.4) [29]. Considering an ensemble of radical pairs in the electronic

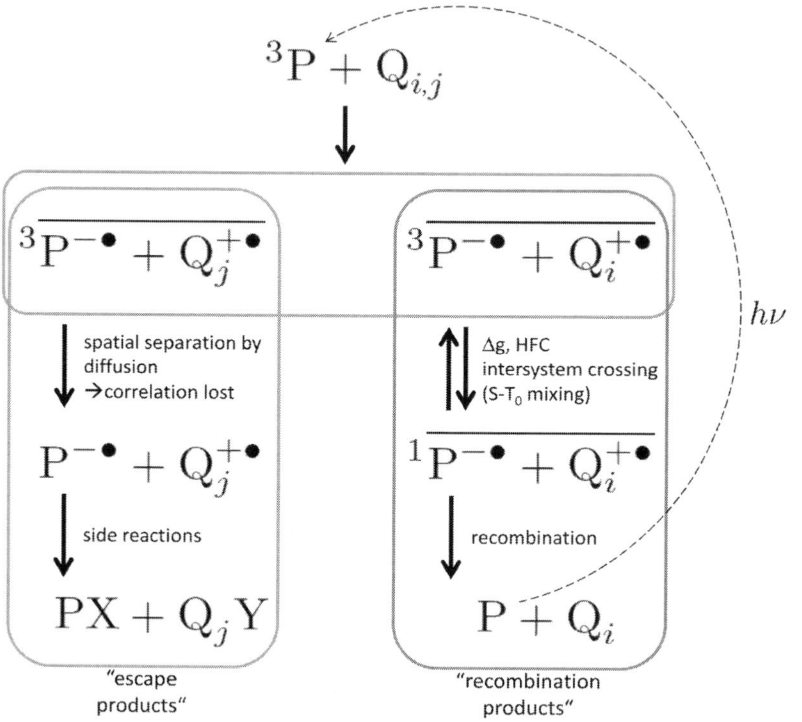

Figure 3.4 Simplified scheme of a RPM for photo-CIDNP. The selective character of both paths with respect to the nuclear spin state of Q results in increased selective polarizations of the recombination (geminate) and escape (transfer) products.

triplet state (gray box), a further differentiation with regard to the nuclear spin states ($i, j \in [\alpha, \beta]$ for a nuclear spin 1/2) into two "branches" is possible. Depending on the difference between the electron precession frequencies (Δg, B_0) and the hyperfine coupling (HFC) constant a (coupling between electronic and nuclear spins), the intersystem crossing (ISC), which causes the conversion between singlet and triplet states (Fig. 3.5), is efficient for only one of the nuclear "branches," here *i* [29]. A recombination is possible only for radical pairs in the singlet state. Therefore, the population of the nuclear state *i*, which is selectively preferred by the ISC, will increase in the recombination products. The nuclear state "branch" *j* is not efficiently converted to singlet state during the lifetime of the correlated pair. In consequence, diffusion separates the radicals and eliminates spin correlation. The free radicals recombine in side reactions and form what is known as escape products, with a highly populated nuclear spin state *j*. If $i = \alpha$ and $j = \beta$ in a conventional MR experiment with a 90° excitation pulse, the recombination products will then show enhanced absorption signals (positive phase, A), while enhanced emission signals (negative phase, E) will be observed for the escape products. Starting with singlet precursors or a more efficient ISC for the other branch, would reverse the signs of these signals [29]. Under certain conditions (low fields, small Δg, large a_k), the signals of a multiplet do not necessarily exhibit the same sign ("net effect," Γ_n), but can show absorption and emission at the same time ("multiplet effect," Γ_m) [38]. These aspects are summarized in the so-called "Kaptein rules," which allow for an understanding of the particular RPM [37–39].

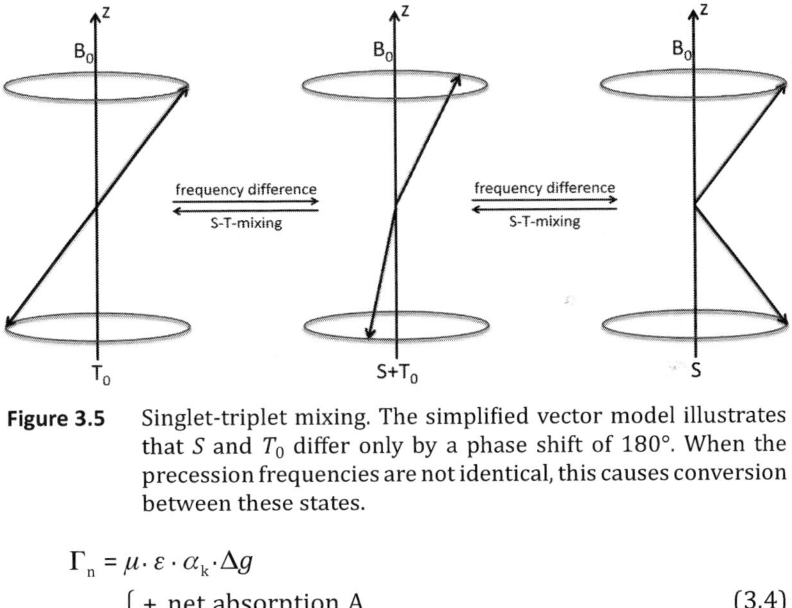

Figure 3.5 Singlet-triplet mixing. The simplified vector model illustrates that *S* and T_0 differ only by a phase shift of 180°. When the precession frequencies are not identical, this causes conversion between these states.

$$\Gamma_n = \mu \cdot \varepsilon \cdot \alpha_k \cdot \Delta g$$
$$\Gamma_n = \begin{cases} + \text{ net absorption A} \\ - \text{ net emission E} \end{cases} \tag{3.4}$$

$$\Gamma_m = \mu \cdot \varepsilon \cdot \alpha_k \cdot \alpha_l \cdot J_{kl} \cdot \sigma_{kl}$$
$$\Gamma_m = \begin{cases} + \text{low field signal(s) E/high field signal(s) A} \\ - \text{low field signal(s) A/high field signal(s) E} \end{cases} \tag{3.5}$$

The particular factors are

$$\mu = \begin{cases} + & \text{triplet precursors} \\ - & \text{singlet precursors} \end{cases}$$

$$\varepsilon = \begin{cases} + & \text{geminate pair reactions} \\ - & \text{transfer/escape products} \end{cases}$$

a = sign of hyperfine coupling constant

Δg = sign of the difference of g factors

J_{kl} = sign of scalar spin coupling between nuclei k and l

$$\sigma_{kl} = \begin{cases} + & \text{nuclei } k \text{ and } l \text{ are part of the same radical fragment} \\ - & \text{nuclei } k \text{ and } l \text{ are part of different radical fragments} \end{cases}$$

In short, CIDNP can be observed for nuclei that exhibit a hyperfine interaction with the unpaired electrons. In a product that contains additional MR active nuclei (without hyperfine coupling), the achieved CIDNP can be transferred among coupled spins [40, 41]. Because the electronic and nuclear precession frequencies are field dependent, the effect of hyperfine couplings and nuclear scalar couplings on the CIDNP and the transfer is also field dependent. For further details, see [41–44].

3.2.3 Parahydrogen-Induced Polarization

PHIP generates the HP by converting the nuclear spin order of parahydrogen into population differences in a particular target molecule. In the early and mid-1980s, Bryndza and Eisenberg made the first experimental observations of the PHIP effect during homogeneous hydrogenation of tricobaltalkylidyne complexes and during the reaction of $Rh_2H_2(CO)_2$(bis(diphenyl phosphino)methane)$_2$ with phenylacetylene under H_2 atmosphere, where enhanced antiphase signals were observed and initially misinterpreted as a result of CIDNP [45, 46]. At the same time, Bowers, Weitekamp, and Pravica predicted that signal enhancements would occur due to hydrogenation with H_2 enriched to a higher fraction of parahydrogen. Here, two experimental procedures were introduced: PASADENA (Parahydrogen And Synthesis Allow Dramatically Enhanced Nuclear Alignment), where the hydrogenation reaction was performed at the high magnetic field of the spectrometer or scanner, and ALTADENA (Adiabatic Longitudinal Transport After Dissociation Engenders Nuclear Alignment), where the reaction took place at Earth's magnetic field [47–49]. In the following years, PHIP became an established term used to summarize all parahydrogen-related HP methods [50]. In the beginning, the PHIP effect was mainly used in chemistry, for instance to investigate and detect kinetics, intermediate states, or stereoselectivity of hydrogenation reactions [51]. In the past decade, the focus has shifted to substrates with biomedical relevance. Consequently, the toxicity of the catalyst, the biocompatibility of the solvent, and the HP of many different classes of target molecules have all become hot topics. A few years ago, the two hydrogenative PHIP protocols were complemented by the non-hydrogenative Signal Amplification By Reversible Exchange (SABRE), which generates the polarization of the target during the lifetime of the complex of catalyst, parahydrogen, and substrate [52]. The largest observed polarizations in PHIP-experiments are up to the order of unity for ^{1}H, and 3% for ^{19}F [16, 53].

As previously indicated, the concept of PHIP relies on enriched parahydrogen (p-H_2) characterized by two spins 1/2 with antisymmetric nuclear spin configuration, i.e., the nuclear singlet state (S_0). Those molecules, which possess triplet state configuration/symmetric nuclear wave functions (T_{-1}, T_0, T_1), are called orthohydrogen (o–H_2). For "normal" hydrogen in thermal equilibrium at room temperature, the ratio o-H_2/p-H_2 is determined by the manifolds of singlet and triplet states and is 3:1. At lower temperatures, an enrichment of p-H_2 (the state with lowest energy) can be achieved. In addition, a proper catalyst such as activated charcoal is required to drive the usually forbidden interconversion of S_0 and T_0 [45, 50]. An enrichment of about 50% (using liquid nitrogen) or up to 98% (using a more sophisticated technical setup with liquid helium) can be achieved [54].

In the case of the hydrogenative protocols ALTADENA and PASADENA, PHIP is generated by a pairwise addition of pH_2 into an unsaturated precursor molecule (Fig. 3.6). The A_2 system of p-H_2 is thereby converted into an AB system, breaking the symmetry of the singlet state and allowing for mixing with the triplet state (Fig. 3.5). Figure 3.7 shows the energy levels of the two-spin system, where the strong non-Boltzmann populations resulting from the hydrogenation are even more evident. In a PASADENA experiment, the sudden transformation of the A_2 system yields a projection of the singlet state $(\alpha\beta - \beta\alpha)/\sqrt{2}$ onto the new eigenstates of the resulting AB system $\alpha\alpha$, $\alpha\beta$, $\beta\alpha$, and $\beta\beta$. This yields a selective population of the states $\alpha\beta$ and $\beta\alpha$, as illustrated in Fig. 3.7b [45, 48, 51]. In comparison to the "standard" spectrum, obtained from Boltzmann polarization, this causes a few changes in the obtained PHIP spectrum (Fig. 3.7a), including (1) increased signal amplitudes due to much higher population differences and (2) typical antiphase multiplets due to the selective population of $\alpha\beta$ and $\beta\alpha$ [45].

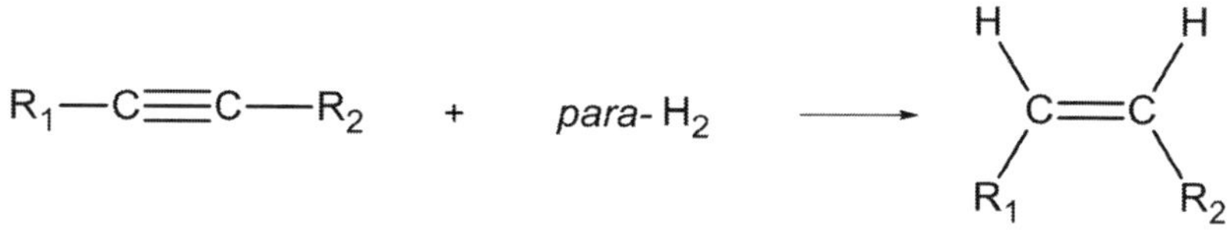

Figure 3.6 The common approach for generating PHIP in a particular target molecule is the hydrogenation of a proper unsaturated precursor.

In the case of an ALTADENA experiment, the hydrogenation is performed at a low magnetic field, where the spins are strongly coupled (i.e., the new eigenstates of the formed product are still of the form S_0, T_{-1}, T_0, and T_1) [49]. Hence, in the product, only the S_0 state of the spin system is populated, as illustrated in Fig. 3.7c. In this case, the transfer into the detection field is adiabatic, meaning that the temporal change of the Hamiltonian is so slow that the initial population of states follows the corresponding eigenstates [55]. Even at high field, only one of the states $\alpha\beta$ and $\beta\alpha$ is strongly populated, resulting in two antiphase signals in the spectrum.

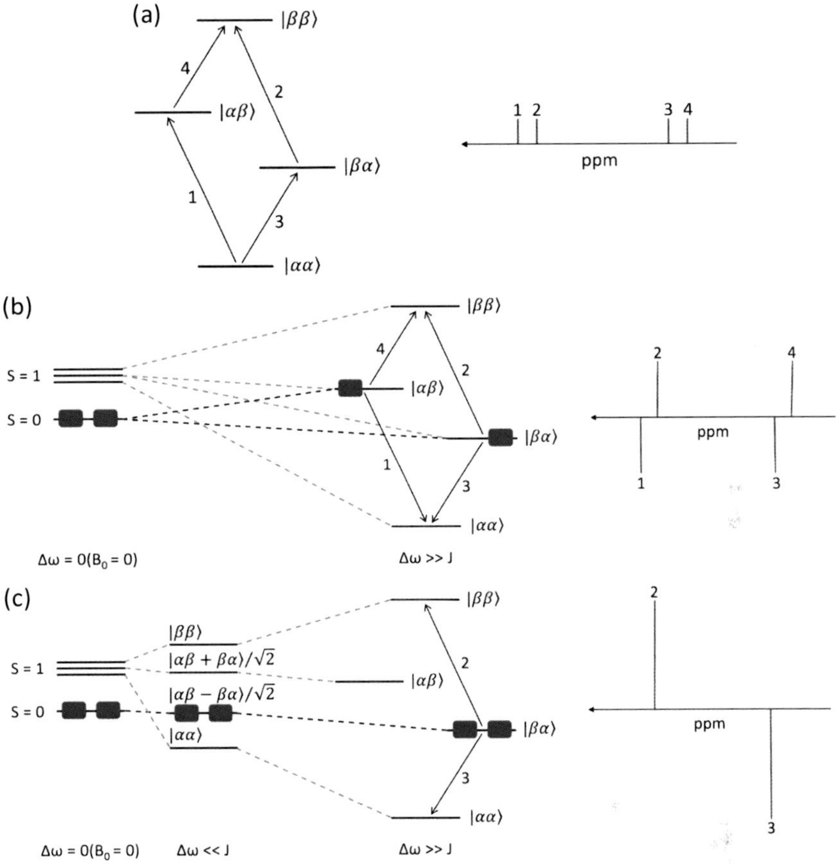

Figure 3.7 Qualitative population models for a system of two spins 1/2 weakly coupled at high field for (a) Boltzmann polarization, (b) PASADENA, and (c) ALTADENA.

In SABRE experiments the procedure is quite different (Fig. 3.8). Here, the symmetry break, which converts the p-H_2 spin order into polarization, is caused by the simultaneous addition of p-H_2 and the target molecule to the Ir-center of the catalyst. Because of the continuous ligand exchange at the catalyst, a repeated HP of the free substrate in solution is possible [52]. For a detailed theoretical description, one needs to consider at least a four-spin system, which is beyond the scope of this review, but for further reading, we recommend the manuscripts of Adams and Pravdivtsev [56, 57]. As with the SABRE approach, the HP generated in a PHIP experiment is not solely limited to the added protons but can be transferred to other MR-active spins (such as other protons or heteronuclei). The following offers a brief overview of the various transfer techniques/mechanisms and suggests further reading.

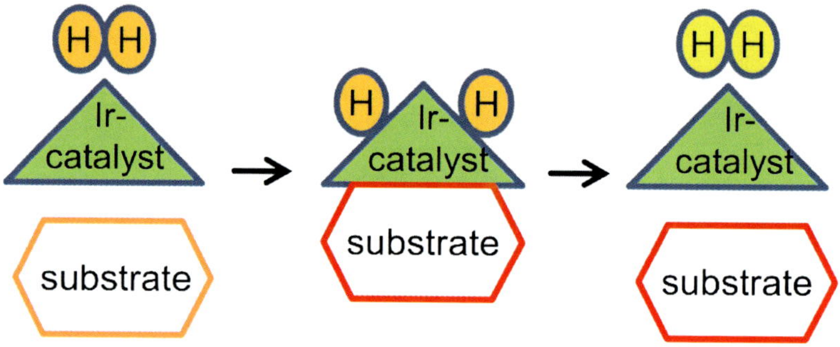

Figure 3.8 SABRE. In the intermediate complex of substrate, Ir-catalyst, and parahydrogen, the singlet symmetry is broken, and the generated polarization is transferred to the target molecule.

- Transfer at high field using INEPT based pulse sequences such as PH-INEPT, PH-INEPT+, INEPT($+\pi/4$) [58].
- Application of a sudden field variation to zero field (field cycling) that causes a projection into the eigenstates of a strongly coupled spin system involving heteronuclei [59].
- Free evolution of the entire spin system under the Hamiltonian at a certain evolution field where, due to scalar couplings and level anti-crossings, coherent transfer takes place [55, 60, 61].
- Application of spin order transfer (SOT) sequences at magnetic fields where heteronuclear interactions are weak and homonuclear couplings are strong [62–65].

3.2.4 Application of HP Methods to MRI

Table 3.1 gives a brief overview of relevant HP techniques and their most important advantages and disadvantages. Aside from investigations of chemical questions, the most common purpose for applying any of these techniques is to significantly increase the signal in MRI experiments. In that context, it is indispensable to take into account the nonconventional properties of these signals that may limit the application of common imaging strategies. For instance, the longitudinal magnetization of a thermal polarized sample, (partly) recovers to the equilibrium value after excitations caused by relaxation processes. Therefore, in "standard" imaging experiments, a stable magnetization prior to the excitation can be assumed. This can be the equilibrium value (repetition time TR > $5T_1$) or a steady state value (T_2 < TR < T_1, achieved after dummy excitations, Fig. 3.9 left). Thus the image-encoding process rests upon well-defined initial conditions. In contrast, HP experiments begin with a much higher initial polarization that irreversibly decays towards the equilibrium value. Furthermore, every excitation diminishes the residual polarization by a factor of cos Θ (Fig. 3.9 right). Therefore, relaxation times are a crucial parameter for any HP investigation and dictate the available temporal window for sample setup, measurement, and data acquisition (e.g., 2D spectra and *k*-space). With respect to ^{19}F MR, where common T_1 values are on the order of seconds, this can be a challenge. For technical or chemical investigations, for example, the problem can be solved by on-site HP generation at the place of investigation, continuous and stable HP delivery, or even repetitively designed measurements with a highly reproducible HP step at the beginning of every loop [66, 67]. However, in biomedical MRI, this is, in many cases, either insufficient or not able to be carried out, requiring a more appropriate experimental design. A first improvement can be achieved with acceleration of the sample transfer, which decreases relaxation losses. For liquids, this acceleration can be realized by an appropriate setup of tubing, electromagnetic solenoid valves, and pumps [68, 69]. Nevertheless, there is an obvious need for faster image encoding, both to maximize the benefit from HP and to minimize modulation of the data due to the decaying signal. Some suitable and proven sequences that allow for rapid *k*-space acquisition are FLASH (Fast Low Angle SHot) [70, 71], RARE (Rapid Acquisition with Refocused Echoes) [70, 72], SSFP (Balanced Steady State Free Precession, TrueFISP/FIESTA) [73, 74], EPI (echo planar imaging) [75], fast MR spectroscopic imaging [76], and TOF (time of flight) [77].

Table 3.1 Brief comparison of the introduced HP methods [16]

Method	Advantages	Disadvantages
Dissolution DNP	• Ideal for nuclei with long T_1 (e.g., ^{13}C or ^{29}Si) • No toxic catalyst • High polarization	• Huge biomolecules cannot be used because of structural modification during thawing • Toxic radicals • Time-consuming thawing step • High equipment complexity and costs
Overhauser DNP	• HP in liquid phases • Study of water dynamics	• Only small molecules • HP process takes a long time • High field → less polarization
CIDNP	• MR studies of radical pair chemical reactions	• Production of radicals is intrinsically required • Only works for a small number of molecules (e.g., peroxides)
Photo-CIDNP	• On-site HP in liquid phase • No metal catalyst	• Less signal enhancement • Usage of radicals • Only for aromatic systems • Often high-power LASERs
PHIP	• Strong polarizations • Fast HP process • Low equipment costs	• Only one-time polarization for each molecule • Unsaturated precursor • Toxic catalyst
SABRE	• Polarization can be repetitively generated • Fast HP process • Low equipment costs • No precursor required	• Less polarization • Toxic catalyst • Limited to special substances

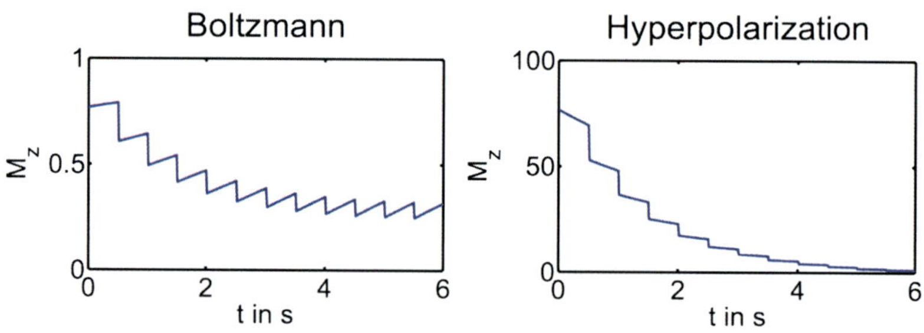

Figure 3.9 Qualitative course of the longitudinal magnetization for repetitive excitations for thermal (left) and HP (right). Simulation parameters: TR = 0.5 s, flip angle $\Theta = 40°$, $T_1 = 5$ s, $T_2 = 0.1$ s.

3.3 Hyperpolarized ^{19}F: Chronological Results

3.3.1 DNP

The first documented results for ^{19}F hyperpolarized compounds achieved with DNP occurred in 1962. Thereafter, the use of ^{19}F DNP was limited for a few decades to experiments in liquids that were based on the Overhauser effect, and it was mainly utilized in basic studies like the detection and characterization of molecular motion, intermolecular interactions, and scalar and dipolar couplings.

Richards and White investigated solutions of 2,4,6-tri-tert-butylphenoxyl radical in semi-fluorinated alcohols, where positive enhanced ^{19}F signals were observed, while ^{1}H exhibited negative signals. In a solution of the same radical in trichlorotrifluoroethane, however, the fluorine resonance was reversed in the normal manner (negative signal) [78]. In the following years, studies were performed to correlate data obtained under different experimental conditions (e.g., magnetic fields) with theoretical models [21, 79–81]. A broad variety of fluorocarbons in different mixtures of solvents and free radicals were examined and revealed the extent of ^{19}F polarization to be dependent on substrates (solvent), substitutes, and radical structure [82–87] (the list of references is not intended to be exhaustive). It could be demonstrated that scalar and dipolar couplings between electrons and ^{19}F nuclei in the examined solutions resulted in positive or negative DNP enhancements.

In the case of dipolar coupling, relaxation occurs through W_2 (Fig. 3.2) and produces an increased population of the higher energy nuclear-spin state, yielding an opposite phase for the enhanced signal compared to the non-enhanced signal (negative enhancement). An enhanced MR signal, with the same phase compared to the signal derived from thermal polarization, is detected when relaxation occurs mainly through W_0, which is the case for dominant scalar coupling [88]. The chemical environment of both the unpaired electron and the receptor nucleus influences the scalar coupling, whereas the dipolar interactions are determined by the geometrical structures of the molecules [89]. Therefore, experimental results depend on the ratio of these interactions. Interestingly, ^{19}F has a much higher tendency toward scalar interactions than ^{1}H [21, 88, 90]. A detailed overview of different theoretical models that calculate the time dependencies that modulate the scalar and dipolar interactions, including a model for translational diffusion, is given in [21, 90].

The field dependence of the coupling factor, which determines the signal enhancement, can be calculated from spin density functions and their contribution towards ρ and σ in (3.3) [20, 21, 91, 92]. Approaching higher fields, the polarization derived from dipolar coupling tends toward zero, whereas the polarization derived from scalar coupling can be high even at higher magnetic fields [91]. A molecular orbital treatment of scalar interaction allows an estimation of the induced spin density at the nucleus. Spin information can be transmitted by the overlapping of the unpaired orbital of the radical with molecular orbitals of the receptor molecule during a collision [90, 93]. In fluorinated systems, the interactions of the ^{19}F nuclei involve always scalar and dipolar contributions [89]. Despite this, a three-spin effect must additionally be considered in molecules which contain other MR active nuclei, such as ^{1}H. After a first treatment of a three-spin system in 1962 [94], Natusch et al. later provided a more general consideration of this phenomenon based on Bloch equations [95]. Using steady-state triple resonance and transient experiments, a good agreement between experimental and calculated data was described.

Since the early 21st century, when bioscientific applications of hyperpolarized fluorine became a relevant focus of research, corresponding ^{19}F DNP (including dissolution DNP) studies were carried out as well. The first MR images of hyperpolarized ^{19}F achieved through the Overhauser effect were documented by Murugesan et al. in 2002. Based on what is known as proton-electron double-resonance imaging (PEDRI), they developed a corresponding technique for ^{19}F called fluorine-electron double-resonance imaging (FEDRI). This was applied to a few fluorinated substrates, such as sodium trifluoro acetate and trifluoroethanol, in the presence of Oxo63 (((tris[8-carboxy-2,2,6,6-tetrakis (2-hydr oxymethyl)benzoyl[1,2-d′]bis(1,3)dithio-4-yl]methyl radical)), to produce signal-enhanced ^{19}F images at the low magnetic field strength of 1.5 mT. FEDRI was also used to determine the samples oxygen concentration enabled by the high sensitivity of fluorine relaxation times towards the presence of paramagnetic impurities

(Fig. 3.10) [96]. Therefore, a higher oxygen concentration leads to a decreased signal intensity. In 2008 Halse and Callaghan used Overhauser DNP to increase the sensitivity of an Earth magnetic field MR experiment and measured 2D ^{19}F-^{1}H COSY spectra of trifluoroethanol using TEMPO as an exogenous agent [97]. In the following year, Peksoz et al. examined the relative contribution of nuclear-electron dipolar and scalar couplings on the ^{19}F DNP enhancements for several fluorinated aromatic and aliphatic solvents in the presence of BDPA (1,3-bisdiphenylene-2-phenylallyl) and GALV (galvinoxyl) at low magnetic field. Based on their results, they concluded that aromatic fluorinated solvents and free radicals are highly suitable for oxymetric measurements in biological samples [89].

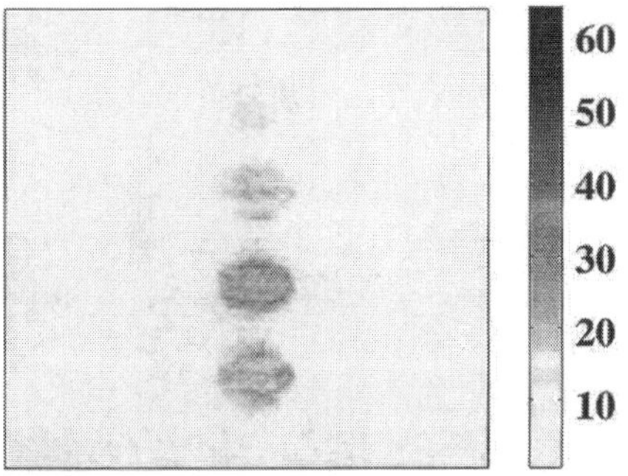

Figure 3.10 FEDRI of 2 M trifluoroethanol solutions containing oxygen between 0% (bottom) and 20% pO_2 (top), which decreases the SNR linearly (right scale). Reprinted from [96] with permission of John Wiley & Sons, Ltd.

In another approach, Lee et al. documented an enhancement ^{19}F of MR signals by factors of up to 3000 in dissolution DNP experiments with a commercial HyperSense DNP polarizer in 2012. The increased sensitivities were then used to study protein-ligand bindings and to quantify corresponding dissociation constants. Furthermore, different fluorinated substrates were used as reporter molecules to monitor the binding to serine protease bovine trypsin [98]. Neudert et al. reported a more detailed study about the estimation of nuclear-electron interactions and saturation factors in DNP experiments with TEMPO for a wide range of radical concentrations using 1,4-difluorobenzene and hexafluorobenzene as solvents. They performed various EPR and nuclear magnetic relaxation dispersion measurements to determine molecular dynamics and interactions [99]. Recently, ^{19}F DNP was used for sensitivity enhanced homonuclear 2D correlation experiments (e.g., COSY, SECSY) of 1,2,4-trifluorobenzene solutions with various exogenous radicals. Making use of the wide chemical shift range of fluorine allowed for the extraction of structural information even at moderate magnetic fields (0.34–3.34 T) [92].

3.3.2 CIDNP

Compared to PHIP and DNP, CIDNP provides lower signal enhancements but at the same time has several advantages, especially for the investigation of biomolecules [16, 100]. Although CIDNP has been widely used to elucidate radical pair processes in chemistry, it can also be applied for biocompatible substrates and solvents [33]. In addition to ^{1}H, other nuclei with even stronger hyperfine coupling constants have been used, involving ^{13}C, ^{15}N, and ^{19}F [101, 102]. Unlike ^{13}C, which has been used in various bioscience applications, ^{19}F has had relatively few applications in the field until recently, despite several advantages such as background-free signal detection. The research using ^{19}F CIDNP can be separated into two categories: (1) analysis of chemical reaction mechanisms of radical pairs and (2) use of fluorinated amino acids as probes in proteins (very few instances) [103].

Soon after the first reported CIDNP results (Section 3.2.2), Rakshys described a ^{19}F CIDNP experiment where the reaction of p-fluorobenzyl chloride and n-butyl lithium dissolved in hexane with a small amount of diethylether yielded positive-and negative-enhanced ^{19}F resonances for the reaction products pp′-difluorobibenzyl and p-fluoropentylbenzene [104]. In a subsequent publication, he showed similar results for the reaction of benzyl halides with sodium naphthalene (Fig. 3.11a) [105]. In 1971, Bargon et al. investigated reactions of free radicals generated by pyrolysis of bis(pentafluorobenzoyl) peroxide using ^{19}F. They examined several substitutes in order to obtain information about the site, where the radical attack occurs. Although no ^{19}F spectra were presented, they reported that fluorinated benzenes produced ^{19}F-CIDNP [106]. Bethell et al. reported in 1972 that polarized 1,2,2-triphenylethyl fluoride was observed in ^{19}F MR spectra when diphenylmethylene reacted with either the solvent benzyl fluoride or with α-bromobenzyl fluoride [107]. In subsequent work, they determined the signs of ^{19}F HFCs and Δg's while investigating the thermal disassociation (120°C) of diphenyldiazomethane dissolved in benzyl chlorides, as well as in fluorinated derivatives, from which (fluorinated) 1,1,2-triphenylchlorides are formed [108].

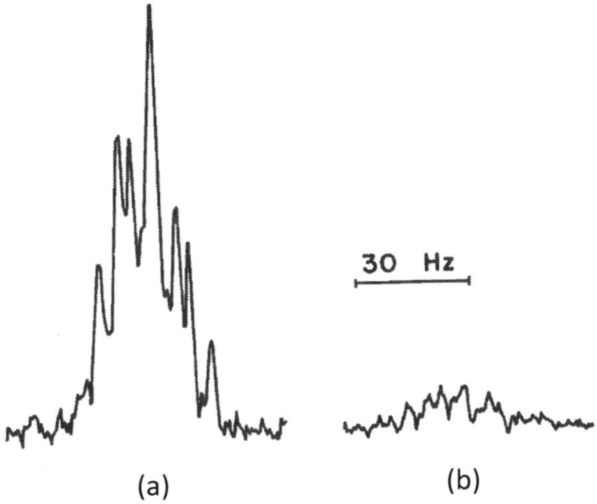

Figure 3.11 ^{19}F CIDNP NMR spectra. Shown are (a) signal enhanced (left) and thermal (right) ^{19}F spectrum of pp′-difluorobibenzyl as one of the first ^{19}F CIDNPs, and (b) ^{19}F photo-CIDNP of 3-fluorotyrosine as a function of duration of laser irradiation. Reprinted from [105] and [36] with permission from Elsevier.

In similar studies by Bargon and Seifert, (o⁻, m⁻, p⁻) monofluorinated benzyl radicals were generated photolytically and exhibit the same behavior of HFC signs independently of the position of fluorine in the molecules [109]. In the same year, Roth and Kaplan also investigated HFC signs, in this case by examining the photolytic dissociation of di(pentafluorobenzoyl) peroxide [110]. Continuing the work of Rakshys, Sagdeev et al. in 1973 studied the magnetic field dependence (5×10^{-5}–2.5 T) of ^{19}F CIDNP for the reaction of ^{19}F substituted benzyl chlorides with n-butyl lithium [111]. In addition to the above-mentioned experiments of Brinkman and Bargon [108, 109], Kaplan et al.

also confirmed the positive and negative HFCs of fluorine in ortho/para and meta position in phenoxyl and anilinyl radicals. These radicals were generated by irradiation with UV-flash lamps in the presence of benzophenone as a photosenzitizer [112]. Bethell et al. (1976) investigated the ^{19}F and ^{1}H signal patterns in CIDNP-spectra that resulted from the thermal decomposition of diaryldiazomethanes and intermediates in the solvent benzyl fluoride. Based on simulations, the authors concluded that the polarization of both nuclei could be interpreted in the framework of CIDNP radical pair theory, but that some characteristics of the spectra might also result from the different relaxation rates of the involved ^{1}H and ^{19}F nuclei [107, 113]. In the same year, Adrian presented a theoretical work that discussed the CIDNP effect of fluorinated quinones (tetrafluoro-1,4-benzoquinone (FQ)) and hydroquinones (tetrafluoro-1,4-hydroquinone (FQH2)) in benzene. That study suggested that the electron polarization was finally transferred to the nuclear spins via the Nuclear Overhauser Effect (NOE), as had earlier been assumed by Seifert and Bargon [114]. However, this view was later discussed in detail by Kuprov, who stated that at high field, this effect should not significantly contribute ^{19}F to CIDNP [115]. In 1977, Thomas et al. proposed stronger CIDNP signals in acidified solutions of α,α,α-trifluoracetophenone (TFA) or p-Cl-TFA and hydroquinonedimethylether in acetonitrile irradiated with UV light [116]. In a later experiment Wagner and Thomas investigated similar solutions involving TFA and cyclohexane [117]. They also documented radical pair mechanisms in reactions of organometallic compounds with xenon-difluoride in organic solvents with ^{1}H and ^{19}F CIDNP [118].

Despite very general aspects like HFC signs or pathways in RPMs, the experiments referred to above did not yet relate to any bioscientific application. This gap was finally bridged by Hore, Khan, Kuprov, and others who published extended theoretical works and protein-engineering methods to use ^{19}F-labeled tyrosine for protein configuration studies. Kuprov et al. (2004) demonstrated the first ^{19}F CIDNP of a fluorinated amino acid. They achieved a 20-fold signal enhancement of 3-fluoro-DL-tyrosine in a D_2O solution containing flavin mononucleotide (FMN) as a photosensitizer after 7 s argon laser irradiation (Fig. 3.11b). In this initial study the authors already speculated that the measured dipolar cross-relaxation rates and NOE ^{19}F magnitudes would still lead to CIDNP-induced ^{1}H-NOEs when 3-fluorotyrosine would be incorporated into larger molecules such as proteins [36]. Shortly after that, the same group ^{19}F presented photo-CIDNP for the first time as a tool to investigate biomolecules. With protein-engineering techniques, green fluorescent protein (GFP) was uniformly labeled with 10 fluorotyrosines.[2] The authors resolved the differences between the undisturbed and the degraded states of the protein via relaxation time measurements and analysis of the spectral characteristics. In that way, the kinetics of protein-folding mechanisms could be examined with respect to sign changes of CIDNP signals [103]. In subsequent work, they substantiated these results with both theoretical considerations and additional experiments elucidating the dependence of the geminate photo-CIDNP effect in fluorinated radicals on nuclear spin relaxation, solvent viscosity, pH, cross-relaxation, and cross-correlation induced by the anisotropy of HFC-and g-tensors [44, 115]. These investigations were motivated by ^{19}F previous observations of CIDNP net polarization in large fluorinated molecules that could not be explained by Kaptein's rules. The developed model succeeded in predicting these sign changes, and the authors concluded that the phase of ^{19}F CIDNP signals from geminate pairs showed a correlation time dependency (affected by temperature, viscosity, etc.). This in turn may allow for a quantitative measurement of side-chain correlation time in proteins labeled with fluorinated tyrosine or tryptophan [44, 103, 115].

3.3.3 PHIP

The successful application of PHIP techniques to ^{19}F increase signals was reported only recently, despite the fact that the technical prerequisites are much less demanding than DNP or CIDNP. The first PHIP experiments aiming at a transfer of HP ^{19}F to were documented by Kuhn et al. in 2006. They investigated monofluorinated styrene and phenylacetylene derivatives under ALTADENA, as well as under PASADENA conditions, experiments which yielded signal enhancements of up to two orders of

[2]GFP is an important fluorescence marker for processes in cells or living systems. Tyrosine is contained in many proteins and is an important precursor of the biosynthesis of, for example dopamine, catecholamines, thyroxine, etc.

magnitude. ALTADENA experiments generated emissive net polarization, while multiplet polarization signals were observed in the case of PASADENA. The observed PASADENA ^{19}F signals were much fainter as compared to ALTADENA due to the very weak ^{1}H-^{19}F-coupling at high field. The highest ^{19}F polarization was achieved in ALTADENA experiments with 3-fluorostyrene because the precursor 3-fluorophenylacetylene exhibits the largest hydrogenation rate. The mechanism of polarization transfer was also examined in PASADENA experiments with different fluoroacetylene derivatives. Here, a quantitative analysis showed a proportional relationship between the achieved polarization and the inverse 6th power of the mean distance "parahydrogenation center—^{19}F." The authors concluded that at low field, the transfer is driven by "through-bond" (coherent transfer) and "through space" (cross relaxation) interactions, while at high field, cross relaxation might dominate [119]. In 2009, Adams et al. presented not only the SABRE approach (Section 3.2.3) for protons, but also proved the feasibility of generating heteronuclear polarization. Instead of pyridine, which is well suited for SABRE experiments because it forms a labile hydrogenated Ir-complex, they used 3-fluoropyridine, which yielded the first SABRE ^{19}F MR spectra [52].

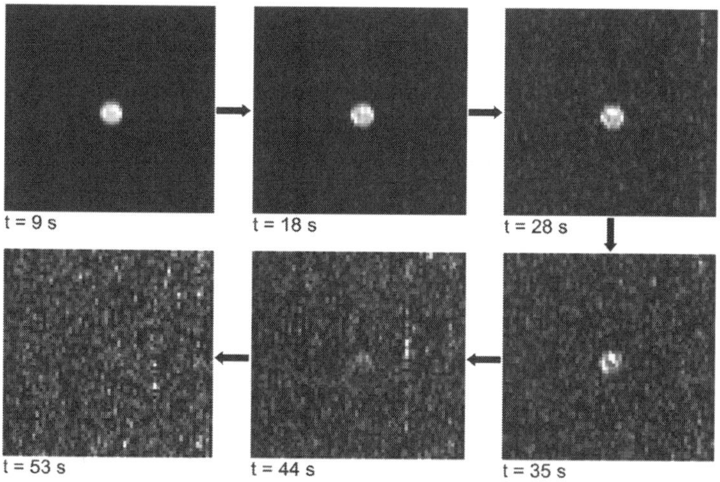

Figure 3.12 ^{19}F FLASH images of PHIP hyperpolarized 3-fluorostyrene. The measured series illustrates the decay of the enhanced signal after HP back to the thermal signal. Adapted from [71] with permission from the PCCP Owner Societies.

The first imaging experiments on hyperpolarized fluorine were performed by Bommerich et al. in 2010. Based on the results of Kuhn mentioned above, who identified 3-fluorostyrene as the derivative with the highest signal enhancement, they used 3-fluorophenylacetylene as the precursor in ALTADENA ^{19}F experiments. In contrast to Kuhn's results, the obtained spectra of hyperpolarized 3-fluorostyrene showed antiphase signals arising from multiplet polarization. To achieve net polarization, which is required for imaging, a diabatic field cycling step was performed immediately after hydrogenation by a fast transfer of the sample tube into a μ-metal cylinder followed by adiabatic transport into the detection field [59]. The generated emissive fluorine signal was utilized to acquire a series of signal-amplified ^{19}F FLASH images (Fig. 3.12) [71].

Enhanced polarizations can also be used to increase sensitivity in low or even in ultra-low magnetic field MR applications [120, 121]. Since the chemical shift differences decrease at low magnetic fields, a simultaneous detection of different types of nuclei becomes feasible, which potentially allows for examination of polarization transfer processes. In 2011, Hamans et al. published the first PHIP-enhanced simultaneous ^{1}H-^{19}F MR detection of ethyl 4,4,4-trifluorocrotonate in mM concentration at Earth's magnetic field. The conversion of the initial spin order into net polarization of proton and fluorine nuclei was achieved with a diabatic-adiabatic field cycling using a Helmholtz pair of coils. Hamans demonstrated a polarization of about 3% for ^{1}H and ^{19}F, enabling the detection of an Earth field spectrum with only a single scan and without further RF shielding [53]. In the following year, Kiryutin et al. presented experiments and theoretical explanations on the transfer of PHIP in scalar coupled spin systems (three monofluorinated styrene derivatives) in a field range between 0.5 mT and 7 T. From discrepancies between simulations and experimental data, they concluded that cross-relaxation probably acts as an additional transfer mechanism. Furthermore, a strong influence of the particular polarization ^{19}F field (where hydrogenation takes place) onto the observed signal was confirmed [122]. In 2013, Plaumann et al. presented the application of PHIP to generate hyperpolarized perfluorocarbons ((perfluoro-n-hexyl)ethene and (perfluoro-n-hexyl)ethane), which are usually biocompatible and potentially enable the extraction of relevant information in vivo. They observed an unexpected transfer of polarization to the most distant fluorines, a process which is not yet fully understood. Nevertheless, this transfer enabled substantial signal-enhanced ^{19}F MRI of hyperpolarized (perfluoron-hexyl)ethene as shown in Fig. 3.13 [72].

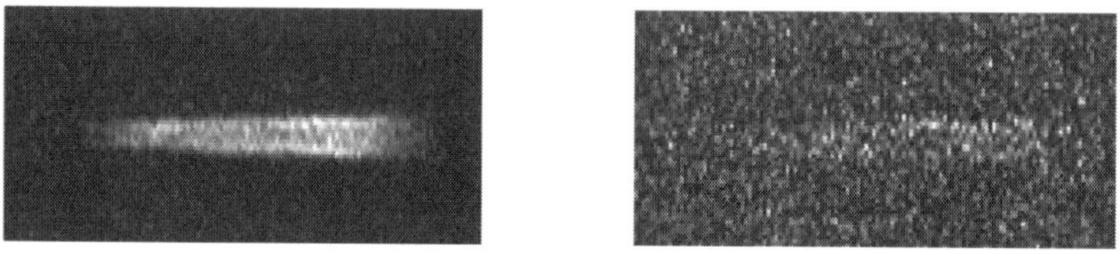

Figure 3.13 ^{19}F RARE MR images of (perfluoro-n-hexyl)ethene in acetone. The left panel shows PHIP hyperpolarized and the right image shows thermal equilibrium. Reproduced from [72] with permission from the Owner Societies.

3.4 Perspectives

Despite the large number of reported HP studies on ^{19}F, there are still very few documented applications of HP for imaging. Before the turn of this century, HP techniques were well established in basic research areas from which they arose (chemistry and physics), but they had not yet been applied to bioscience and biomedical imaging. Although important steps have already been taken to address this gap, in particular with regard to ^{1}H and ^{13}C, spatially ^{19}F resolved measurements with are still in the early stages of development. This is surprising, given that ^{19}F exhibits favorable properties such as its large chemical shift range and the absence of background signal. There are, of course, obstacles that apply to all HP methods, like the small number of biomedical relevant molecules addressed so far, the toxicity of involved substances, and the short lifetimes of the generated non-equilibrium polarizations. However, recent developments may pave the way for biocompatible HP agents, with additional purification steps, as well as the application of sophisticated reaction procedures [15]. A smart approach to get rid of these toxic compounds is the heterogeneous catalysis for PHIP of gases, which could be extended to fluorinated derivatives, and which might allow for a substitution of expensive hyperpolarized ^{3}He imaging in the near future [123].

The most fundamental aspect yet to be worked out is the prolongation of relaxation times for ^{19}F hyperpolarized samples. At this point, the fact that ^{19}F has a magnetic moment comparable to ^{1}H is a double-edged sword. On the one hand, this indicates a similar experimental behavior, which simplifies sequence design and leads to comparable sensitivities and spatial resolution in MRI experiments. On the other hand it is also be associated with comparably short relaxation times, which are often much shorter than for low-γ nuclei such as ^{13}C. To some degree, an extension of the HP lifetimes can be achieved using deuterated target molecules/solvents (smaller magnetic moment). A more sophisticated approach would be the application of long living states (LLS), in which the quantum mechanical stability of singlet states is used to prevent polarization from relaxation processes [124, 125]. Since ^{19}F has a spin 1/2, it can also form singlet states, as has already been documented for other nuclei [126]. Additionally, image acceleration techniques like compressed sensing could reduce relaxation losses and capitalize on the benefits of HP to improve image quality [123].

Aside from these general aspects, there are some recent developments in the HP methods that may support new applications in life science. In particular, the latest advances in ^{13}C dissolution DNP demonstrate the ability to monitor the metabolism of glucose and pyruvate in vivo, which is an important step for MR based molecular imaging [15]. ^{19}F PHIP, in particular SABRE, could be a valuable tool, producing MR-markers, which are structurally related to ^{18}F-labeled PET-tracers and would allow for a wider and less expensive implementation. Recent developments in new non-metal-based hydrogenation catalysts might also promote the biomedical application [127]. Reducing the technical and financial expenditure involved with HP techniques will enable their application at more sites and therefore stimulate new implementations. An interesting new advance in the field is the use of LEDs instead of lasers for photo-CIDNP experiments, which significantly simplifies the experimental setup and safety issues [128]. Furthermore, initial spatially resolved CIDNP experiments described by Trease et al. may allow for bioimaging applications similar to DNP or PHIP [129]. Currently, the above-mentioned obstacles are being addressed by many HP research groups and collaboration networks, which frequently generate innovative approaches and fascinating progress. In light of its advantages, such as its lack of background signal in biological tissues and its high chemical shift sensitivity, the future of ^{19}F hyperpolarization is quite promising.

References

1. Ross, B. D., Bhattacharya, P., Wagner, S., Tran, T., and Sailasuta, N. (2010). Hyperpolarized MR imaging: Neurologic applications of hyperpolarized metabolism, *Am. J. Neuroradiol.,* **31**(1), pp. 24–33.
2. Overhauser, A. (1953). Polarization of nuclei in metals, *Phys. Rev.,* **92**(2), pp. 411–415.
3. Carver, T., and Slichter, C. (1953). Polarization of nuclear spins in metals, *Phys. Rev.,* **92**(1), pp. 212–213.
4. Abragam, A. (1955). Overhauser effect in nonmetals, *Phys. Rev.*, **98**(6), pp. 1729–1735.
5. Hurd, R. E., Yen, Y. F., Chen, A., and Ardenkjaer-Larsen, J. H. (2012). Hyperpolarized 13C metabolic imaging using dissolution dynamic nuclear polarization, *J. Magn. Reson. Imaging,* **36**(6), pp. 1314–1328.
6. Pound, R. (1950). Nuclear electric quadrupole interactions in crystals, *Phys. Rev.*, **79**(4), pp. 685–702.
7. Jeffries, C. (1957). Polarization of nuclei by resonance saturation in paramagnetic crystals, *Phys. Rev.*, **106**(1), pp. 164–165.
8. Kessenikh, A. V., Lushchikov, V. I., Manenkov, A. A., and Taran, Y. V. (1963). Relaxation and dynamical polarization of protons in polyethylenes, *Sov. Phys. Solid State,* **5**, pp. 321–329.
9. Hwang, C., and Hill, D. (1967). Phenomenological model for the new effect in dynamic polarization, *Phys. Rev. Lett.,* **19**(18), pp. 1011–1014.
10. Wind, R. A., Duijvestijn, M. J., van der Lugt, C., Manenschijn, A., and Vriend, J. (1985). Applications of dynamic nuclear polarization in 13C NMR in solids, *Prog. Nucl. Magn. Reson. Spectrosc.,* **17**, pp. 33–67.
11. Gerfen, G. J., Becerra, L. R., Hall, D. A., Griffin, R. G., Temkin, R. J., and Singel, D. J. (1995). High frequency (140 GHz) dynamic nuclear polarization: Polarization transfer to a solute in frozen aqueous solution, *J. Chem. Phys.*, **102**(24), p. 9494.
12. Hu, K. N. (2011). Polarizing agents and mechanisms for high-field dynamic nuclear polarization of frozen dielectric solids, *Solid State Nucl. Magn. Reson.*, **40**(2), pp. 31–41.

13. Potapov, A., Yau, W. M., and Tycko, R. (2013). Dynamic nuclear polarization-enhanced 13C NMR spectroscopy of static biological solids, *J. Magn. Reson.*, **231**, pp. 5–14.

14. Ardenkjaer-Larsen, J. H., Fridlund, B., Gram, A., Hansson, G., Hansson, L., Lerche, M. H., Servin, R., Thaning, M., and Golman, K. (2003). Increase in signal-to-noise ratio of 10,000 times in liquid-state NMR, *Proc. Natl. Acad. Sci. U. S. A.*, **100**(18), pp. 10158–10163.

15. Rodrigues, T. B., Serrao, E. M., Kennedy, B. W. C., Hu, D. E., Kettunen, M. I., and Brindle, K. M. (2013). Magnetic resonance imaging of tumor glycolysis using hyperpolarized 13C-labeled glucose, *Nat. Med.*, **20**(1), pp. 93–97.

16. Lee, J. H., Okuno, Y., and Cavagnero, S. (2014). Sensitivity enhancement in solution NMR: Emerging ideas and new frontiers, *J. Magn. Reson.*, **241**, pp. 18–31.

17. Maly, T., Debelouchina, G. T., Bajaj, V. S., Hu, K. N., Joo, C. G., Mak-Jurkauskas, M. L., Sirigiri, J. R., van der Wel, P C A, Herzfeld, J., Temkin, R. J., and Griffin, R. G. (2008). Dynamic nuclear polarization at high magnetic fields, *J. Chem. Phys.*, **128**(5), p. 052211.

18. Serra, S. C., Rosso, A., and Tedoldi, F. (2012). Electron and nuclear spin dynamics in the thermal mixing model of dynamic nuclear polarization, *Phys. Chem. Chem. Phys.*, **14**(38), p. 13299.

19. Song, C., Hu, K. N., Joo, C. G., Swager, T. M., and Griffin, R. G. (2006). TOTAPOL: A biradical polarizing agent for dynamic nuclear polarization experiments in aqueous media, *J. Am. Chem. Soc.*, **128**(35), pp. 11385–11390.

20. Günther, U. L. (2013). Dynamic nuclear hyperpolarization in liquids, *Top. Curr. Chem.*, **335**, 23–69 (Springer).

21. Hausser, K. H., and Stehlik, D. (1968). Dynamic nuclear polarization in liquids, in: *Advances in Magnetic and Optical Resonance*, vol. 3 (Academic Press Inc.).

22. Abragam, A., and Goldman, M. (1978). Principles of dynamic nuclear polarisation, *Rep. Prog. Phys.*, **41**(3), pp. 395–467.

23. Goldman, M. (1970). *Spin Temperature and Nuclear Magnetic Resonance in Solids* (Clarendon Press, Oxford).

24. Batel, M. (2013). Dissolution dynamic-nuclear-polarization and its combination with cross polarization, dissertation ETH Zurich, URL http://e-collection.library.ethz.ch/view/eth:8392?q=8392.

25. Bargon, J., Fischer, H., and Johnsen, U. (1967). Kernresonanz-Emissionslinien während rascher Radikalreaktionen. I. Aufnahmeverfahren und Beispiele. *Z. Naturforsch. A*, **22**, pp. 1551–1555.

26. Ward, H. R., and Lawler, R. G. (1967). Nuclear magnetic resonance emission and enhanced absorption in rapid organometallic reactions, *J. Am. Chem. Soc.*, **89**(21), pp. 5518–5519.

27. Bargon, J., and Fischer, H. (1967). Kernresonanz-Emissionslinien während rascher Radikalreaktionen. II. Chemisch induzierte dynamische Kernpolarisation, *Z. Naturforsch. A*, **22**, pp. 1556–1562.

28. Lawler, R. G. (1967). Chemically induced dynamic nuclear polarization, *J. Am. Chem. Soc.*, **89**(21), pp. 5519–5521.

29. Kuhn, L. (2013). *Hyperpolarization Methods in NMR Spectroscopy*, Topics in Current Chemistry, vol. **338** (Springer, Berlin).

30. Kaptein, R., and Oosterhoff, L. J. (1969). Chemically induced dynamic nuclear polarization III (anomalous multiplets of radical coupling and disproportionation products), *Chem. Phys. Lett.*, **4**(4), pp. 214–216.

31. Closs, G. L. (1969). Mechanism explaining nuclear spin polarizations in radical combination reactions, *J. Am. Chem. Soc.*, **91**(16), pp. 4552–4554.

32. Ward, H. R. (1972). Chemically induced dynamic nuclear polarization (CIDNP). I. Phenomenon, examples, and applications, *Acc. Chem. Res.*, **5**(1), pp. 18–24.

33. Kaptein, R., Dijkstra, K., and Nicolay, K. (1978). Laser photo-CIDNP as a surface probe for proteins in solution, *Nature*, **274**(5668), pp. 293–294.

34. Morozova, O. B., Kiryutin, A. S., Sagdeev, R. Z., and Yurkovskaya, A. V. (2007). Electron transfer between guanosine radical and amino acids in aqueous solution. 1. Reduction of guanosine radical by tyrosine, *J. Phys. Chem. B*, **111**(25), pp. 7439–7448.

35. Matysik, J., Alia, Gast, P., van Gorkom, H. J., Hoff, A. J., and de Groot, H. J. M. (2000). Photochemically induced nuclear spin polarization in reaction centers of photosystem II observed by ^{13}C-solid-state NMR reveals a strongly asymmetric electronic structure of the P680(+) primary donor chlorophyll, *Proc. Natl. Acad. Sci. U. S. A.*, **97**(18), pp. 9865–9870.

36. Kuprov, I., and Hore, P. J. (2004). Chemically amplified ^{19}F–^{1}H nuclear Overhauser effects, *J. Magn. Reson.*, **168**(1), pp. 1–7.

37. Seifert, K. G. (1976). Chemisch induzierte dynamische Kernspin-Polarisation (CIDNP), *Chemie in unserer Zeit*, **10**(3), pp. 84–93.

38. Pine, S. H. (1972). Chemically induced dynamic nuclear polarization, *J. Chem. Educ.*, **49**(10), p. 664.

39. Hutton, R. S., Roth, H. D., and Bertz, S. H. (1983). Nuclear spin polarization effects in systems with large hyperfine couplings. Limitations of Kaptein's rules, *J. Am. Chem. Soc.*, **105**(21), pp. 6371–6377.

40. de Kanter, F. J. J., and Kaptein, R. (1979). CIDNP transfer via nuclear dipolar relaxation and spin-spin coupling, *Chem. Phys. Lett.*, **62**(3), pp. 421–426.

41. Ivanov, K. L., Miesel, K., Yurkovskaya, A. V., Korchak, S. E., Kiryutin, A. S., and Vieth, H. M. (2006). Transfer of CIDNP among coupled spins at low magnetic field, *Appl. Magn. Reson.*, **30**(3)–4, pp. 513–534.

42. Grosse, S., Gubaydullin, F., Scheelken, H., Vieth, H. M., and Yurkovskaya, A. V. (1999). Field cycling by fast NMR probe transfer: Design and application in field-dependent CIDNP experiments, *Appl. Magn. Reson.*, **17**(2–3), pp. 211–225.

43. Grosse, S., Yurkovskaya, A. V., Lopez, J., and Vieth, H. M. (2001). Field dependence of chemically induced dynamic nuclear polarization (CIDNP) in the photoreaction of N-acetyl histidine with 2,2′-dipyridyl in aqueous solution, *J. Phys. Chem. A*, **105**(26), pp. 6311–6319.

44. Kuprov, I. (2005). Chemically induced dynamic nuclear polarization of 19F nuclei, DPhil thesis University of Oxford, URL http://arxiv.org/abs/physics/0604156.

45. Natterer, J., and Bargon, J. (1997). Parahydrogen induced polarization, *Prog. Nucl. Magn. Reson. Spectrosc.*, **31**(4), pp. 293–315.

46. Eisenberg, R. (1991). Parahydrogen-induced polarization: A new spin on reactions with molecular hydrogen, *Acc. Chem. Res.*, **24**(4), pp. 110–116.

47. Bowers, C. R., and Weitekamp, D. P. (1986). Transformation of symmetrization order to nuclear-spin magnetization by chemical reaction and nuclear magnetic resonance, *Phys. Rev. Lett.*, **57**(21), pp. 2645–2648.

48. Bowers, C. R., and Weitekamp, D. P. (1987). Parahydrogen and synthesis allow dramatically enhanced nuclear alignment, *J. Am. Chem. Soc.*, **109**(18), pp. 5541–5542.

49. Pravica, M. G., and Weitekamp, D. P. (1988). Net NMR alignment by adiabatic transport of parahydrogen addition products to high magnetic field, *Chem. Phys. Lett.*, **145**(4), pp. 255–258.

50. Duckett, S. B., and Mewis, R. E. (2012). Application of parahydrogen induced polarization techniques in NMR spectroscopy and imaging, *Acc. Chem. Res.*, **45**(8), pp. 1247–1257.

51. Duckett, S. B., and Wood, N. J. (2008). Parahydrogen-based NMR methods as a mechanistic probe in inorganic chemistry, *Coord. Chem. Rev.*, **252**(21)–22, pp. 2278–2291.

52. Adams, R. W., Aguilar, J. A., Atkinson, K. D., Cowley, M. J., Elliott, P. I. P., Duckett, S. B., Green, G. G. R., Khazal, I. G., Lopez-Serrano, J., and Williamson, D. C. (2009a). Reversible interactions with para-hydrogen enhance NMR sensitivity by polarization transfer, *Science,* **323**(5922), pp. 1708–1711.

53. Hamans, B. C., Andreychenko, A., Heerschap, A., Wijmenga, S. S., and Tessari, M. (2011). NMR at earth's magnetic field using para-hydrogen induced polarization, *J. Magn. Reson.*, **212**(1), pp. 224–228.

54. Hövener, J. B., Bär, S., Leupold, J., Jenne, K., Leibfritz, D., Hennig, J., Duckett, S. B., and von Elverfeldt, D. (2013). A continuous-flow, high-throughput, high-pressure parahydrogen converter for hyperpolarization in a clinical setting, *NMR Biomed.*, **26**(2), pp. 124–131.

55. Ivanov, K. L., Yurkovskaya, A. V., and Vieth, H. M. (2008). Coherent transfer of hyperpolarization in coupled spin systems at variable magnetic field, *J. Chem. Phys.*, **128**(15), p. 154701.

56. Adams, R. W., Duckett, S. B., Green, R. A., Williamson, D. C., and Green, G. G. R. (2009b). A theoretical basis for spontaneous polarization transfer in non-hydrogenative parahydrogen-induced polarization, *J. Chem. Phys.,* **131**(19), p. 194505.

57. Pravdivtsev, A. N., Yurkovskaya, A. V., Vieth, H. M., Ivanov, K. L., and Kaptein, R. (2013). Level anti-crossings are a key factor for understanding para-hydrogen-induced hyperpolarization in SABRE experiments, *Chemphyschemistry,* **14**(14), pp. 3327–3331.

58. Haake, M., Natterer, J., and Bargon, J. (1996). Efficient NMR pulse sequences to transfer the parahydrogen-induced polarization to hetero nuclei, *J. Am. Chem. Soc.,* **118**(36), pp. 8688–8691.

59. Jóhannesson, H., Axelsson, O., and Karlsson, M. (2004). Transfer of para-hydrogen spin order into polarization by diabatic field cycling, *C. R. Phys.*, **5**(3), pp. 315–324.

60. Miesel, K., Ivanov, K. L., Yurkovskaya, A. V., and Vieth, H. M. (2006). Coherence transfer during field-cycling NMR experiments, *Chem Phys. Lett.*, **425**(1–3), pp. 71–76.

61. Kiryutin, A. S., Yurkovskaya, A. V., Kaptein, R., Vieth, H. M., and Ivanov, K. L. (2013). Evidence for Coherent transfer of para-hydrogen-induced polarization at low magnetic fields, *J. Phys. Chem. Lett.*, **4**(15), pp. 2514–2519.

62. Goldman, M., and Jóhannesson, H. (2005). Conversion of a proton pair para order into 13C polarization by rf irradiation, for use in MRI, *C. R. Phys.*, **6**(4–5), pp. 575–581.

63. Kadlecek, S., Emami, K., Ishii, M., and Rizi, R. R. (2010). Optimal transfer of spin-order between a singlet nuclear pair and a heteronucleus, *J. Magn. Reson.*, **205**(1), pp. 9–13.

64. Bär, S., Lange, T., Leibfritz, D., Hennig, J., von Elverfeldt, D., and Hövener, J. B. (2012). On the spin order transfer from parahydrogen to another nucleus, *J. Magn. Reson.*, **225**, pp. 25–35.

65. Cai, C., Coffey, A. M., Shchepin, R. V., Chekmenev, E. Y., and Waddell, K. W. (2013). Efficient transformation of parahydrogen spin order into heteronuclear magnetization, *J. Phys. Chem. B*, **117**(5), pp. 1219–1224.

66. Roth, M., Kindervater, P., Raich, H. P., Bargon, J., Spiess, H. W., and Münnemann, K. (2010). Continuous 1H and 13C signal enhancement in NMR spectroscopy and MRI using parahydrogen and hollow-fiber membranes, *Angew. Chem. Int. Ed. Engl.*, **49**(45), pp. 8358–8362.

67. Lloyd, L. S., Adams, R. W., Bernstein, M., Coombes, S., Duckett, S. B., Green, G. G. R., Lewis, R. J., Mewis, R. E., and Sleigh, C. J. (2012). Utilization of SABRE-derived hyperpolarization to detect low-concentration analytes via 1D and 2D NMR methods, *J. Am. Chem. Soc.*, **134**(31), pp. 12904–12907.

68. Hövener, J. B., Chekmenev, E. Y., Harris, K. C., Perman, W. H., Robertson, L. W., Ross, B. D., and Bhattacharya, P. (2009). PASADENA hyperpolarization of 13C biomolecules: Equipment design and installation, *MAGMA*, **22**(2), pp. 111–121.

69. Reynolds, S., Bucur, A., Port, M., Alizadeh, T., Kazan, S. M., Tozer, G. M., and Paley, M. N. J. (2014). A system for accurate and automated injection of hyperpolarized substrate with minimal dead time and scalable volumes over a large range, *J. Magn. Reson.*, **239**, pp. 1–8.

70. Golman, K., Axelsson, O., Jóhannesson, H., Månsson, S., Olofsson, C., and Petersson, J. S. (2001). Parahydrogen-induced polarization in imaging: Subsecond (13)C angiography, *Magn. Reson. Med.*, **46**(1), pp. 1–5.

71. Bommerich, U., Trantzschel, T., Mulla-Osman, S., Buntkowsky, G., Bargon, J., and Bernarding, J. (2010). Hyperpolarized ^{19}F-MRI: Parahydrogen-induced polarization and field variation enable ^{19}F-MRI at low spin density, *Phys. Chem. Chem. Phys.*, **12**(35), p. 10309.

72. Plaumann, M., Bommerich, U., Trantzschel, T., Lego, D., Dillenberger, S., Sauer, G., Bargon, J., Buntkowsky, G., and Bernarding, J. (2013).

Parahydrogen-induced polarization transfer to ^{19}F in perfluorocarbons for ^{19}F NMR spectroscopy and MRI, *Chem. Eur. J.*, **19**(20), pp. 6334–6339.

73. Magnusson, P., Johansson, E., Månsson, S., Petersson, J. S., Chai, C. M., Hansson, G., Axelsson, O., and Golman, K. (2007). Passive catheter tracking during interventional MRI using hyperpolarized ^{13}C, *Magn. Reson. Med.*, **57**(6), pp. 1140–1147.

74. Bhattacharya, P., Harris, K., Lin, A. P., Mansson, M., Norton, V. A., Perman, W. H., Weitekamp, D. P., and Ross, B. D. (2005). Ultra-fast three dimensional imaging of hyperpolarized ^{13}C in vivo, *MAGMA*, **18**(5), pp. 245–256.

75. Saam, B., Yablonskiy, D. A., Gierada, D. S., and Conradi, M. S. (1999). Rapid imaging of hyperpolarized gas using EPI, *Magn. Reson. Med.*, **42**(3), pp. 507–514.

76. Larson, P. E. Z., Bok, R., Kerr, A. B., Lustig, M., Hu, S., Chen, A. P., Nelson, S. J., Pauly, J. M., Kurhanewicz, J., and Vigneron, D. B. (2010). Investigation of tumor hyperpolarized [1-13 C]pyruvate dynamics using time-resolved multiband RF excitation echo-planar MRSI, *Magn. Reson. Med.*, **63**(3), pp. 582–591.

77. Telkki, V. V., Zhivonitko, V. V., Ahola, S., Kovtunov, K. V., Jokisaari, J., and Koptyug, I. V. (2010). Microfluidic gas-flow imaging utilizing parahydrogen-induced polarization and remote-detection NMR, *Angew. Chem. Int. Ed. Engl.*, **49**(45), pp. 8363–8366.

78. Richards , R. E., and White, J. W. (1962). Relative couplings between free radicals and hydrogen and fluorine nuclei by the Overhauser effect, *Proc. Chem. Soc.*, pp. 119–120.

79. Dwek, R. A., Kenworthy, J. G., Natusch, D. F. S., Richards, R. E., and Shields, D. J. (1966). Nuclear electron double resonance and scalar interactions in solutions, *Proc. R. Soc. Lond. A*, **291**(1427), pp. 487–499.

80. Müller-Warmuth, W. (1966). Skalare Kern-Elektronen-Wechselwirkungen zwischen verschiedenen Molekülen in Lösungen, *Z. Naturforsch. A*, **21**, pp. 153–157.

81. Potenza, J. (1972). Measurement and applications of dynamic nuclear polarization, *Adv. Mol. Relaxation Processes*, **4**(3–4), pp. 229–354.

82. Thompson, D., and Poindexter, E. (1965). Differential dynamic polarization of fluorine and hydrogen nuclei, *Phys. Rev. Lett.*, **15**(7), pp. 281–282.

83. Stewart, J. R., Poindexter, E. H., and Potenza, J. A. (1967). Dynamic enhancement of fluorine nuclear magnetic resonance signals. Effects of chemical environment, *J. Am. Chem. Soc.*, **89**(24), pp. 6017–6021.

84. Müller-Warmuth, W., van Steenwinkel, R., and Yalçiner, A. (1971). Intermolecular nuclear-electron interaction and scalar spectral intensities in solutions of 1,3,5-trifluorobenzene and substituted fluorobenzenes containing free radicals, *Mol. Phys.*, **21**(3), pp. 449–459.

85. Bates, R. D., Wagner, B. E., and Poindexter, E. H. (1976). Dynamic fluorine-19 polarization in fluorinated strained cyclic alkanes and alkenes, *J. Phys. Chem.*, **80**(3), pp. 320–323.

86. Chandrakumar, N., and Narasimhan, P. T. (1982). ^{19}F dynamic nuclear polarization studies on the system $CF_2Cl–CFCl_2$/tritertiary butyl phenoxyl radical: Evidence for indirect chlorine participation in enhancing fluorine scalar rates, *J. Chem. Phys.*, **77**(5), p. 2697.

87. Franklin, G., and Bates, R. D. (1983). Dynamic 1H and 19F polarization studies of interactions of fluorohaloethanes with free radical solutes, *J. Magn. Reson.*, **53**(3), pp. 503–508.

88. Lingwood, M. D., and Han, S. (2011). Solution-state dynamic nuclear polarization, *Annu. Rep. NMR Spectrosc*, **73** pp. 83–126.

89. Peksoz, A., Yalciner, A., and Cimenoglu, M. A. (2009). A low field fluorine-electron double resonance study for GALV and BDPA in some aliphatic and aromatic solvents, *Z. Naturforsch. A*, **64**, pp. 477–484.

90. Müller-Warmuth, W., and Meise-Gresch, K. (1983). Molecular motions and interactions as studied by dynamic molecular motions and interactions as studied by dynamic nuclear polarization (DNP) in free radical solutions, *Adv. Magn. Reson.*, **11**, pp. 1–45.

91. Loening, N. M., Rosay, M., Weis, V., and Griffin, R. G. (2002). Solution-state dynamic nuclear polarization at high magnetic field, *J. Am. Chem. Soc.*, **124**(30), pp. 8808–8809.

92. George, C., and Chandrakumar, N. (2014). Chemical-shift-resolved ^{19}F NMR spectroscopy between 13.5 and 135 MHz: Overhauser-DNP-enhanced diagonal suppressed correlation spectroscopy, *Angew. Chem.*, **126**(32), pp. 8581–8584.

93. Tripathi, A., and Narasimhan, P. T. (1985). Molecular orbital treatment of differential Overhauser enhancements of solvent nuclei, *Mol. Phys.*, **54**(6), pp. 1415–1422.

94. Hausser, K. H., and Reinbold, F. (1962). Dynamic polarisation in a three-spin system, *Phys. Lett.*, **2**(1), pp. 53–54.

95. Natusch, D. F. S., Richards, R. E., and Taylor, D. (1966). Steady state and transient Overhauser effect in systems of three spins, *Mol. Phys.*, **11**(5), pp. 421–435.

96. Murugesan, R., English, S., Reijnders, K., Yamada, K., Cook, J. A., Mitchell, J. B., Subramanian, S., and Krishna, M. C. (2002). Fluorine electron double resonance imaging for 19F MRI in low magnetic fields, *Magn. Reson. Med.*, **48**(3), pp. 523–529.

97. Halse, M. E., and Callaghan, P. T. (2008). A dynamic nuclear polarization strategy for multi-dimensional Earth's field NMR spectroscopy, *J. Magn. Reson.*, **195**(2), pp. 162–168.

98. Lee, Y., Zeng, H., Ruedisser, S., Gossert, A. D., and Hilty, C. (2012). Nuclear magnetic resonance of hyperpolarized fluorine for characterization of protein–ligand interactions, *J. Am. Chem. Soc.*, **134**(42), pp. 17448–17451.

99. Neudert, O., Mattea, C., Spiess, H. W., Stapf, S., and Münnemann, K. (2013). A comparative study of ^{1}H and ^{19}F Overhauser DNP in fluorinated benzenes, *Phys. Chem. Chem. Phys.*, **15**(47), p. 20717.

100. Janssen, G. J., Daviso, E., Son, M., Groot, H. J. M., Alia, A., and Matysik, J. (2010). Observation of the solid-state photo-CIDNP effect in entire cells of cyanobacteria Synechocystis, *Photosynth. Res.*, **104**(2–3), pp. 275–282.

101. Diller, A., Roy, E., Gast, P., van Gorkom, H. J., de Groot, H. J. M., Glaubitz, C., Jeschke, G., Matysik, J., and Alia, A. (2007). 15N photochemically induced dynamic nuclear polarization magic-angle spinning NMR analysis of the electron donor of photosystem II, *Proc. Natl. Acad. Sci. U. S. A.*, **104**(31), pp. 12767–12771.

102. Prakash, S., Alia, A., Gast, P., de Groot, H. J. M., Jeschke, G., and Matysik, J. (2007). ^{13}C chemical shift map of the active cofactors in photosynthetic reaction centers of Rhodobacter sphaeroides revealed by photo-CIDNP MAS NMR, *Biochemistry*, **46**(31), pp. 8953–8960.

103. Khan, F., Kuprov, I., Craggs, T. D., Hore, P. J., and Jackson, S. E. (2006). 19F NMR studies of the native and denatured states of green fluorescent protein, *J. Am. Chem. Soc.*, **128**(33), pp. 10729–10737.

104. Rakshys, J. W. (1970). Chemically-induced dynamic polarization of 19F from spin-delocalized radicals, *J. Chem. Soc. D*, **10**, p. 578.

105. Rakshys, J. W. (1971). Use of 19F CIDNP in reaction mechanism studies. Evidence for the intermediacy of benzyl anion in the

reaction of benzyl halides with sodium naphthalene, *Tetrahedron Lett.*, **12**(49), pp. 4745–4748.

106. Bargon, J. (1971). Chemically induced dynamic nuclear polarization and substituent effects in some homolytic aromatic substitution reactions, *J. Am. Chem. Soc.*, **93**(18), pp. 4630–4632.

107. Bethell, D., Brinkman, M. R., and Hayes, J. (1972). Chemically induced dynamic nuclear spin polarisation in products of reactions of diphenylmethylene with benzyl fluorides: Novel effects arising from intramolecular interactions of polarised ^{1}H and ^{19}F nuclei, *J. Chem. Soc. Chem. Commun.*, **8**, p. 475.

108. Brinkman, M. R., Bethell, D., and Hayes, J. (1973). Determination of the absolute signs of ^{19}F hyperfine coupling constants by chemically induced dynamic nuclear spin polarization (CIDNP) in fluorinated aromatic systems, *J. Chem. Phys.*, **59**(6), p. 3431.

109. Bargon, J., and Seifert, K. G. (1973). Determination of the signs of the fluorine hyperfine coupling constants of monosubstituted benzyl radicals, *J. Phys. Chem.*, **77**(24), pp. 2877–2879.

110. Roth, H. D., and Kaplan, M. L. (1973). Chemically induced polarization of fluorine-19 nuclei. Fluorine-substituted phenyl radicals, *J. Am. Chem. Soc.*, **95**(1), pp. 262–264.

111. Sagdeev, R. Z., Molin, Y. N., Salikhov, K. M., Leshina, T. V., Kamha, M. A., and Shein, S. M. (1973). CIDNP effects in various magnetic fields during the reaction of substituted benzyl chlorides with n-butyllithium, *Org. Magn. Reson.*, **5**(12), pp. 599–602.

112. Kaplan, M. L., Manion, M. L., and Roth, H. D. (1974). Chemically induced polarization of fluorine-19 nuclei. Fluorinated phenoxyl and anilinyl radicals, *J. Phys. Chem.*, **78**(18), pp. 1837–1839.

113. Bethell, D., Brinkman, M. R., Hayes, J., and McDonald, K. (1976). Intermediates in the decomposition of aliphatic diazo-compounds. Part XII. Chemically induced dynamic polarisation of fluorine-19 and proton nuclei in the insertion of diarylmethylenes into benzyl fluoride, *J. Chem. Soc., Perkin Trans.*, **2**(8), p. 966.

114. Adrian, F. J., Vyas, H. M., and Wan, J. K. S. (1976). Magnetic field and concentration dependence of CIDNP in some quinone photolyses: Further evidence for an Overhauser mechanism, *J. Chem. Phys.*, **65**(4), pp. 1454–1461.

115. Kuprov, I., Craggs, T. D., Jackson, S. E., and Hore, P. J. (2007). Spin relaxation effects in photochemically induced dynamic nuclear

polarization spectroscopy of nuclei with strongly anisotropic hyperfine couplings, *J. Am. Chem. Soc.*, **129**(29), pp. 9004–9013.

116. Thomas, M. J., and Wagner, P. J. (1977). Acid catalyzed CIDNP chemically induced dynamic nuclear polarization during photoinduced electron transfer, *J. Am. Chem. Soc.*, **99**(11), pp. 3845–3847.

117. Wagner, P. J., and Thomas, M. J. (1980). 1H, 13C, and 19F nuclear polarizations in the photochemistry of α,α,α-trifluoroacetophenone, *J. Am. Chem. Soc.*, **102**(12), pp. 4173–4182.

118. Sadykov, R. A., Yakovlev, V. N., Bulgakov, R. G., Kazakov, V. P., Tolstikov, G. A., and Kruppa, A. I. (1989). A study of the mechanism of the reaction of $(C_2H_5O)_2AlC_2H_5$ with XeF_2 by ^1H and ^{19}F CIDNP, *J. Organomet. Chem.*, **377**(1), pp. 1–8.

119. Kuhn, L. T., Bommerich, U., and Bargon, J. (2006). Transfer of parahydrogen-induced hyperpolarization to ^{19}F, *J. Phys. Chem. A*, **110**(10), pp. 3521–3526.

120. Colell, J., Türschmann, P., Glöggler, S., Schleker, P., Theis, T., Ledbetter, M., Budker, D., Pines, A., Blümich, B., and Appelt, S. (2013). Fundamental aspects of parahydrogen enhanced low-field nuclear magnetic resonance, *Phys. Rev. Lett.*, **110**(13), p. 137602.

121. Theis, T., Ganssle, P., Kervern, G., Knappe, S., Kitching, J., Ledbetter, M. P., Budker, D., and Pines, A. (2011). Parahydrogen-enhanced zero-field nuclear magnetic resonance, *Nat. Phys.*, **7**(7), pp. 571–575.

122. Kiryutin, A. S., Ivanov, K. L., Yurkovskaya, A. V., Kaptein, R., and Vieth, H. M. (2012). Transfer of parahydrogen induced polarization in scalar coupled systems at variable magnetic field, *Z. Phys. Chem.*, **226**(11–12), pp. 1343–1362.

123. Kovtunov, K. V., Barskiy, D. A., Coffey, A. M., Truong, M. L., Salnikov, O. G., Khudorozhkov, A. K., Inozemtseva, E. A., Prosvirin, I. P., Bukhtiyarov, V. I., Waddell, K. W., Chekmenev, E. Y., and Koptyug, I. V. (2014). High-resolution 3D proton MRI of hyperpolarized gas enabled by parahydrogen and Rh/TiO_2 heterogeneous catalyst, *Chem. Eur. J.*, **20**(37), pp. 11636–11639.

124. Carravetta, M., and Levitt, M. H. (2005). Theory of long-lived nuclear spin states in solution nuclear magnetic resonance. I. Singlet states in low magnetic field, *J. Chem. Phys.*, **122**(21), p. 214505.

125. Feng, Y., Davis, R. M., and Warren, W. S. (2012). Accessing long-lived nuclear singlet states between chemically equivalent spins without breaking symmetry, *Nat. Phys.*, **8**(11), pp. 831–837.

126. Tayler, M. C. D. (2012). Theory and practice of singlet nuclear magnetic resonance, Dissertation University of Southampton, URL http://www.southampton.ac.uk/~mhl/publications/theses/2012/MCDT/MCDT-PhD-thesis-centered_subm.pdf.

127. Zhivonitko, V. V., Telkki, V. V., Chernichenko, K., Repo, T., Leskelä, M., Sumerin, V., and Koptyug, I. V. (2014). Tweezers for parahydrogen: A metal-free probe of nonequilibrium nuclear spin states of H_2 molecules, *J. Am. Chem. Soc.*, **136**(2), pp. 598–601.

128. Feldmeier, C., Bartling, H., Riedle, E., and Gschwind, R. M. (2013). LED based NMR illumination device for mechanistic studies on photochemical reactions: Versatile and simple, yet surprisingly powerful, *J. Magn. Reson.*, **232**, pp. 39–44.

129. Trease, D., Bajaj, V. S., Paulsen, J., and Pines, A. (2011). Ultrafast optical encoding of magnetic resonance, *Chem. Phys. Lett.*, **503**(4–6), pp. 187–190.

Part 2

^{19}F Imaging Agents

Chapter 4

Active Targeting of Perfluorocarbon Nanoemulsions

Sebastian Temme,[a] Christoph Grapentin,[b] Tuba Güden-Silber,[a] and Ulrich Flögel[a]

[a]*Department of Molecular Cardiology, Heinrich Heine University, Universitätsstraße 1, 40225, Düsseldorf, Germany*
[b]*Department of Pharmaceutical Technology and Biopharmacy, Albert Ludwig University, Hermann-Herder-Strasse 9, 79104 Freiburg, Germany*

sebastian.temme@uni-duesseldorf.de

4.1 A Short Introduction to Perfluorocarbons and Perfluorocarbon Nanoemulsions

Perfluorocarbons (PFCs) are fluorinated alkanes in which all hydrogen atoms have been substituted by fluorine atoms. Low-molecular-weight perfluorocarbons (up to four carbon atoms) are gaseous at 20°C, whereas larger perfluorocarbons are color- and odorless liquids or solids [1]. Since fluorine atoms are bigger than hydrogen atoms, the perfluoroalkyl chains are bulkier than the corresponding hydrocarbons [2]. In addition, effective overlapping of orbitals as well as fluorine's high electronegativity render the

Fluorine Magnetic Resonance Imaging
Edited by Ulrich Flögel and Eric Ahrens

ISBN 978-981-4745-31-4 (Hardcover), 978-981-4745-32-1 (eBook)
www.panstanford.com

C–F bond one of the most stable single bonds in organic chemistry [2, 3]. PFCs are characterized by low cohesion forces resulting in low intermolecular van der Waals forces and thus low surface tension, low viscosity, high density and anti-friction properties. Of note, PFCs have a high density which can reach up to 2 g/mL.

The dense electron cloud, in combination with the strong C–F bond, leads to a remarkable stability of the perfluorinated chains against reactive agents. These physicochemical properties explain why PFCs are physiologically inert upon in vivo application in animals as well as humans. PFCs are not metabolized and thus absolutely biocompatible [2, 4]. Therefore, no adverse effects have been observed for PFCs in a range of 460–520 Da [2].

PFCs are neither lipo- nor hydrophilic but exceptionally fluorophilic and, therefore, not mixible with water or hydrophobic substances. Thus, for intravenous in vivo application, they have to be emulsified to obtain PFC nanoemulsions (PFC-NE).[1] The most appropriate way to obtain stable nanoemulsions with narrow size distribution is high-pressure homogenization or microfluidization [5–7]. Other techniques like extrusion or ultrasonication often result in nanoemulsions with very broad size distributions [5, 6].

Since phospholipids (from either egg yolk or soy bean) exhibit an excellent safety profile for intravenous application, they are most often used as emulsifiers. The obtained PFC-NE consist of a perfluorocarbon core surrounded by a lipid monolayer, which is different to liposomes that are formed by lipid bilayers enclosing an aqueous phase. Nanoemulsion droplets typically have a particle size in the range of 100 to 200 nm depending on factors described below.

Size and its distribution are, besides charge or storage stability, important characteristics of PFC-NE that are influenced by several parameters like the ratio of PFC to emulsifier, the PFC(s) itself, the emulsifier composition and the preparation technique (e.g., process pressure and homogenization time in high-pressure homogenization or microfluidization). Size can be manipulated by the ratio of phospholipid to PFC by increasing the lipid to PFC ratio which leads to a smaller average size of nanoemulsion droplets and vice versa [5, 6]. However, higher amounts of phospholipid favor the formation of liposomes, which

[1]A detailed overview about PFC-NE is given in Chapter 14 of this book.

do not contain PFCs, and these compete with PFC-NE for cellular uptake and impact on storage stability [8]. Size distribution is determined by the preparation process, and PFC-NE with a very narrow size variation can be obtained by subsequent size exclusion chromatography [5, 6].

The PFC or mixture of PFCs also influence the stability of PFC-NE [9, 10]. Stability of PFC-NE is limited by Ostwald ripening and coalescence (for a brief overview see Ref. [11]). Ostwald ripening is a process that leads to the generation of larger particles due to transfer of individual PFCs from one particle to another. In contrast, coalescence is a process where nanoemulsion droplets merge resulting in bigger particles. Attempts to stabilize PFC-NE make use of reagents minimizing Ostwald ripening, e.g., addition of perfluorodecyl bromide (PFDB) which has a higher vapor pressure [9] or the incorporation of semifluorinated alkanes acting as molecular dowels [12].

4.2 Generation of Targeted Perfluorocarbon Nanoemulsions

Up to now, PFC-NE in combination with ^{19}F MRI have been widely used for oximetry (see Chapter 5) and, more recently, for inflammation imaging. Upon intravenous injection, PFC-NE particles are predominantly taken up by blood monocytes and to a lesser extent by B-cells and, neutrophil granulocytes, which can be tracked as "^{19}F hot spots" after infiltration of these cells into inflammatory foci [13–18]. Lately, we found that specific epicardium-derived stem cells in the heart also internalize PFC-NE after myocardial infarction [19]. However, the accumulation of PFCs at the site of injury, or within the reticuloendothelial system, can also occur by "passive targeting" through a leaky endothelial barrier. The liver sinusoids have a fenestrated endothelium with pores of about 100 nm, the spleen has gaps of several hundred nm, and in the endothelium of inflamed tissues and tumors, pores with 400 nm diameter are found; thus these pores permit nanoparticles to leave the blood stream and diffuse into the tissue. Since tumors as well as inflamed tissue are characterized by an increased inflow from the blood, and a decreased drain towards the lymphatic system, nanoparticles transiently accumulate. This effect is termed "enhanced permeability and retention" (EPR) and has been exploited for both therapeutics and imaging [20, 21], specifically for visualization of tumor tissue by ^{19}F MRI via deposited PFCs [21, 22].

For a more targeted tracking of cellular structures, specific individual phagocytic immune cells (inflammatory vs. reparative monocytes) or T-cells (which do not take up PFC-NE under normal conditions) an "active targeting" approach is required. Active targeting makes use of specific ligands attached to the nanoparticle surface to facilitate the direct interaction of particle and target structure. Several distinct ligands like antibodies, antibody fragments, proteins, hormones, etc. have been exploited for active targeting approaches. Besides specificity, an additional advantage could be elevated sensitivity, which is a key concern of ^{19}F MRI. Compared to other imaging techniques, like PET, MRI is insensitive [23] and local ^{19}F concentration at monitored sites is intrinsically low. Therefore, specific targeting of PFC-NE can increase the fluorine concentration within target cell types and, hence, increase sensitivity.

4.2.1 Targeting Ligands

As pointed out before, targeting of PFC-NE to distinct cell types is based on the attachment of specific ligands to the particle surface. For the various imaging modalities, like PET, SPECT, MRI, ultrasound, etc., numerous different ligands for specific targeting of nanoparticles for diagnosis and also therapeutics have been evaluated. These include targeting moieties from antibody-related structures (mAbs, Fab', $F(ab)_2$, nanobodies, scFvs, diabodies, microbodies) to peptides, carbohydrates, aptamers, and others [24–26]. Due to the large amount of literature concerning targeted nanoparticles, we primarily refer to studies using PFC-NE as a scaffold.

4.2.1.1 Antibodies and antibody derivatives

The most widely used targeting ligands are monoclonal IgG antibodies.[2] IgGs have a molecular weight of about 150–160 kDa and are composed of four subunits linked by disulfide bridges

[2]Monoclonal antibodies: Antibodies obtained from a hybridoma which is obtained by fusion of IgG secreting B-cells with a myeloma cell line. All antibodies generated by one specific hydridoma are identical in structure, binding site and isotype.

(2 × light chain = 2 × 25 kDa; 2 × heavy chain = 2 × 50 kDa; see Fig. 4.1). Monoclonal antibodies (mAb) are characterized by high affinity and specificity towards their antigen. Limitations of mAbs are the high molecular weight and the Fc part that facilitates effector functions like binding to Fc-receptors of macrophages and neutrophils, antibody dependent cytotoxic reactions (ADCC)[3] or complement activation. In addition, xenogenic IgGs provoke immune reactions mainly caused by the Fc region.

Unwanted side effects of IgGs can be eliminated by removal of the Fc part by enzymatic digestion. Pepsin cleaves underneath the hinge region to generate $F(ab)_2$ fragments whereas papain cleavage results in two Fab fragments linked by one disulfide bridge (Fig. 4.1a). Importantly, Fab/$F(ab)_2$ fragments retain their binding properties but have a lower weight and show reduced side effects.

In contrast to Fab fragments, a single chain antibody (single chain variable fragment; scFv) is not a fragment of an antibody obtained by enzymatic digestion but a fusion protein of the variable regions of the heavy (V_H) and light (V_L) chains of immunoglobulins connected by a short linker peptide (Fig. 4.1b) [27].[4] These proteins are generated by cloning of the variable domains of the heavy and light chains from antibody producing hybridomas, splenocytes or B-cells by polymerase chain reaction. Subsequently, both chains are connected by an artificial linker and expressed in *E. coli*, insect, yeast or eukaryotic cells [28].

A further class of compounds comprises single domain antibodies (sdAbs) which consist of a singular monomeric variable domain. These fragments were initially found in camelids or sharks [29]. In 2003, the Belgian company Ablynx (www.ablynx.com) established the term "nanobody" for single domain antibodies derived from camelids (Fig. 4.1c). Nanobodies have a low molecular weight of about 10–15 kDa and are composed of about 100 amino acids. In addition, they lack the Fc part, have a shorter plasma half-life and display better tissue penetration. Some nanobodies are extremely resistant to heat up

[3]ADCC: Binding of antibodies to target cells (normally tumor cells or infected cells) that are recognized by NK-cells via their Fc-receptors. This results in the killing of the target cell by the NK-cell.

[4]The V_H and V_L domains can associate noncovalently (Fv), by a peptide linker (scFv), by disulfide bond (dsFv) or both (sc-dsFv) [27].

to 90°C [30]. Interestingly, in contrast to conventional IgGs nanobodies possess an extended CDR3 loop which can facilitate binding to hidden target structures like active sites of enzymes [29]. Moreover, administration of camelid nanobodies to humans should be of low immunogenicity, since the variable regions between camelids and humans are very similar and, therefore, do not provoke adverse immune reactions [31].

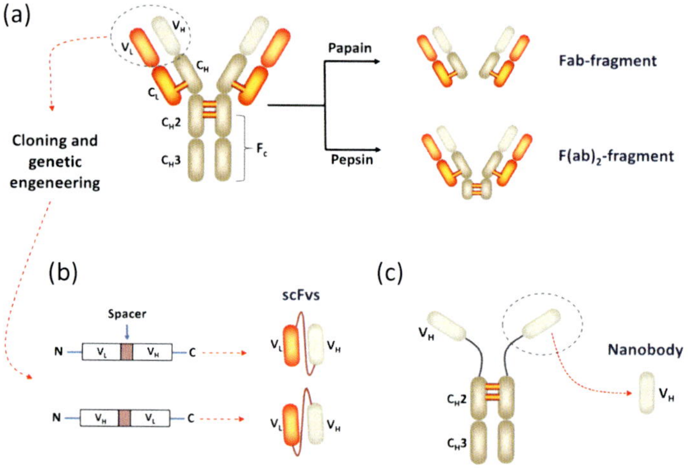

Figure 4.1 Antibody related structures which can be exploited for active targeting of PFC-NE. (a) Schematic drawing of an conventional IgG antibody (150 kDa) which consists of two heavy (light beige) and two light chains (orange) that are cross-linked by disulfide-linkages (orange lines). The heavy and light chains are composed of variable (*V*) and constant regions (*C*). The heavy chains contain two additional constant domains (C_H2, C_H3) which form the constant region of the mAb. Fc = Fc-region (Fc = fragment of crystallization). Digestion of the IgG with papain or pepsin generates two Fab or one $F(ab)_2$ fragments, respectively (Fab = fragment of antigen-binding). (b) Single chain antibodies (scFv = short chain variable fragment) are synthetic constructs which are generated by cloning of the variable antigen-binding regions of the antibody (V_L and V_H) from the antibody-secreting B-cells and connecting both chains via short peptide linker (red line). The orientation of the V_L/V_H chains can be N-terminal or C-terminal. (c) Nanobodies are the variable antigen-binding domains of camelid or shark antibodies. These antibodies have the advantage of solely consisting of heavy chains and the antigen-binding fragment is therefore intrinsically stable.

4.2.1.2 Peptides and other targeting ligands

Peptides are sequences of amino acids with less than 50 residues that can be linear or cyclic [24]. A variety of peptides has been generated for targeting of distinct cell types or structures like thrombi, inflamed endothelium or tumor vasculature [24, 32]. Peptides are very convenient because of their low molecular weight (approximately 1–5 kDa), easy handling, high stability, and that they can be obtained custom-made at a relatively low price from multiple suppliers. Moreover, peptides can be easily modified during the manufacturing process by incorporation of additional cysteines, lysines or biotin to enable the binding to bind nanoparticles (see Section 4.2.2). Some of the most commonly used peptides are from the RGD peptide-family which bind to $\alpha_v\beta_3$ integrins which are upregulated during inflammation and are also found on tumor endothelium [24, 33]. Other peptides are known to bind to fibrin or collagen [34–36], or to tumor cells [35, 36]. An overview of peptides of interest can be found in the PepBank database (www.pepbank.mgh.harvard.edu).

Besides antibodies and peptides, targeting of nanoparticles can be performed with a multitude of other ligands, like aptamers [37], carbohydrates [38], lectins [39], folate [40], biotin [41, 42], transferrin [43], and many more.

4.2.2 Coupling of Targeting Ligands to PFC-NE

Targeting ligands are usually linked to the phospholipid surface of PFC-NE via an anchor-molecule (e.g., fatty acids, phospholipids or a sterol derivative) containing a reactive group at the distal end of a polyethylene glycol (PEG) chain which acts as a spacer between phospholipid surface and targeting ligand. Additionally, the PEG chains impair opsonization by serum proteins and consequently slow nonspecific uptake by phagocytic cells like monocytes and macrophages [44].

4.2.2.1 Functional groups for coupling reactions

Several functional groups of the targeting ligands can be exploited for attachment of the ligand to the phospholipid shell of PFC-NE. Each method has advantages and disadvantages that must be critically evaluated with regard to the specifics of the individual targeting approach. To get a comprehensive overview about coupling techniques, we refer the reader to excellent summaries published elsewhere [45–47]. Among the functional groups used for linking of proteins or peptides to nanoparticles are amines (ε-amino group on lysines or the N-terminal α-amino group), sulfhydryl groups (–SH), carboxylate residues (C-terminus or aspartic acid/glutamic acid) or modified carbohydrates.

Primary amines (in particular lysine residues) are abundantly found in antibodies, Fab-fragments, scFvs or nanobodies and are uniformly distributed over the entire antibody surface. Efficient coupling to amines can be conducted using the NHS ester (*N*-hydroxysuccinimide ester) system which generates an amide bond with NH_2-groups. Conjugation is simple, but occurs randomly, and may impact binding properties, since lysines are also found within the antigen-binding site [48]. Furthermore, the orientation of the antibody conjugate might not be ideal and lead to a decreased binding affinity.

The C_H2-domain of the Fc-part contains carbohydrates which can also be used for coupling. Oxidation with sodium periodate generates aldehyde groups that can be further functionalized with hydrazine derivatives leading to a stable hydrazone bond. Additionally, the aldehyde residues can also react with primary amines to generate imine bonds, but in contrast to hydrazone bonds, imines are prone to hydrolysis under acidic conditions. Principally, coupling to carbohydrates ensures the correct orientation of the antibody to maintain an antibody conjugate with high affinity towards the target structure. However, it has to be taken into account that some antibodies may be glycosylated near the antigen-binding site and that modification of these sites can impair antigen-binding. Additionally, some monoclonal Abs (mAbs) produced by hybridomas and all mAbs produced by bacteria are not glycosylated at all [46] and, therefore, cannot be functionalized by hydrazine derivatives. In contrast, some nanobodies and variable domains of IgG are indeed glycosylated [49, 50] which makes them suitable for coupling via the hydrazine technique.

The sulfhydryl group is a popular target in many modification strategies, since the reaction with maleimide and thiol (–SH) is very efficient and leads to stable thioethers. However, free sulfhydryl groups are rare or absent in proteins and have to be generated by mild reduction of disulfide bonds or introduced via a thiolating agent [46, 48]. Since sulfhydryl groups are sensitive to oxidation and rapidly form disulfide bonds, it is mandatory that oxygen and metals are absent during coupling reactions. The approach can be conveniently used for modification of peptides, in that a cysteine is attached to one terminus of the peptide for coupling to the maleimide anchor.

For functionalization purposes, antibodies contain several disulfide bonds which are essential for maintaining the structure of the molecule. The most sensitive disulfide bridges for reducing agents are found in the hinge regions; this property enables the cleavage of the antibody into monovalent halves, without disrupting the structure, disulfide bonds in the variable domains, and antibody binding properties [48]. To this end, the antibody is first digested with pepsin and subsequently, the $F(ab)_2$ fragments are reduced by 2-MEA (2-mercaptoethylamine) leading to free SH groups. Importantly, this elegant technique also leads to the proper orientation of the antigen-binding site and has already been used to couple Fab-fragments to DSPE-PEG-maleimide [51].

Since nanobodies and scFvs do not possess any accessible sulfhydrylgroups, the coupling via a maleimide reaction might be conducted by thiolating agents (e.g., 2-iminothiolane) or by adding a cysteine residue to the C- or N-terminus of the protein by genetic engineering. One example is a scFv directed to the activated form of the gpIIb/IIIa receptor on platelets, which has been further modified with an additional cysteine to enable the coupling to maleimide groups [5, 6].

Another way to generate functionalized nanoparticles is the biotin-avidin (or streptavidin) system. The popularity of this system is based on the extreme stability of the avidin-biotin complex with a dissociation constant of 1.3×10^{-15} mol/L [52]. For coupling of ligands to nanoparticles using this approach, the nanoparticles have to be equipped with streptavidin and incubated with biotinylated ligands or vice versa. Of note, streptavidin decorated nanoparticles can be labeled with a multitude of biotinylated ligands. For example, Wang et al. used biotinylated scFvs to label streptavidin-microbubbles for detection of thrombus formation in murine carotid arteries [53]. Recently, liposomes [54] and also PFC-NE [41] have been reported to be successfully labeled with streptavidin and incubated with biotinylated ligands to achieve an active targeting.

4.2.2.2 Generation of targeted PFC-NE

In general, three ways to generate targeted PFC-NE modified with an anchor-ligand conjugate are feasible (see Fig. 4.2).

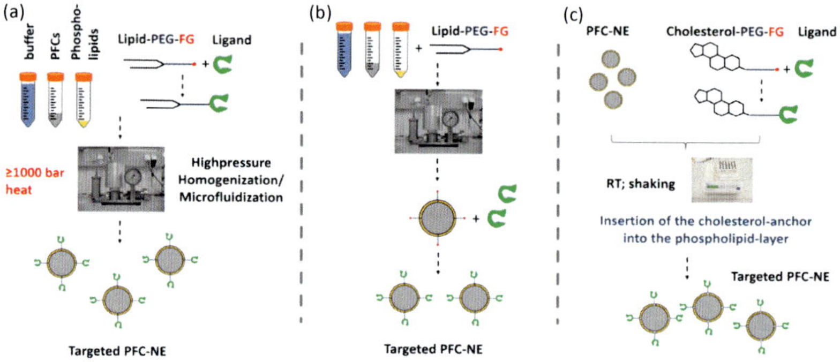

Figure 4.2 Generation of targeted PFC-NE. (a) The targeting ligand is linked to a phospholipid and subjected to high-pressure homogenization or microfluidization resulting in targeted PFC-NE. (b) A lipid with a reactive group (Lipid-PEG-FG) is added to the main lipids used for the generation of PFC-NE, and the ligand is coupled to the reactive group after formation of the PFC-NE. (c) In terms of sterol-based post-insertion, the ligand is coupled to a cholesterol anchor in a separate step. The cholesterol-PEG-conjugate is then added to pre-formed PFC-NE which results in a spontaneous insertion of the cholesterol moiety into the lipid shell of the PFC-NE. PEG = polyethylene glycol; FG = functional group; RT = room temperature.

In the first approach (Fig. 4.2a), the ligand is conjugated to the lipid anchor, the conjugate is added to phospholipids to a small degree (usually three to five mol%) and this mixture is coprocessed via high-pressure homogenization (Fig. 4.2a). However, during high-pressure homogenization, harsh conditions like shear-stress and local heating can lead to degradation or denaturation of labile ligands. Therefore, this method is restricted to small, robust molecules [55].

In the second method (Fig. 4.2b), PFC-NE with reactive groups on the surface are generated first, and the targeting ligand is attached to the droplet surface in a second step. In general, phospholipids (DSPE) equipped with a functional group (maleimide, NHS) are incorporated into the lipid layer of PFC-NE. Of note, when phospholipids with maleimide or NHS are used, a partial deactivation of the reactive groups during the manufacturing process might occur (e.g., due to hydrolysis which is particularly relevant for the NHS group).

In the third method (Fig. 4.2c), a cholesterol-ligand or phospholipid-ligand conjugate is generated and subsequently inserted into the phospholipid monolayer of preformed PFC-NE by post-insertion (Fig. 4.2c). Post-insertion using DSPE-PEG-moieties is routinely used to modify liposomes [56, 57] but has not been tested for PFC-NE. This approach requires elevated temperatures (approximately 60°C) that might be incompatible with thermodynamically labile ligands, and the low vapor pressure of some PFCs may lead to droplet growth by Ostwald ripening. In contrast, sterol-based post-insertion is a method in which the preformed PFC-NE is incubated with a cholesterol ligand conjugate under very mild conditions (i.e., room temperature and gentle agitation). The cholesterol part inserts into the phospholipid layer leading to functionalized PFC-NE. Another major advantage of the post-insertion strategy is that it is also suitable for very small volumes. Commercial high-pressure homogenizers usually need a minimum volume of several milliliters, whereas post-insertion can be conducted with few microliters, which allows the routine screening of multiple targeting ligands. A potential disadvantage of the sterol-anchor might be its lower stability within the lipid monolayer compared to the DSPE-anchor.

4.3 Applications Using Actively Targeted PFC-NE

Multiple studies have applied perfluorocarbon nanocarriers modified with distinct targeting ligands for a variety of imaging modalities like X-ray [58], ultrasound [59], near infrared microscopy [60] as well as MRI [41, 42, 55, 61–68]. However, for most of the MRI-studies, the surface of PFC-NE was decorated with a targeting ligand together with a high payload of gadolinium to generate T1-weighted contrast for read-out by conventional ^{1}H MRI [41, 55, 62, 63, 67, 68]. Surprisingly, up to now only few studies have taken advantage of the negligible physiological background in ^{19}F MRI and measured the distribution of the targeted PFC-NE directly via the ^{19}F nucleus [42, 61, 63, 65, 66].

4.3.1 Inflammation

Inflammation is initially characterized by a local activation of cells–mostly tissue macrophages, fibroblasts and the endothelium –leading to redness, heat and swelling due to vessel dilatation, permeabilization of the endothelium and increased fluid entry into the inflamed area. Subsequently, immune cells from the surrounding tissue and the blood migrate into the inflamed region to fight invaded pathogens, remove debris or aid in the healing processes. In many cases, neutrophil granulocytes are the first cells that are recruited, followed by monocyte subsets with distinct functions (e.g., inflammatory vs. reparative) and after a few days T-cells will also enter the inflamed area.

4.3.1.1 Imaging immune cells

Neutrophil granulocytes, macrophages and monocytes ingest pathogens and cellular debris to remove the infectious agents and to clean up the site of injury. Due to their phagocytic properties, these cells also endocytose PFC-NE, which has recently been applied to visualize inflammation in a variety of clinically relevant disease models [13, 14, 16–18, 69–73]. However, using conventional PFCs, all phagocytic cells are labeled and it is difficult to differentiate between individual immune cell types. In studies where PFCs were injected after myocardial infarction (MI) [18, 74], myocarditis [16], heart transplantation [13], or upon pulmonary infection [17], predominantly macrophages were loaded with PFCs. In contrast, in a collagen-induced arthritis model [14] or at an early time-point after LPS-induced inflammation [15] the majority of the ^{19}F signal was found in neutrophil granulocytes. Interestingly, if PFCs are injected at day 4 after MI in a rat model of ischemia-reperfusion, epicarduim derived stem cells of the heart also showed strong PFC uptake [19].

Up to now, there have been no studies demonstrating that PFC-NE can be actively targeted to specific immune cell populations like distinct monocyte subsets, M1/M2 macrophages or even T-cells for in vivo ^{19}F MRI. Nevertheless, there are interesting studies with ex vivo labeled immune cells. Srinivas et al. performed ex vivo labeling of T-cells and followed their homing to local lymph nodes and pancreas in a murine diabetes model

[75, 76].[5] Waiczies et al. and Ahrens et al. labeled dendritic cells and followed their migration to the draining lymph nodes [77, 78]. Although these studies did not rely on an active targeting approach, they demonstrate the potentiality of loading T-cells and DCs with PFC payloads and their subsequent visualization by noninvasive in vivo ^{19}F MRI.

The feasibility of specific immune cell targeting of PFC-NE has recently been demonstrated by Mayenfels who used sterol-based post-insertion to target the macrophage cell line J774 by coupling an anti-CD204 antibody to a Chol-PEG$_{1300}$-NHS anchor. CD204 is the macrophage scavenger receptor 1 (MSR1) which has been found on macrophages in atherosclerotic lesions and also in tumors. Although J774 cells endocytose conventional PFC-NE, PEGylation of PFC-NE nearly completely abolished the uptake by J774 cells which could be rescued by coupling a CD204 mAb to the PEGylated particle surface (Fig. 4.3). Interestingly, CD204 expression on tumor-associated macrophages has been associated with a more malignant tumor phenotype and therefore may offer prognostic value in diagnosis of certain tumors [79–81].

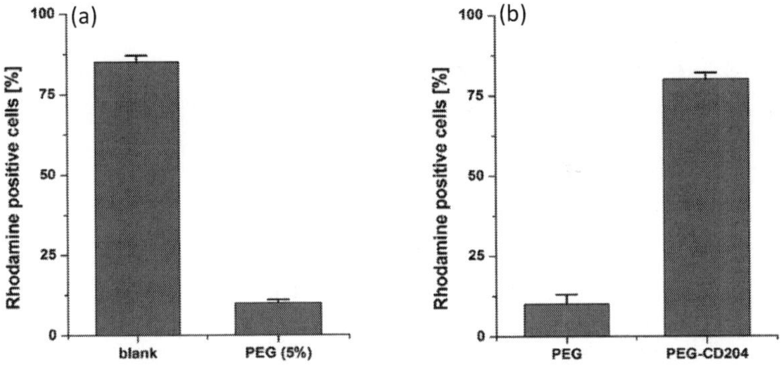

Figure 4.3 Cell uptake of PFC-NE having different compositions. (a) Conventional PFC-NE is strongly taken up by J774 macrophages (a, left bar), whereas PEGylation of the PFC-NEs strongly decreases the uptake (a, right bar). (b) Coupling of CD204 antibodies (which are directed against a scavenger receptor of macrophages) to the surface of PEGylated PFC-NE reestablished the uptake by J774 macrophages. Adapted with permission from Ref. [6].

4.3.1.2 Visualization of the activated endothelium

One of the earliest events during inflammation is the activation of the endothelium by danger signals and cytokines. This activation results in the formation of small gaps between the otherwise tightly sealed endothelial barrier and the expression of adhesion molecules which permit the binding, rolling and transmigration of immune cells into the inflamed area. Among these adhesion molecules are selectins, Ig-like proteins or integrins. Since upregulation of VCAM-1 or other integrins is an early hallmark of inflamed tissue, these adhesion molecules represent attractive target structures for imaging purposes.

Of note, VCAM-1 is not only upregulated by inflamed endothelium, but also in atherosclerotic lesions, and in many types of cancer. To visualize VCAM-1 expression in these settings, Pan et al. used a fusion-peptide consisting of a VCAM-1 targeting peptide and a melittin derived membrane-integrating sequence to post-functionalize PFC-NE (20% w/v PFCE or PFOB) [82]. This approach will be described in more detail below (Section 4.3.4).

Interestingly, alterations in regional VCAM-1 expression are also associated with kidney disease and found in patients with type II diabetes. Here, a strong inverse correlation between the glomerular filtration rate and plasma VCAM-1 levels was observed. Furthermore, plasma VCAM-1 levels were also increased in patients with chronic renal failure. To investigate whether a proatherogenic diet impacts on renal VCAM-1 expression, Southworth et al. used PFCE-NE equipped with anti-VCAM peptides to visualize alterations in VCAM-1 levels in the kidneys of ApoE-deficient mice exposed to western diet for 35 weeks [66]. ^{19}F MR spectroscopy and imaging indeed revealed increased fluorine signal in excised kidneys of $ApoE^{-/-}$ under diet compared to controls. Importantly, the specificity of this approach was proven by binding of targeted vs. nontargeted NE to the mouse endothelial cell line 2F-2B expressing high levels of VCAM-1.

4.3.1.3 Inflammation-associated angiogenesis

The formation of blood vessels is rarely observed in normal adult tissue except during wound healing or within the female reproductive cycle [83]. Angiogenesis within the normal wound healing process restores the tissue-supply with nutrients and oxygen and is therefore an important factor for efficient healing

[84]. However, aberrant vascularization or formation of blood vessels in chronic inflammatory diseases can have detrimental effects leading to increased tissue destruction as shown for inflammatory bowel disease, retinopathy or cancer [85, 86].

In this context, application of noninvasive ^{19}F MRI was recently reported for the visualization of inflammation-associated formation of new blood vessels [65]. To this end, an $\alpha_v\beta_3$-targeting ligand (quinolone nonpeptide vitronectin antagonist, described below) was incorporated into NE containing PFOB with a diameter of about 210 nm. Targeted and control emulsions were intravenously injected in a rat model of lung ischemia three days after surgery. Simultaneous ^{1}H/^{19}F MRI revealed marked signals in the left lung, whereas lower signals were found in the right lobe which was not affected by ischemia or in animals which were treated with control emulsions. However, strong signals also arouse from the liver and spleen, organs that are known to be major sites of nanoparticle deposition.

4.3.2 Cancer

Cancer, characterized by unlimited proliferation of tissue cells or blood cells, is one of the four major causes of death worldwide. Here, early diagnosis is of particular importance since many types of cancer develop metastases over time which leads to the spreading of the tumor to distant tissues like the lung or the brain. Interestingly, there are already some observations in very early PFC-NE studies which found deposition of PFC-NE in a variety of experimentally induced tumors [87–89]. Today it is known that conventional PFC-NE (and other nanoparticles) passively accumulate within tumors due to the EPR effect (see Section 4.2) and by the uptake of tumor macrophages. For a more specific active targeting of nanoparticles, the focus can be (i) tracking the tumor cells, (ii) binding to the activated endothelium, (iii) the immune cells within the tumor, or (iv) the detection of neoangiogenesis, which is another important feature of tumor growth, since the formation of blood vessels is crucial for supply of the tumor tissue with nutrients and oxygen.

In a study to directly label the tumor cells, Mayenfels coupled cholesterol-PEG$_{1300}$-NHS or -PEG$_{2000}$-maleimide anchors to anti-GD2 antibodies and used sterol-based post-insertion to functionalize PFC-NE [6]. The disialoganglioside GD2 is highly expressed in several tumors especially in the brain, whereas the levels in normal tissue are low. Thus, GD2 antibodies are an interesting target for detection and treatment of neuroblastoma and have already been used in clinical trials. The specific binding to the GD2 overexpressing neuroblastoma cell line *Kelly* was assessed by flow cytometry, while coupling of nonspecific IgG or incubation of cells displaying no GD2 expression did not lead to active targeting of GD2-NE.

As mentioned above, tumor tissue is characterized by rapid proliferation of tumor cells. One important factor for cell proliferation is folate (or folic acid) implicating an upregulation of its receptor in a variety of tumors. This has been exploited to detect folate-receptor positive tumors by i.v. administration of folate-labeled PFC-NE as shown in Fig. 4.4. Subsequent ^{19}F MRI actually confirmed specific ^{19}F signals in implanted tumors besides some deposition of PFC-NE in the liver [61]. With enhanced proliferation, supply of the increased tumor mass with nutrients and oxygen becomes essential and boosts angiogenesis which is characterized by the expression of the $\alpha_v\beta_3$-integrin on the luminal side of the sprouting endothelial cells [90, 91]. Contrast agents targeted to these integrins have already been used to visualize tumor neoangiogenesis by optical methods, ultrasound, nuclear techniques and MRI [92]. Moreover, antagonists of $\alpha_v\beta_3$-integrin have been applied to inhibit tumor angiogenesis and metastasis.

One widely used targeting structure that binds to $\alpha_v\beta_3$-integrin is the RGD peptide sequence (Arg-Gly-Asp). This tripeptide is extensively investigated since many proteins of the extracellular matrix, from blood or the cell surface contain RGD motifs which are recognized by several integrins [93]. As a consequence, many applications of RGD peptides for imaging or drug-delivery purposes have been explored [26, 94, 95]. However, a key concern using RGD peptides as a targeting ligand is that integrins display a very broad expression level. The $\alpha_v\beta_3$-integrin for example is expressed by proliferating cells, immature endothelial cells, cancer cells and many immune cells [60]. To overcome the difficulties with the wide integrin expression, a recent study used RGD-labeled PFC-NE of about 250 nm in diameter to visualize newly formed blood vessels in a murine glioblastoma model [96]. Essentially, the size of these particles should be too large to passively leave the blood stream and therefore avoid unspecific distribution within the tissue. Indeed, the authors found higher PFC signals with the RGD-targeted NE compared to control NE within the first 90 min after administration. Nevertheless, they also observed considerable signal with untargeted PFC-NE, which was suggested to be caused by circulating PFC-NE in the tumor vasculature. However, it has to be taken into account that tumors display an abnormal blood vessel system with gaps of several hundreds of nanometers between endothelial cells and that several integrins are also upregulated on cancer cells which could lead to the diffusion of nanoparticles into the tissue and subsequent binding to the cancer cells.

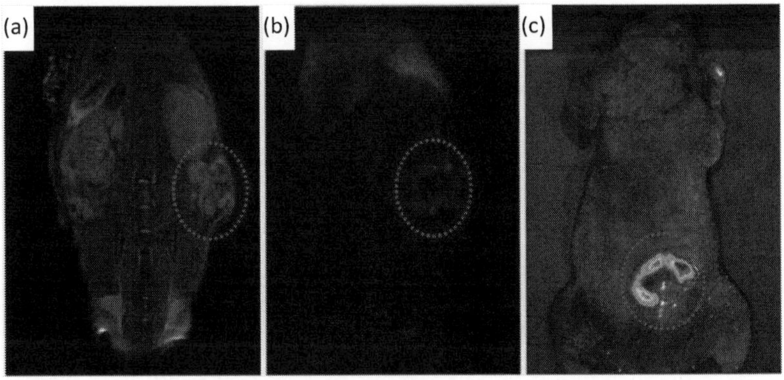

Figure 4.4 Detection of folate-receptor positive tumors using folate-labeled PFC-nanoemulsions. (a) ^{1}H MR image that displays the anatomy. Spinal cord and kidneys are visible. The location of the tumor is indicated by the red dashed line. (b) 19F MRI image showing the fluorine signal in the tumor as well as the liver. (c) Fluorescence image highlighting the localization of accumulated rhodamine labeled folate-PFC-NEs. Reprinted with permission from Ref. [61].

Since RGD-peptides display several disadvantages, nonpeptide targeting ligands (indazole- and quinolone-based peptidomimetics) to $\alpha_v\beta_3$-integrin have been developed by Bristol-Myers Squibb Medical Imaging (PCT/US1998/012010) [97]. The quinolone ligand has also been utilized in several studies in combination with PFC-NE [58, 60, 65, 68]. For example, Akers and coworkers used the near infrared dye cypate and PFOB emulsions as a scaffold equipped with the quinolone $\alpha_v\beta_3$ nonpeptide integrin ligand as a specific targeting moiety leading to targeted NE which should reside in the vessel lumen and bind only to endothelial $\alpha_v\beta_3$ integrins [60]. With this approach, the authors analyzed angiogenesis of implanted tumors within the ear by near-infrared microscopy and found that targeted PFC-NE remained in the tumor vasculature whereas smaller targeting ligands diffused into the tumor tissue. These results show that NEs restrained in the vasculature have specificity for cancer-induced angiogenesis. Other studies also used PFOB-NE to visualize neoangiogenesis in Vx-2 rabbit tumors by Gd-enhanced MRI or ^{111}In-based nuclear imaging [58, 68]. Unfortunately, none of these studies measured the fluorine content and solely used the nanoemulsions as a scaffold.

As already pointed out above, VCAM-1 is another molecule which is upregulated on the endothelium of the tumor vasculature and therefore has been used to image neoangiogenesis within cancer tissues [82].

4.3.3 Thrombosis

Thrombi are the causative agents of stroke, myocardial infarction or deep venous thrombosis and subsequent pulmonary embolism. Moreover, in recent years it has become evident that thrombosis is also a severe cofactor for morbidity and mortality in tumor patients [98]. Thrombi develop due to vessel injury, increased coagulativity or blood stasis (Virchow triad) which subsequently leads to activation and aggregation of platelets and the formation of a dense fibrin network which seals the site of injury and impairs blood loss or the spread of pathogens. Since activated platelets or fibrin are the main components of thrombi, many imaging-approaches for thrombus detection have focused on the targeting of fibrin or platelets [99, 100].

Fibrin forms a unique dense meshwork which stabilizes the clot. About one hour after thrombus formation, a process called clot retraction occurs which further tightens the fibrin network. Clot retraction is mediated by the active form of FXIIIa which is a transglutaminase that cross-links the different fibrin strands [101]. Interestingly, the half-life of FXIIIa is quite short which is thought to be due to the oxidation of cysteine residues within the catalytic center. FXIIIa also cross-links α2-antiplasmin (among several other proteins) to the clot which blocks the activity of the serin protease plasmin and impairs clot lysis [102, 103].

To visualize fibrin within thrombi, the group of Wickline and Lanza has conducted several studies using fibrin antibodies to target PFOB-NE to blood clots ex vivo or in vivo [41, 42, 62]. Winter as well as Flacke and coworkers used biotinylated and gadolinium-decorated PFOB-NE together with biotinylated anti-fibrin mAbs for in vitro detection of human clots by MRI [41, 62]. Subsequent high resolution MRI scans revealed that the nanoparticles bound almost exclusively to the surface of the thrombi. Since complete antibodies have some disadvantages for in vivo applications (see also Section 4.2.1.1), Flacke et al. also generated fibrin specific Fab-fragments to obtain targeted PFOB-NE for visualization of thrombi by T1-weighted ^{1}H MRI ex vivo as well as in vivo in dogs. Another study from the same group used a similar thrombus detection strategy, but directly measured the fluorine content of the biotinylated PFC-NEs which contained either PFCE or PFOB [42].

Fibrin antibodies detect early as well as chronic thrombi, therefore it is not possible to differentiate between fresh and organized thrombi. However, fresh clots are more treatable with fibrinolytic therapy than older and organized thrombi. A probe that specifically detects early thrombi could help to assess fibrinolytic lysis schemata to avoid severe bleeding risks. Several years ago, a small peptide probe was described which is covalently cross-linked to the fibrin network by active factor XIIIa during the first hour after thrombus initiation [104]. The targeting peptide was derived from α2-antiplasmin (see Section 4.3.3)—a serum protein that is incorporated into the thrombus and confers clot resistance to endogenous lysis. This FXIIIa peptide-probe has also been used for imaging purposes by NI imaging to detect experimental thrombi in the superficial femoral artery. More recently, Miserus and coworkers have generated a Gd-based α2-antiplasmin probe for detection of carotid artery thrombi by MRI [105].

Based on these investigations, our group has used this peptide to functionalize PFC-NE for specific detection of deep venous thrombi which are difficult to identify by conventional ultrasound techniques. With this approach we were able to detect

$FeCl_3$-induced thrombi in the inferior vena cava by noninvasive ^{19}F MRI with high SNR (>70) for thrombi with less than 1 mm in diameter (Fig. 4.5). A control peptide which is less efficiently cross-linked to the fibrin network showed more than 10-fold lower ^{19}F signals. Moreover, since we added the peptide to the distal end of a polyethylene glycol-chain, we achieved a strongly reduced PFC-uptake by blood monocytes and a shorter half-life in the blood. As a consequence, we found strong ^{19}F signals in the thrombus but not in the area of the surgery, which resulted in one single ^{19}F spot at the location of the thrombus (see Fig. 4.5b, top middle) [106].

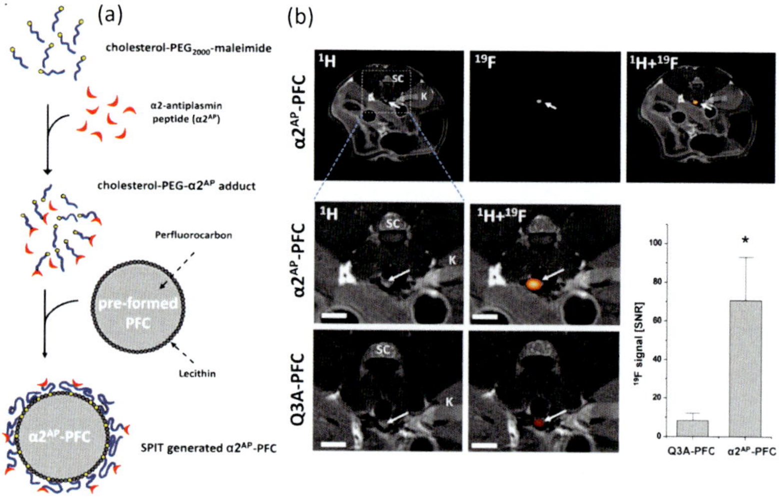

Figure 4.5 Specific detection of thrombi using $\alpha 2$-antiplasmin peptide labeled PFC-NE. (a) Scheme showing the principle of sterol-based post-insertion. First, a peptide derived from $\alpha 2$-antiplasmin is linked to a cholesterol-PEG-maleimide. Next, the $\alpha 2^{AP}$-PEG-cholesterol conjugate is added to preformed PFC-NE which leads to the spontaneous insertion of the cholesterol-moiety into the lipid layer of the NE. (b) Detection of $FeCl_3$-induced deep venous thrombi by in vivo $^1H/^{19}F$ MRI. Images of ^1H and ^{19}F (top) and the merged $^1H/^{19}F$ image. Dashed lines show the magnified areas surrounding the vena cava. The lower panels display ^1H (left) and merged (middle) images of animals receiving $\alpha 2^{AP}$-PFCs (top) or Q3A-PFCs (bottom). Reprinted with permission from Ref. [106].

Additionally, we were able to visualize small thrombi in the lungs after experimental induction of pulmonary embolism. Since α2-antiplasmin probes do not accumulate in older thrombi, these agents are of particular interest to differentiate early thrombi from organized clots which are often resistant to fibrinolysis. On the other hand, this approach could help to detect acute thrombus formation as complication during implantation of devices like stents or valves. Nevertheless, it has to be taken into account that the time frame of the α2-antiplasmin probes is restricted due to the relatively short enzymatic half-life of FXIIIa of about 30 min [107].

Besides fibrin, platelets are another major player for the formation of thrombi. Platelets bind to the activated endothelium and also actively participate in the formation of the blood clot as well as in clot retraction. The most abundant protein expressed on activated platelets is gpIIb/IIIa, also known as CD61/CD41. About 80,000 copies of gpIIb/IIIa are found on the surface of one resting platelet [108]. Upon activation, gpIIb/IIIa changes its conformation which facilitates a tight binding of the platelets to fibrin. The group of Prof. Peter generated a single chain antibody (scFv anti-LIBS) which exclusively binds to the activated form of gpIIb/IIIa and has been successfully used to detect activated platelets by ultrasound [53], PET [109] and MRI via coupled iron oxide nanoparticles [110]. Recently, Mayenfels labeled PFC-NE with the scFv anti-LIBS and demonstrated the feasibility of directing the targeted NE to the activated form of gpIIb/IIIa in clon3 cells expressing high levels of this receptor [6].

4.3.4 Atherosclerotic Plaques and Restenosis

The formation of an atherosclerotic plaque is a multi-step process. Initially, endothelial dysfunction leads to infiltration of leukocytes into the subendothelial space and accumulation of LDL-cholesterol followed by proliferation of vascular smooth muscle cells. Infiltrated macrophages develop into lipid-rich macrophages (foam cells) which accumulate within the vascular intima. Thereafter, the plaque gradually evolves into an advanced and complex lesion characterized by a lipid core and a fibrous cap [111]. Again, the recruitment of inflammatory cells from the circulation is predominantly mediated by adhesion molecules, which are expressed on the vascular endothelium. Two important adhesion molecules which are found in atherosclerotic regions are ICAM-1 and VCAM-1. These proteins are not expressed under normal conditions but become upregulated in atherosclerotic lesions and other inflammatory states (see Section 4.3.1). Both proteins are involved in the tight binding of leukocytes to the vessel wall; while VCAM-1 expression is restricted to lesions and lesion-prone areas, ICAM-1 expression is also found in more distal regions of the plaque [112]. Thus, VCAM-1 has been exploited as a targeting structure to visualize atherosclerotic lesions [113–115].

To target VCAM-1, Pan and coworkers used a fusion-peptide consisting of VCAM-1 targeting peptide and a melittin-derived membrane-integrating sequence to post-functionalize PFC-NE [82]. Melittin is a cytolytic peptide which comprises about 50% of the venom of the honeybee *apis mellifera* and has been observed to spontaneously integrate into the lipid layer of PFC-NE. Interestingly, whereas melittin peptide destroys the lipid layers of liposomes and cells, it does not destabilize the lipid layer of PFC-NE. The authors modified the peptide to remove the cytolytic part and used the membrane-integrating part to functionalize the nanoemulsions. Targeted nanocarriers self-assembled by mixing NEs with the peptide. These nanoemulsions were evaluated in vivo in models of ApoE-deficient mice (atherosclerosis model) and a breast cancer mouse model with STAT-1-deficient tumor cells; both models displayed increased binding of the targeted NE to the affected endothelium as assessed by ex vivo ^{19}F MR measurements.

Current treatment of atherosclerotic plaques involves the insertion of stents for stable dilation of the stenotic blood vessel which is accompanied by strong anticoagulation and antiplatelet therapy to avoid restenosis. Nevertheless, the rate of restenosis with conventional metal stents is about 20%, whereas drug-eluting stents (releasing paclitaxel or sirolimus) can reduce the rate to less than 10% [116]. One major factor contributing to restenosis are vascular smooth muscle cells (VSMCs) which proliferate and immigrate into the affected area as part of the normal wound healing process and display tissue factor on their surface that can induce the blood coagulation cascade. Impairing the proliferation and migration of VSMCs is thus a potential means to reduce restenosis. Therefore, Lanza and coworkers aimed to target nanoparticles to VSMCs for concomitant delivery of doxorubicine or paclitaxel to impair proliferation [63]. To this end, both drugs were passively absorbed to the NE surface and additionally a high payload of Gd was bound to the NE surface. For directing the drug-loaded NE towards VSMCs, cells were first incubated with biotinylated tissue factor mAbs, then with streptavidin and finally with biotinylated PFOB-NE. High resolution ^{1}H MRI revealed an T1-weighted (T1w) enhancement of VSMCs exposed to the targeted NE, while ^{19}F MR spectroscopy was used to quantify the PFOB signal for determination of the carried drug payloads. A similar approach was conducted with isolated smooth muscle cells from pig aortas that were targeted using tissue factor antibodies to PFC-NE [64]. In this study, it was shown that even sparse cellular epitopes can be sufficient to generate Gd-enhanced T1w contrast at 1.5 T—even with picomolar concentrations of PFCs with very high payloads of Gd-complexes bound to the particle surface (92,000 Gd/particle).

4.3.5 Targeting of Stem Cells

Stem cells play an imminent role in many biological processes. The visualization of these cell populations to monitor their fate and function is of major interest. Several groups have employed an ex vivo labeling strategy, using SPIOs or PFCs, for MRI tracking of stem cells that were grafted to reconstitute destroyed tissue function [117–120]. In situ labeling of endogenous stem cells by intravenous delivery of contrast agents is much more complicated, since most stem cells are found only in trace amounts in tissue and exhibit limited phagocytic properties to incorporate the administered contrast agents.

Interestingly, it has recently been suggested that nearly all tissues contain small numbers of dormant mesenchymal stem cells (MSC) that can be activated and migrate to the site of tissue injury to perform repair functions. Although these endogenous MSCs are found in low numbers only, they can undergo tremendous proliferation upon activation. For example, upon myocardial infarction (MI) a mesenchymal-like cell population massively proliferates within day 3–7 post surgery to form an epicardial layer covering the entire site of myocardial injury [121]. Importantly, it has been reported that a sub-population of these distinct cardiac stem cells has the ability to migrate into the damaged area to facilitate repair functions and to differentiate into fibroblasts, smooth muscle cells or cardiomyocytes [122]. Our group has recently observed that this cell type internalizes PFC-NE in vivo when administered on day 4 after MI, a time point where monocyte infiltration into the injured heart has already declined [19]. Isolation and ex vivo incubation confirmed that these cells are indeed capable of internalizing PFC-NE and cellular debris. However, it should be noted that the delivery of conventional PFC-NE on day 4 post MI does not exclusively label this cell population, but also macrophages present within the infarcted area. Nevertheless, these data indicate that under certain circumstances, a targeted labeling of stem cells may be possible if a significant amount of this cell type is present and circulating PFC-NE can reach and enter these cells from the blood stream via adequate vascularization.

MSC populations are attractive targets, since this cell type seems to be generally involved in repair processes after injury in many different organs [123]. One step towards targeting of MSCs was again performed by Mayenfels. MSCs are characterized by several distinct cell surface markers like CD73, CD90 and CD105. In a first step, Mayenfels labeled PFC-NE (20% perfluorodecaline) with complete anti-CD90 mAbs (or nonspecific control IgGs) via cholesterol-PEG_{1300}-NHS. The CD90-PFC-NE resulted in a strong binding to human MSCs, whereas nonspecific IgG-PFC-NE did not result in any targeting as demonstrated by flow cytometry. Interestingly, our group has revealed that epicardium derived stem cells (see previous paragraph) also express CD90, CD73, and CD105 [19]. Although CD90 is also present on T-cells, this might pave the way for a specific labeling of epicardial stem cells by CD90 since the T-cell count in the heart, even after MI, is quite low.

4.4 Summary and Outlook

Fluorine MRI has gained a lot of attraction within the past years. ^{19}F is nearly absent from the human body and therefore fluorinated contrast agents generate an unequivocal "positive" contrast. Intravenously administered PFC-NE is endocytosed by immune cells which leads to the accumulation of ^{19}F signal at the site of inflammation. However, the ^{19}F MR imaging of inflammation represents the combination of signals from different immune cells (monocytes and macrophages, neutrophil granulocytes, etc.). As a consequence, a strong ^{19}F signal, for example in the heart, can be derived from a high number of neutrophils with potential harmful properties or a large amount of M2 macrophages with beneficial functions for healing processes. However, in the case of tumors it might be vice versa where M2 macrophages are detrimental since they facilitate neoangiogenesis and impair immune responses. A further very attractive cell population are T-cells, since they are key players in immune responses against viruses, tumors and play a critical role in autoimmune diseases and transplant rejections.

Although numerous studies have been performed using active targeting of different kinds of nanoparticles, overall, specific targeting approaches for PFC-NE are still limited. Further progress in directing PFC-NE to distinct cell types or acellular structures will strongly enhance the versatility of the ^{19}F MRI technique. Moreover, the visualization of individual immune cell populations will enable the assessment of the quality of the immune response or help monitoring therapeutic interventions.

Another interesting option for fluorine MRI will be the parallel imaging of different PFC-NE. The fluorine atoms in diverse PFCs display a relatively large range of chemical shifts that could enable the simultaneous imaging of distinct PFC-NE. Initial studies have been performed several years ago by combining PFCE and PFOB to visualize different cell populations or to label thrombi [42, 120]. In combination with a specific active targeting approach it might be possible to perform a simultaneous imaging of thrombi and macrophages, or neutrophils and T-cells, which would offer a noninvasive option to characterize immune responses and healing processes in more detail. In particular, in the case of personalized medicine, this technique could aid in the diagnosis and help to select and monitor specific individual treatment regiments to optimize the outcome for the patient and reduce cost of therapies.

References

1. Faude F., Wolf S., and Wiedemann P. Flüssige Perfluorkarbone in der Glaskörper-und Netzhautchirurgie, 2000; **97**, 652–661.
2. Krafft M. P. Fluorocarbons and fluorinated amphiphiles in drug delivery and biomedical research. *Adv. Drug Deliv. Rev.*, 2001; **47**(2–3), 209–228.
3. O'Hagan D. Understanding organofluorine chemistry. An introduction to the C-F bond. *Chem. Soc. Rev.*, 2008; **37**(2), 308–319.
4. Riess J. G. Understanding the fundamentals of perfluorocarbons and perfluorocarbon emulsions relevant to in vivo oxygen delivery. *Artif. Cells Blood Substit. Immobil. Biotechnol.*, 2005; **33**(1), 47–63.
5. Grapentin C., Mayenfels F., Barnert S., Süss R., Schubert R., Temme S., Jacoby C., Schrader J., Flögel U. Optimization of perfluorocarbon nanoemulsions for molecular imaging by ^{19}F MRI. In: Seifalian A, de Mel A, Kalaskar DM eds. (Hrsg): Nanomedicine One Central Press, 2014; 268–286.
6. Mayenfels F. Fluorcarbonhaltige Nanoemulsionen zur Anwendung in der ^{1}H/ ^{19}F-Magnetresonanztomographie Albert-Ludwigs Universität, Freiburg i. Br.; Thesis, 2012.
7. Riess, J. G. Oxygen carriers ('blood substitutes')—raison d'etre, chemistry, and some physiology. *Chem. Rev.*, 2001, 101, 2797–2919.
8. Krafft, M. P. Detrimental Effect of Excess Lecithin on the Stability of Fluorocarbon Lecithin Emulsions. *J. Phys. Chem.*, 1991, **95**(14), 5673–5676.
9. Arlauskas. Analysis of Ostwald Ripening in Fluorocarbon Emulsions by Sedimentation Field-Flow Fractionation. *Artif Cells Blood Substit Immobil Biotechnol.*, 1994, **22**(4), 1317–23.
10. Weers J. G., Ni Y., Tarara T. E., Pelura T. J., Arlauskas R. A. The effect of molecular-diffusion on initial particle-size distributions in phospholipid-stabilized fluorocarbon emulsions. *Colloids Surf A Physicochem Eng Aspects*, 1994; **84**(1), 81–87.
11. Grapentin C., Barnert S., Schubert R. Monitoring the stability of perfluorocarbon nanoemulsions by Cryo-TEM image analysis and dynamic light scattering. *PLoS One*, 2015; **10**(6), e0130674.
12. Bertilla S. M., Thomas J. L., and Krafft, M. P. Cosurfactant effect of a semifluorinated alkane at a fluorocarbon/water interface: Impact on the stabilization of fluorocarbon-in-water emulsions. *Langmuir*, 2004; **20**, 3920–3924.

13. Flögel U., Su S., Kreideweiß I., Ding Z., Galbarz L., Fu J., Jacoby C., Witzke O., Schrader J. Noninvasive detection of graft rejection by in vivo (19)F MRI in the early stage. *Am. J. Transplant.*, 2011; **11**(2), 235–244.

14. Flögel U., Burghoff S., van Lent P. L., Temme S., Galbarz L., Ding Z., El-Tayeb A., Huels S., Bönner F., Borg N., Jacoby C., Muller C. E., van den Berg W. B., Schrader J. Selective activation of adenosine A2A receptors on immune cells by a CD73-dependent prodrug suppresses joint inflammation in experimental rheumatoid arthritis. *Sci. Transl. Med.*, 2012; **4**(146), 146ra108.

15. Temme S., Jacoby C., Ding Z., Bönner F., Borg N., Schrader J., Flögel U. Technical advance: monitoring the trafficking of neutrophil granulocytes and monocytes during the course of tissue inflammation by noninvasive ^{19}F MRI. *J. Leukoc. Biol.*, 2014; **95**(4), 689–97.

16. Jacoby C., Borg N., Heusch P., Sauter M., Bönner F., Kandolf R., Klingel K., Schrader J., Flogel U. Visualization of immune cell infiltration in experimental viral myocarditis by ^{19}F MRI in vivo. *MAGMA*, 2014; **27**(1), 101–106.

17. Ebner B., Behm P., Jacoby C., Burghoff S., French B. A., Schrader J., Flögel U. Early assessment of pulmonary inflammation by ^{19}F MRI in vivo. *Circ. Cardiovasc. Imaging*, 2010; **3**(2), 202–210.

18. Flögel U., Ding Z., Hardung H., Jander S., Reichmann G., Jacoby C., Schubert R., Schrader J. In vivo monitoring of inflammation after cardiac and cerebral ischemia by fluorine magnetic resonance imaging. *Circulation,* 2008; **118**(2), 140–148.

19. Ding Z., Temme S., Quast C., Friebe D., Jacoby C., Zanger K., Bidmon H. J., Grapentin C., Schubert R., Flögel U., Schrader J. Epicardium-Derived Cells Formed After Myocardial Injury Display Phagocytic Activity Permitting In Vivo Labeling and Tracking. *Stem Cells Transl. Med.*, 2016; **5**(5), 639–50.

20. Kobayashi H., Turkbey B., Watanabe R., Choyke P. L. Cancer drug delivery: considerations in the rational design of nanosized bioconjugates. *Bioconjug. Chem.*, 2014; **25**(12), 2093–100.

21. Schleich N., Po C., Jacobs D., Ucakar B., Gallez B., Danhier F., Preat V. Comparison of active, passive and magnetic targeting to tumors of multifunctional paclitaxel/SPIO-loaded nanoparticles for tumor imaging and therapy. *J. Control. Release,* 2014; **194**, 82–91.

22. Diou O., Tsapis N., Giraudeau C., Valette J., Gueutin C., Bourasset F., Zanna S., Vauthier C., Fattal E. Long-circulating perfluorooctyl bromide nanocapsules for tumor imaging by ^{19}FMRI. *Biomaterials,* 2012; **33**(22), 5593–5602.

23. Ahrens E. T., Bulte J. W. Tracking immune cells in vivo using magnetic resonance imaging. *Nat. Rev. Immunol.*, 2013; **13**(10), 755–763.

24. Bertrand N., Wu J., Xu X., Kamaly N., Farokhzad O. C. Cancer nanotechnology: The impact of passive and active targeting in the era of modern cancer biology. *Adv. Drug Deliv. Rev.*, 2014; **66**, 2–25.

25. Holliger P., Hudson P. J. Engineered antibody fragments and the rise of single domains. *Nat. Biotechnol.*, 2005; **23**(9), 1126–1136.

26. Zhao G., Rodriguez B. L. Molecular targeting of liposomal nanoparticles to tumor microenvironment. *Int. J. Nanomed.*, 2013; **8**, 61–71.

27. Wörn A., Pluckthun A. Stability engineering of antibody single-chain Fv fragments. *J. Mol. Biol.*, 2001; **305**(5), 989–1010.

28. Ahmad Z. A., Yeap S. K., Ali A. M., Ho W. Y., Alitheen N. B., Hamid M. scFv antibody: Principles and clinical application. *Clin. Dev. Immunol.*, **2012**; 2012, 980250.

29. Siontorou C. G. Nanobodies as novel agents for disease diagnosis and therapy. *Int. J. Nanomed.*, 2013; **8**, 4215–4227.

30. van der Linden R. H., Frenken L. G., de G. B., Harmsen M. M., Ruuls R. C., Stok W., de R. L., Wilson S., Davis P., Verrips C. T. Comparison of physical chemical properties of llama VHH antibody fragments and mouse monoclonal antibodies. *Biochim. Biophys. Acta*, 1999; 1431(**1**), 37–46.

31. Chakravarty R., Goel S., Cai W. Nanobody: The "magic bullet" for molecular imaging? *Theranostics*, 2014; **4**(4), 386–398.

32. Vives E., Schmidt J., Pelegrin A. Cell-penetrating and cell-targeting peptides in drug delivery. *Biochim. Biophys. Acta*, 2008; **1786**(2), 126–138.

33. Kelly C., Jefferies C., Cryan S. A. Targeted liposomal drug delivery to monocytes and macrophages. *J. Drug Deliv.*, **2011**; 2011, 727241.

34. Caravan P., Das B., Dumas S., Epstein F. H., Helm P. A., Jacques V., Koerner S., Kolodziej A., Shen L., Sun W. C., Zhang Z. Collagen-targeted MRI contrast agent for molecular imaging of fibrosis. *Angew. Chem. Int. Ed. Engl.*, 2007; **46**(43), 8171–8173.

35. Fogal V., Zhang L., Krajewski S., Ruoslahti E. Mitochondrial/cell-surface protein p32/gC1qR as a molecular target in tumor cells and tumor stroma. *Cancer Res.*, 2008; **68**(17), 7210–7218.

36. Konkalmatt P. R., Deng D., Thomas S., Wu M. T., Logsdon C. D., French B. A., Kelly K. A. Plectin-1 targeted AAV vector for the molecular imaging of pancreatic cancer. *Front. Oncol.*, 2013; **3**, 84.

37. Meyer C., Hahn U., Rentmeister A. Cell-specific aptamers as emerging therapeutics. *J. Nucleic Acids*, **2011**; 2011, 904750.

38. Irache J. M., Salman H. H., Gamazo C., Espuelas S. Mannose-targeted systems for the delivery of therapeutics. *Expert. Opin. Drug Deliv.*, 2008; **5**(6), 703–724.

39. Bies C., Lehr C. M., Woodley J. F. Lectin-mediated drug targeting: History and applications. *Adv. Drug Deliv. Rev.*, 2004; **56**(4), 425–435.

40. Low P. S., Kularatne S. A. Folate-targeted therapeutic and imaging agents for cancer. *Curr. Opin. Chem. Biol.*, 2009; **13**(3), 256–262.

41. Winter P. M., Caruthers S. D., Yu X., Song S. K., Chen J., Miller B., Bulte J. W., Robertson J. D., Gaffney P. J., Wickline S. A., Lanza G. M. Improved molecular imaging contrast agent for detection of human thrombus. *Magn. Reson. Med.*, 2003; **50**(2), 411–416.

42. Morawski A. M., Winter P. M., Yu X., Fuhrhop R. W., Scott M. J., Hockett F., Robertson J. D., Gaffney P. J., Lanza G. M., Wickline S. A. Quantitative "magnetic resonance immunohistochemistry" with ligand-targeted ^{19}F nanoparticles. *Magn. Reson. Med.*, 2004; **52**(6), 1255–1262.

43. Daniels T. R., Bernabeu E., Rodriguez J. A., Patel S., Kozman M., Chiappetta D. A., Holler E., Ljubimova J. Y., Helguera G., Penichet M. L. The transferrin receptor and the targeted delivery of therapeutic agents against cancer. *Biochim. Biophys. Acta*, 2012; **1820**(3), 291–317.

44. Amoozgar Z., Yeo Y. Recent advances in stealth coating of nanoparticle drug delivery systems. *Wiley Interdiscip. Rev. Nanomed. Nanobiotechnol.*, 2012; **4**(2), 219–233.

45. Hermanson, G. T. Antibody modification and conjugation. In *Bioconjugate Techniques*, 3rd ed. Academic Press, London, 2013.

46. Manjappa A. S., Chaudhari K. R., Venkataraju M. P., Dantuluri P., Nanda B., Sidda C., Sawant K. K., Murthy R. S. Antibody derivatization and conjugation strategies: Application in preparation of stealth immunoliposome to target chemotherapeutics to tumor. *J. Control. Release*, 2011; **150**(1), 2–22.

47. Nobs L., Buchegger F., Gurny R., Allemann E. Current methods for attaching targeting ligands to liposomes and nanoparticles. *J. Pharm. Sci.*, 2004; **93**(8), 1980–1992.

48. Hermanson, G. T. Antibody modification and conjugation. In *Bioconjugate Techniques*, 3rd ed. Academic Press, London, 2013. pp. 867–920.

49. Harmsen M. M., van Solt C. B., Fijten H. P. Enhancement of toxin- and virus-neutralizing capacity of single-domain antibody fragments by N-glycosylation. *Appl. Microbiol. Biotechnol.*, 2009; **84**(6), 1087–1094.

50. Mimura Y., Ashton P. R., Takahashi N., Harvey D. J., Jefferis R. Contrasting glycosylation profiles between Fab and Fc of a human IgG protein studied by electrospray ionization mass spectrometry. *J. Immunol. Methods*, 2007; **326**(1–2), 116–126.

51. Beduneau A., Saulnier P., Hindre F., Clavreul A., Leroux J. C., Benoit J. P. Design of targeted lipid nanocapsules by conjugation of whole antibodies and antibody Fab' fragments. *Biomaterials*, 2007; **28**(33), 4978–4990.

52. Hermanson, G. T. 2013. (Strept)avidin-biotin systems. In *Bioconjugate Techniques*, 3rd ed., Academic Press, London, pp. 465–505.

53. Wang X., Hagemeyer C. E., Hohmann J. D., Leitner E., Armstrong P. C., Jia F., Olschewski M., Needles A., Peter K., Ahrens I. Novel single-chain antibody-targeted microbubbles for molecular ultrasound imaging of thrombosis: Validation of a unique noninvasive method for rapid and sensitive detection of thrombi and monitoring of success or failure of thrombolysis in mice. *Circulation*, 2012; **125**(25), 3117–3126.

54. Chen M. H., Soda Y., Izawa K., Kobayashi S., Tani K., Maruyama K., Tojo A., Asano S. A versatile drug delivery system using streptavidin-tagged pegylated liposomes and biotinylated biomaterials. *Int. J. Pharm.*, 2013; **454**(1), 478–485.

55. Winter P. M., Cai K., Caruthers S. D., Wickline S. A., Lanza G. M. Emerging nanomedicine opportunities with perfluorocarbon nanoparticles. *Expert. Rev. Med. Devices*, 2007; 4(**2**), 137–145.

56. Ishida T., Iden D. L., Allen T. M. A combinatorial approach to producing sterically stabilized (Stealth) immunoliposomal drugs. *FEBS Lett.*, 1999; **460**(1), 129–133.

57. Moreira J. N., Ishida T., Gaspar R., Allen T. M. Use of the post-insertion technique to insert peptide ligands into pre-formed stealth liposomes with retention of binding activity and cytotoxicity. *Pharm. Res.*, 2002; **19**(**3**), 265–269.

58. Hu G., Lijowski M., Zhang H., Partlow K. C., Caruthers S. D., Kiefer G., Gulyas G., Athey P., Scott M. J., Wickline S. A., Lanza G. M. Imaging of Vx-2 rabbit tumors with alpha(nu)beta3-integrin-targeted ^{111}In nanoparticles. *Int. J. Cancer*, 2007; **120**(9), 1951–1957.

59. Wang C. H., Kang S. T., Lee Y. H., Luo Y. L., Huang Y. F., Yeh C. K. Aptamer-conjugated and drug-loaded acoustic droplets for ultrasound theranosis. *Biomaterials*, 2012; **33**(6), 1939–1947.

60. Akers W. J., Zhang Z., Berezin M., Ye Y., Agee A., Guo K., Fuhrhop R. W., Wickline S. A., Lanza G. M., Achilefu S. Targeting of alpha(nu)beta(3)-integrins expressed on tumor tissue and neovasculature using fluorescent small molecules and nanoparticles. *Nanomedicine* (Lond.), 2010; **5**(5), 715–726.

61. Bae P. K., Jung J., Lim S. J., Kim D., Kim S. K., Chung B. H. Bimodal perfluorocarbon nanoemulsions for nasopharyngeal carcinoma targeting. *Mol. Imaging Biol.*, 2013; **15**(4), 401–410.

62. Flacke S., Fischer S., Scott M. J., Fuhrhop R. J., Allen J. S., McLean M., Winter P., Sicard G. A., Gaffney P. J., Wickline S. A., Lanza G. M. Novel MRI contrast agent for molecular imaging of fibrin: Implications for detecting vulnerable plaques. *Circulation*, 2001; **104**(11), 1280–1285.

63. Lanza G. M., Yu X., Winter P. M., Abendschein D. R., Karukstis K. K., Scott M. J., Chinen L. K., Fuhrhop R. W., Scherrer D. E., Wickline S. A. Targeted antiproliferative drug delivery to vascular smooth muscle cells with a magnetic resonance imaging nanoparticle contrast agent: Implications for rational therapy of restenosis. *Circulation*, 2002; **106**(22), 2842–2847.

64. Morawski A. M., Winter P. M., Crowder K. C., Caruthers S. D., Fuhrhop R. W., Scott M. J., Robertson J. D., Abendschein D. R., Lanza G. M., Wickline S. A. Targeted nanoparticles for quantitative imaging of sparse molecular epitopes with MRI. *Magn Reson. Med.*, 2004; **51**(3), 480–486.

65. Schmieder A. H., Wang K., Zhang H., Senpan A., Pan D., Keupp J., Caruthers S. D., Wickline S. A., Shen B., Wagner E. M., Lanza G. M. Characterization of early neovascular response to acute lung ischemia using simultaneous $^{19}F/^{1}H$ MR molecular imaging. *Angiogenesis*, 2014; **17**(1), 51–60.

66. Southworth R., Kaneda M., Chen J., Zhang L., Zhang H., Yang X., Razavi R., Lanza G., Wickline S. A. Renal vascular inflammation induced by Western diet in ApoE-null mice quantified by ^{19}F NMR of VCAM-1 targeted nanobeacons. *Nanomedicine*, 2009; **5**(3), 359–367.

67. Winter P. M., Morawski A. M., Caruthers S. D., Fuhrhop R. W., Zhang H., Williams T. A., Allen J. S., Lacy E. K., Robertson J. D., Lanza G. M.,

Wickline S. A. Molecular imaging of angiogenesis in early-stage atherosclerosis with alpha(v)beta3-integrin-targeted nanoparticles. *Circulation* 2003; **108**(18), 2270–2274.

68. Winter P. M., Caruthers S. D., Kassner A., Harris T. D., Chinen L. K., Allen J. S., Lacy E. K., Zhang H., Robertson J. D., Wickline S. A., Lanza G. M. Molecular imaging of angiogenesis in nascent Vx-2 rabbit tumors using a novel alpha(nu)beta3-targeted nanoparticle and 1.5 tesla magnetic resonance imaging. *Cancer Res.*, 2003; **63**(18), 5838–5843.

69. Balducci A., Helfer B. M., Ahrens E. T., O'Hanlon C. F., III, Wesa A. K. Visualizing arthritic inflammation and therapeutic response by fluorine-19 magnetic resonance imaging (^{19}F MRI). *J. Inflamm.* (Lond.), 2012; **9**(1), 24.

70. Balducci A., Wen Y., Zhang Y., Helfer B. M., Hitchens T. K., Meng W. S., Wesa A. K., Janjic J. M. A novel probe for the non-invasive detection of tumor-associated inflammation. *Oncoimmunology*, 2013; **2**(2), e23034.

71. Hitchens T. K., Ye Q., Eytan D. F., Janjic J. M., Ahrens E. T., Ho C. ^{19}F MRI detection of acute allograft rejection with in vivo perfluorocarbon labeling of immune cells. *Magn. Reson. Med.*, 2011; 65(**4**), 1144–1153.

72. Kadayakkara D. K., Ranganathan S., Young W. B., Ahrens E. T. Assaying macrophage activity in a murine model of inflammatory bowel disease using fluorine-19 MRI. *Lab Invest.*, 2012; 92(**4**), 636–645.

73. van Heeswijk R. B., De B. J., Kania G., Gonzales C., Blyszczuk P., Stuber M., Eriksson U., Schwitter J. Selective in vivo visualization of immune-cell infiltration in a mouse model of autoimmune myocarditis by fluorine-19 cardiac magnetic resonance. *Circ. Cardiovasc. Imaging*, 2013; **6**(2), 277–284.

74. Bönner F., Jacoby C., Temme S., Borg N., Ding Z., Schrader J., Flogel U. Multifunctional MR monitoring of the healing process after myocardial infarction. *Basic Res. Cardiol.*, 2014; **109**(5), 430.

75. Srinivas M., Morel P. A., Ernst L. A., Laidlaw D. H., Ahrens E. T. Fluorine-19 MRI for visualization and quantification of cell migration in a diabetes model. *Magn. Reson. Med.*, 2007; **58**(4), 725–734.

76. Srinivas M., Turner M. S., Janjic J. M., Morel P. A., Laidlaw D. H., Ahrens E. T. In vivo cytometry of antigen-specific t cells using ^{19}F MRI. *Magn. Reson. Med.*, 2009; **62**(3), 747–753.

77. Ahrens E. T., Flores R., Xu H., Morel P. A. In vivo imaging platform for tracking immunotherapeutic cells. *Nat. Biotechnol.*, 2005; **23**(8), 983–987.

78. Waiczies H., Lepore S., Janitzek N., Hagen U., Seifert F., Ittermann B., Purfurst B., Pezzutto A., Paul F., Niendorf T., Waiczies S. Perfluorocarbon particle size influences magnetic resonance signal and immunological properties of dendritic cells. *PLoS One*, 2011; **6**(7), e21981.

79. Ichimura T., Morikawa T., Kawai T., Nakagawa T., Matsushita H., Kakimi K., Kume H., Ishikawa S., Homma Y., Fukayama M. Prognostic significance of CD204-positive macrophages in upper urinary tract cancer. *Ann. Surg. Oncol.*, 2014; **21**(6), 2105–2112.

80. Kelley J. L., Ozment T. R., Li C., Schweitzer J. B., Williams D. L. Scavenger receptor-A (CD204): A two-edged sword in health and disease. *Crit. Rev. Immunol.*, 2014; **34**(3), 241–261.

81. Sugimoto M., Mitsunaga S., Yoshikawa K., Kato Y., Gotohda N., Takahashi S., Konishi M., Ikeda M., Kojima M., Ochiai A., Kaneko H. Prognostic impact of M2 macrophages at neural invasion in patients with invasive ductal carcinoma of the pancreas. *Eur. J. Cancer*, 2014; **50**(11), 1900–1908.

82. Pan H., Myerson J. W., Hu L., Marsh J. N., Hou K., Scott M. J., Allen J. S., Hu G., San R. S., Lanza G. M., Schreiber R. D., Schlesinger P. H., Wickline S. A. Programmable nanoparticle functionalization for in vivo targeting. *FASEB J.*, 2013; **27**(1), 255–264.

83. Granger, D. N., Senchenkova E. Inflammation and the microcirculation. San Rafael (CA): Morgan & Claypool Life Sciences; 2010.

84. Tonnesen M. G., Feng X., Clark R. A. Angiogenesis in wound healing. *J. Invest. Dermatol. Symp. Proc.*, 2000; **5**(1), 40–46.

85. Chidlow J. H., Jr., Shukla D., Grisham M. B., Kevil C. G. Pathogenic angiogenesis in IBD and experimental colitis: New ideas and therapeutic avenues. *Am. J. Physiol. Gastrointest. Liver Physiol.*, 2007; **293**(1), G5–G18.

86. Timar J., Dome B., Fazekas K., Janovics A., Paku S. Angiogenesis-dependent diseases and angiogenesis therapy. *Pathol. Oncol. Res.*, 2001; **7**(2), 85–94.

87. Long D. M., Multer F. K., Greenburg A. G., Peskin G. W., Lasser E. C., Wickham W. G., Sharts C. M. Tumor imaging with x-rays using macrophage uptake of radiopaque fluorocarbon emulsions. *Surgery*, 1978; **84**(1), 104–112.

88. Patronas N. J., Hekmatpanah J., Doi K. Brain-tumor imaging using radiopaque perfluorocarbon. A preliminary report. *J. Neurosurg.*, 1983; **58**(5), 650–653.

89. Young S. W., Enzmann D. R., Long D. M., Muller H. H. Perfluoroctylbromide contrast enhancement of malignant neoplasms: Preliminary observations. *Am. J. Roentgenol.*, 1981; **137**(1), 141–146.

90. Clark R. A., Tonnesen M. G., Gailit J., Cheresh D. A. Transient functional expression of alphaVbeta 3 on vascular cells during wound repair. *Am. J. Pathol.*, 1996; **148**(5), 1407–1421.

91. Max R., Gerritsen R. R., Nooijen P. T., Goodman S. L., Sutter A., Keilholz U., Ruiter D. J., De Waal R. M. Immunohistochemical analysis of integrin alpha vbeta3 expression on tumor-associated vessels of human carcinomas. *Int. J. Cancer*, 1997; **71**(3), 320–324.

92. Schottelius M., Laufer B., Kessler H., Wester H. J. Ligands for mapping alphavbeta3-integrin expression in vivo. *Acc. Chem. Res.*, 2009; **42**(7), 969–980.

93. Ruoslahti E. RGD and other recognition sequences for integrins. *Annu. Rev. Cell Dev. Biol.*, 1996; **12**, 697–715.

94. Mukherjee B., Satapathy B. S., Mondal L., Dey N. S., Maji R. Potentials and challenges of active targeting at the tumor cells by engineered polymeric nanoparticles. *Curr. Pharm. Biotechnol.*, 2013; **14**(15), 1250–1263.

95. Svensen N., Walton J. G., Bradley M. Peptides for cell-selective drug delivery. *Trends Pharmacol. Sci.*, 2012; **33**(4), 186–192.

96. Giraudeau C., Geffroy F., Meriaux S., Boumezbeur F., Robert P., Port M., Robic C., Le B. D., Lethimonnier F., Valette J. ^{19}F molecular MR imaging for detection of brain tumor angiogenesis: In vivo validation using targeted PFOB nanoparticles. *Angiogenesis*, 2013; **16**(1), 171–179.

97. Harris T. D., Kalogeropoulos S., Nguyen T., Liu S., Bartis J., Ellars C., Edwards S., Onthank D., Silva P., Yalamanchili P., Robinson S., Lazewatsky J., Barrett J., Bozarth J. Design, synthesis, and evaluation of radiolabeled integrin alpha v beta 3 receptor antagonists for tumor imaging and radiotherapy. *Cancer Biother. Radiopharm.*, 2003; **18**(4), 627–641.

98. Lauw M. N., van Doormaal F. F., Middeldorp S., Buller H. R. Cancer and venous thrombosis: Current comprehensions and future perspectives. *Semin. Thromb. Hemost.*, 2013; **39**(5), 507–514.

99. Ciesienski K. L., Caravan P. Molecular MRI of thrombosis. *Curr. Cardiovasc. Imaging Rep.*, 2010; **4**(1), 77–84.

100. Houshmand S., Salavati A., Hess S., Ravina M., Alavi A. The role of molecular imaging in diagnosis of deep vein thrombosis. *Am. J. Nucl. Med. Mol. Imaging*, 2014; **4**(5), 406–425.

101. Kasahara K., Souri M., Kaneda M., Miki T., Yamamoto N., Ichinose A. Impaired clot retraction in factor XIII A subunit-deficient mice. *Blood*, 2010; **115**(6), 1277–1279.

102. Greenberg C. S., Birckbichler P. J., Rice R. H. Transglutaminases: Multifunctional cross-linking enzymes that stabilize tissues. *FASEB J.*, 1991; **5**(15), 3071–3077.

103. Munday A. D., Lopez J. A. Factor XIII: Sticking it to platelets. *Blood*, 2013; **122**(19), 3246–3247.

104. Jaffer F. A., Tung C. H., Wykrzykowska J. J., Ho N. H., Houng A. K., Reed G. L., Weissleder R. Molecular imaging of factor XIIIa activity in thrombosis using a novel, near-infrared fluorescent contrast agent that covalently links to thrombi. *Circulation*, 2004; **110**(2), 170–176.

105. Miserus R. J., Herias M. V., Prinzen L., Lobbes M. B., Van Suylen R. J., Dirksen A., Hackeng T. M., Heemskerk J. W., van Engelshoven J. M., Daemen M. J., van Zandvoort M. A., Heeneman S., Kooi M. E. Molecular MRI of early thrombus formation using a bimodal alpha2-antiplasmin-based contrast agent. *JACC. Cardiovasc. Imaging*, 2009; **2**(8), 987–996.

106. Temme S., Grapentin C., Quast C., Jacoby C., Grandoch M., Ding Z., Owenier C., Mayenfels F., Fischer J. W., Schubert R., Schrader J., Flogel U. Noninvasive imaging of early venous thrombosis by ^{19}F magnetic resonance imaging With targeted perfluorocarbon nanoemulsions. *Circulation*, 2015; **131**(16), 1405–1414.

107. Robinson B. R., Houng A. K., Reed G. L. Catalytic life of activated factor XIII in thrombi. Implications for fibrinolytic resistance and thrombus aging. *Circulation* 2000; **102**(10), 1151–1157.

108. Wagner C. L., Mascelli M. A., Neblock D. S., Weisman H. F., Coller B. S., Jordan R. E. Analysis of GPIIb/IIIa receptor number by quantification of 7E3 binding to human platelets. *Blood*, 1996; **88**(3), 907–914.

109. Alt K., Paterson B. M., Ardipradja K., Schieber C., Buncic G., Lim B., Poniger S. S., Jakoby B., Wang X., O'Keefe G. J., Tochon-Danguy H. J., Scott A. M., Ackermann U., Peter K., Donnelly P. S., Hagemeyer C. E. Single-chain antibody conjugated to a cage amine chelator and labeled with positron-emitting copper-64 for diagnostic imaging of activated platelets. *Mol. Pharm.*, 2014; **11**(8), 2855–2863.

110. von Elverfeldt D., Maier A., Duerschmied D., Braig M., Witsch T., Wang X., Mauler M., Neudorfer I., Menza M., Idzko M., Zirlik A., Heidt T., Bronsert P., Bode C., Peter K., von Zur M. C. Dual-contrast molecular imaging allows noninvasive characterization of myocardial ischemia/reperfusion injury after coronary vessel occlusion in mice by magnetic resonance imaging. *Circulation*, 2014; **130**(8), 676–687.

111. Preiss D. J., Sattar N. Vascular cell adhesion molecule-1: A viable therapeutic target for atherosclerosis? *Int. J. Clin. Pract.*, 2007; **61**(4), 697–701.

112. Cybulsky M. I., Iiyama K., Li H., Zhu S., Chen M., Iiyama M., Davis V., Gutierrez-Ramos J. C., Connelly P. W., Milstone D. S. A major role for VCAM-1, but not ICAM-1, in early atherosclerosis. *J. Clin Invest.*, 2001; **107**(10), 1255–1262.

113. Nahrendorf M., Keliher E., Panizzi P., Zhang H., Hembrador S., Figueiredo J. L., Aikawa E., Kelly K., Libby P., Weissleder R. 18F-4V for PET-CT imaging of VCAM-1 expression in atherosclerosis. *JACC Cardiovasc. Imaging*, 2009; **2**(10), 1213–1222.

114. Orbay H., Hong H., Zhang Y., Cai W. Positron emission tomography imaging of atherosclerosis. *Theranostics*, 2013; **3**(11), 894–902.

115. Inaba Y., Lindner J. R. Molecular imaging of disease with targeted contrast ultrasound imaging. *Transl. Res.*, 2012; **159**(3), 140–148.

116. Marx S. O., Totary-Jain H., Marks A. R. Vascular smooth muscle cell proliferation in restenosis. *Circ. Cardiovasc. Interv.*, 2011; **4**(1), 104–111.

117. Zhou R., Acton P. D., Ferrari V. A. Imaging stem cells implanted in infarcted myocardium. *J. Am. Coll. Cardiol.*, 2006; **48**(10), 2094–2106.

118. Boehm-Sturm P., Aswendt M., Minassian A., Michalk S., Mengler L., Adamczak J., Mezzanotte L., Lowik C., Hoehn M. A multi-modality platform to image stem cell graft survival in the naive and stroke-damaged mouse brain. *Biomaterials*, 2014; **35**(7), 2218–2226.

119. Boehm-Sturm P., Mengler L., Wecker S., Hoehn M., Kallur T. In vivo tracking of human neural stem cells with ^{19}F magnetic resonance imaging. *PLoS One*, 2011; **6**(12), e29040.

120. Partlow K. C., Chen J., Brant J. A., Neubauer A. M., Meyerrose T. E., Creer M. H., Nolta J. A., Caruthers S. D., Lanza G. M., Wickline S. A. ^{19}F magnetic resonance imaging for stem/progenitor cell tracking with multiple unique perfluorocarbon nanobeacons. *FASEB J.*, 2007; **21**(8), 1647–1654.

121. van W. B., Gunst Q. D., Moorman A. F., van den Hoff M. J. Cardiac regeneration from activated epicardium. *PLoS One*, 2012; **7**(9), e44692.

122. Epstein J. A. Franklin H. Epstein Lecture. Cardiac development and implications for heart disease. *N. Eng. J. Med.*, 2010; **363**(17), 1638–1647.

123. Shi Y., Su J., Roberts A. I., Shou P., Rabson A. B., Ren G. How mesenchymal stem cells interact with tissue immune responses. *Trends Immunol.*, 2012; 33(**3**), 136–143.

Chapter 5

Responsive Probes for ^{19}F MRS/MRI

Aneta Keliris,[a] Klaus Scheffler,[a,b] and Jörn Engelmann[a]

[a]*High-Field Magnetic Resonance Centre, Max Planck Institute for Biological Cybernetics, Tübingen, Germany*
[b]*Department for Biomedical Magnetic Resonance, University of Tübingen, Tübingen, Germany*

aneta.keliris@tuebingen.mpg.de

5.1 Introduction

Fluorine (^{19}F) magnetic resonance imaging (MRI) and spectroscopy (MRS) are increasingly gaining interest and relevance in biomedical and clinical research [1–5]. The 100% natural abundance of ^{19}F, high intrinsic NMR sensitivity (83% relative to ^{1}H) and the large chemical shift range (>300 ppm as compared to ~10 ppm in ^{1}H NMR) as well as the high sensitivity to changes in the molecular and chemical microenvironment make this nuclide an ideal candidate for a wide range of MRI/MRS applications. Conveniently, the ^{19}F resonance frequency is close to that of ^{1}H, allowing the use of most ^{1}H NMR/MRI instrumentation for ^{19}F measurements. Further, the concentration of detectable endogenous mobile fluorine in the body is well below the detection limit of ^{19}F MRI. Higher concentrations of fluorine are immobilized in

Fluorine Magnetic Resonance Imaging
Edited by Ulrich Flögel and Eric Ahrens

ISBN 978-981-4745-31-4 (Hardcover), 978-981-4745-32-1 (eBook)
www.panstanford.com

bone matrix and teeth but the corresponding ^{19}F signals are not visible to conventional MRI due to the very short T_2 relaxation times. Thus, exogenously administered fluorinated reporter molecules or drugs can be visualized without any interference from endogenous background signals. This is a great advantage of ^{19}F in comparison to ^{1}H MRI that frequently suffers from low contrast due to the presence of natural background signal from intrinsic proton molecules, leading to difficulties in unambiguous signal interpretation in acquired images. However, the correct anatomical localization of the ^{19}F signal requires the acquisition of proton ^{1}H MR images (consecutively or simultaneously).

In 1985, McFarland and coworkers performed the first in vivo ^{19}F MRI studies in rats [6]. Since then, a continuous development has taken place in the field of ^{19}F MRI/MRS, constantly broadening the range of applications (e.g., monitoring of drug metabolism, cell tracking, plaque detection, sensing microenvironmental changes, etc.). Historically, (per)fluorinated molecules (e.g., perfluorocarbons and highly fluorous gases) have been used for ^{19}F MRI/MRS, however, in recent years new ^{19}F probes have been developed. There have been several promising reports on fluorinated micelles, dendrimers, hyperbranched polymers, as well as responsive and multimodal probes. These fluorinated agents are generally categorized into two groups: (a) targeted and (b) responsive depending on their mode of action. Accordingly, targeted ^{19}F-reporters function by accumulating at the site of interest via binding to specific targets, for instance cell surface epitopes, or are used for tracking of macrophages, immune and stem cells [3, 7]. Whereas for targeted agents the detected signal is dependent basically on their local concentration, for responsive ^{19}F probes a designated modulation of the fluorine signal takes place in situ only in response to specific stimuli in the local biochemical environment. Hence, responsive agents have the potential for high specificity and may provide deeper insights into biochemical processes of interest. Accordingly, several promising ^{19}F agents have been reported that proved to be responsive to key physiological processes and diagnostically relevant parameters such as enzymatic activity, changes in metal ion concentration, pH, and pO_2 [5].

In this chapter, we aim to provide a brief overview on the recent developments in responsive ^{19}F reporters, the mechanisms underlying their "responsiveness," as well as the obstacles related to detection sensitivity and efficient utilization of these agents in vivo.

5.2 Response Mechanisms

Responsive ^{19}F probes primarily function by alternating their ^{19}F chemical shift and/or relaxation rates ($R_1 = 1/T_1$ and $R_2 = 1/T_2$) via stimuli driven reversible (e.g., ion sensing) or irreversible (e.g., sensing of enzymatic activity) changes in molecular structure (Fig. 5.1). In some responsive probes, interactions with the biological marker of interest induce alterations of the relaxation parameters, albeit without changing the structure of the fluorinated agent (e.g., pO_2 mapping by perfluorocarbons).

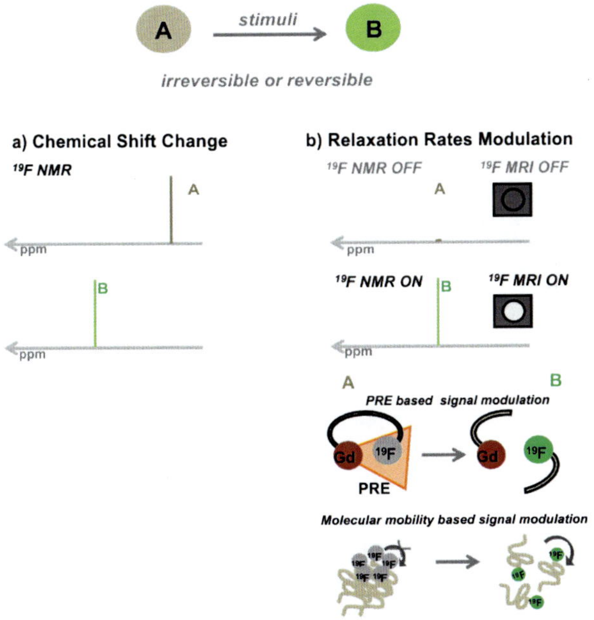

Figure 5.1 Schematic representation of the main response mechanisms.

The most frequently exploited mechanism of activation so far involves chemical shift changes observed upon interaction of low molecular weight fluorinated agents (e.g., amino acid derivatives and pharmaceutical drugs) with their target.

Another way of controlling the ^{19}F signal is based on the modulation of probe ^{19}F relaxation times via paramagnetic relaxation enhancement (PRE). The presence of a paramagnetic metal ion in close proximity to the fluorine induces a strong relaxation enhancement and a chemical shift change, depending on the metal ion used. The ^{19}F signal is modulated in response to changes in the microenvironment of the probe by either alteration of the distance between fluorine and the paramagnetic center or perturbation in local electron density in a reversible or irreversible way [8–10]. Relaxation times can be shortened to such extent that the ^{19}F signal becomes undetectable by common ^{19}F MR sequences in the absence of target. This has been achieved either via utilizing the PRE effect of strongly paramagnetic moieties such as gadolinium or by steadily restricting the molecular dynamic motion of fluorinated agents (via supramolecular self-assembly or immobilization, see Fig. 5.1). The stimulus triggered activation of such agents leads to prolonged ^{19}F T_2 values, and thus recovery of the attenuated signal in ^{19}F NMR spectra and MR images. Various "off-on" probes for sensing microenvironmental changes employ this principle.

A recent alternative approach, named iCEST (ion chemical exchange saturation transfer), is based on chemical exchange saturation transfer (CEST) combined with ^{19}F MRI. CEST, extensively used in ^{1}H MRI, exploits the gain in sensitivity by the specific magnetization transfer from a low concentrated to a bulk molecular species [11].

5.3 Classes of ^{19}F Responsive Probes

5.3.1 pH-Activatable ^{19}F Probes

A tight regulation of pH homoeostasis in the body is of essential importance. A large deviation from the physiological pH like in acidosis and alkalosis is accompanied by substantial alterations of normal biochemical processes (e.g., enzyme kinetics). For instance, tissue acidosis is a characteristic hallmark of inflammation processes and tumors. Local changes in pH of affected regions compared to normal tissue (e.g., pH 7.4 versus 6.8 in cancer lesions) can be measured by using properly functionalized ^{19}F sensors. This generally involves incorporation of organic groups capable of undergoing protonation/deprotonation at defined pH that causes structural changes within the ^{19}F sensor and consequently leads to the desired alternation of fluorine signal. Deutsch et al. have pioneered the development of pH sensitive ^{19}F MR chemical shift indicators by using a series of fluorinated alanine derivatives (chemical shift range $\Delta\delta \sim 2$ ppm, Fig. 5.2) for the assessment of intra- and extra-cellular pH [12].

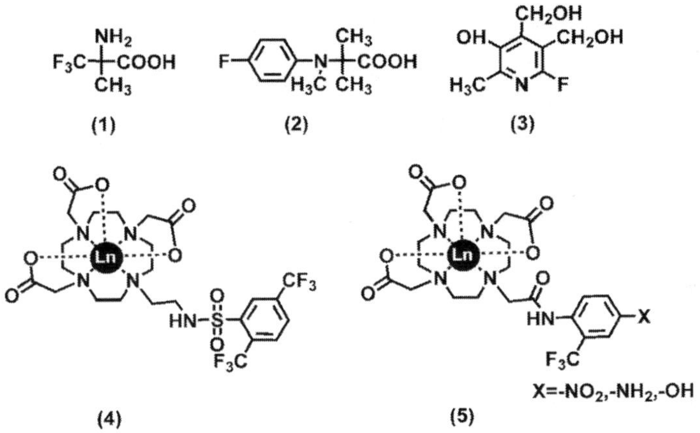

Figure 5.2 Examples of pH-responsive ^{19}F probes. Displayed are (1) α-trifluororomethyl alanine, (2) *N,N*-(methyl-2-carboxyisopropyl)-4-fluoroaniline, (3) 6-fluoropyridoxol, and (4) and (5) are macrocyclic lanthanide (III) complexes with different fluorinated pH-responsive moieties (Ln, lanthanide ion).

Further improvement was achieved by using aromatic molecules with a larger chemical shift response, for instance fluorine-substituted aniline derivatives (~5–15 ppm acid/base shift, Fig. 5.2) [13] and the vitamin B6 analog 6-fluoropyridoxol (~10 ppm acid/base shift, Fig. 5.2) [14]. More recent studies introduced a new class of pH-sensitive ^{19}F chemical shift agents, paramagnetic fluorine-labeled lanthanide complexes. The main benefits of incorporating lanthanide (III) metal ions into ^{19}F probes are to reduce the relaxation times of the fluorine nuclei, facilitating shorter acquisition times, and amplifying ^{19}F chemical shift sensitivity to small structural changes [15]. In this regard, Senanayake et al. reported fluorinated macrocyclic lanthanide (III) complexes with a sulfonamide moiety that undergoes reversible coordination to the metal ion center controlled by its pH-dependent protonation states (Fig. 5.2) [10]. This structural reorganization of the sulfonamide-chelating arm within the complex caused a significant shift in the resonance frequency of the CF_3 groups. In another study, a *para*-substituent (i.e., $-NO_2$, $-OH$ or $-NH_2$) was introduced as a functional unit for regulating protonation/deprotonation of the amide hydrogen and thereby the pH sensitivity of fluorinated lanthanide (III) complexes (Fig. 5.2) [16]. The nitro-substituted holmium complexes, for instance, showed a pH-dependent shift of the CF_3 group from −55.1 to −36.8 ppm.

In the past few years, there have been several reports on promising ^{19}F nanoprobes that can be "turned on" from an "off state" through the regulation of their ^{19}F relaxation times in a pH-dependent manner [17–19]. Accordingly, a series of PEGylated nanogels composed of cross-linked poly[2-(*N*,*N*-diethylamino)ethyl methacrylate]-copoly(2,2,2-trifluoroethyl methacrylate) (PEAMA-co-PTFEMA) gel core were obtained that showed an off-on regulation of ^{19}F MR signals by a relaxation enhancement mechanism upon adjusting the extracellular pH from 7.4 to 6.5 [19]. The change was driven by "swelling" of the gel core upon protonation of the amino groups in PEAMA at acidic pH, resulting in a 8.7-fold larger hydrodynamic volume of the nanogel. This enhanced the molecular motion of ^{19}F moieties within the probe and consequently led to an increase of ^{19}F T_2 relaxation times (T_2 = 56.8 ms vs. ~100 μs at pH 7.4) resulting in a "turning on" of the initially quenched ^{19}F signal.

Recently, Huang et al. reported on a series of multichromatic ^{19}F MR nanoprobes with remarkable binary (off/on) response to narrow pH transitions ($\Delta pH_{OFF/ON} \approx 0.25$ pH unit) [18]. Here, a suppression of ^{19}F signals at pH > pK_a was achieved by the self-assembly of hydrophobic micelles that effectively restricted motion of the fluorinated polymer chains. Upon pH activation (pH < pK_a), micelles disassembled to unimers with reappearance of the ^{19}F MR signal. Using diblock copolymers consisting of different pH responsive segments (with different pK_a values) and ^{19}F reporter moieties (different ^{19}F chemical shifts), the authors have shown the feasibility of the proposed approach as a qualitative measure of pH differences in the microenvironment (i.e., pH = 7.4, 6.5, 5.5) by ^{19}F MR techniques [18].

In another recent study, a nanosized ^{19}F on/off system consisting of Au-fluorescent mesoporous silica nanoparticles (Au-FMSNPs), was used for pH-triggered ^{19}F MR/fluorescence imaging of cancer cells [17]. The ^{19}F agent was sealed inside the nanopores of FMSNPs by covalently attaching gold nanoparticles to their surface via acid-cleavable hydrazone linkages. These Au NPs were functionalized with folic acid, a targeting moiety of high affinity towards folate receptors expressed on cancer cells. Upon incubation of normal human lung fibroblast cells (MRC-5) with Au-FMSNPs, no signal in ^{19}F MR images was observed as the ^{19}F agent remained entrapped inside nanoparticles and Au-FMSNPs localized outside of the plasma membrane of MRC-5 cells. In the case of folate receptor positive human lung cancer cells (A549), the folate-functionalized nanoparticles efficiently entered the cells, the hydrazone bonds were cleaved in the intracellular acidic environment (pH 5–6), and the uncapped ^{19}F agent was released resulting in a detectable intracellular ^{19}F MR signal.

Further, Gianolio et al. have demonstrated pH mapping by utilizing a ratiometric $^{19}F/^{1}H$ MRI approach based on poly-β-cyclodextrin. Although the ^{19}F signal was not directly modulated by pH, it was used to normalize the relaxation enhancement values generated by a pH-responsive Gd-based contrast agent [20].

5.3.2 Metal Ion Responsive ^{19}F Sensors

Metal ions such as Ca^{2+}, Zn^{2+}, or Fe^{2+} play an essential role in numerous biological processes (i.e., signal transduction, gene expression, catalytic action, and muscle contraction). Their excess or deficiency may cause severe diseases. Hence, ^{19}F sensors that can specifically report on fluctuations in ion concentrations in real time are of great importance for understanding of various physiological processes. The majority of ^{19}F probes for ion detection reported so far explored changes in chemical shift of the ^{19}F signal, occurring upon binding of the ion of interest to its chelator. In this regard, fluorinated NMR sensors have been demonstrated that reported on $[Mg^{2+}]$ [21] and $[Na^{+}]$ [22] in biological systems, as well as a series of fluoro-cryptands and crown ethers with structures optimized for hosting of various other metal ions [23].

To date, among the physiological relevant metal ions, the detection of [Ca^{2+}] has been utilized primarily. Calcium ions play an essential role in numerous signaling pathways and cellular processes such as neuronal transmission, muscle contraction, cell growth and regulation of enzyme activity. Thus, the specific detection of its fluctuations by means of ^{19}F MR techniques would be of great value, particularly for in vivo studies. Accordingly, the fluorine-labeled Ca^{2+} indicator 5F-BAPTA (5,5′-difluoro-1,2-bis(O-aminophenoxy)ethane-*N*,*N*,*N*′,*N*′-tetraacetic acid) has been developed and showed a ^{19}F chemical shift of 6 ppm upon reversible binding of Ca^{2+} [24]. This probe has been employed to measure [Ca^{2+}] in cell culture, perfused brain slices and perfused beating heart (i.e., rat and ferret) and in vivo [25–28]. Besides, 5F-BAPTA can also be used for the detection of other metal ions (e.g., Cd^{2+}, Zn^{2+}) by observing different chemical shift changes depending on the nature of the chelated metal ion [27, 29].

Bar-Shir et al. have demonstrated a novel application of the well-known 5F-BAPTA molecule for MRI. Their promising iCEST (ion chemical exchange saturation transfer) approach combined ^{19}F and CEST MRI for the specific detection of Ca^{2+} ions. This method involved radiofrequency labeling at the ^{19}F resonance frequency of low concentrated Ca^{2+}-5F-BAPTA and subsequent detection of the magnetization transfer to the ^{19}F frequency of bulk free 5F-BAPTA, with a large signal enhancement with Ca^{2+} binding. Using the proposed iCEST strategy, a low concentration of Ca^{2+} (~500 nM) could be efficiently detected in vitro (Fig. 5.3) with high specificity over other divalent cations (Zn^{2+}, Mg^{2+}) due to their different chemical shifts [11]. The authors reported that the best performance was observed at a Ca^{2+}/5F-BAPTA ratio of 1:1000 to 1:2000. The obtained signal change in the iCEST image depended on the detectability of the bulk 5F-BAPTA. Thus, even Ca^{2+}-concentrations below 500 nM might be detectable. However, the applicability of this approach to visualize intracellular calcium concentrations (typically ~10–100 nM) needs to be proven.

More recently, Harvey et al. demonstrated a paramagnetic ^{19}F chemical shift probe in which a CF_3 group was locked in a stable position within an arylamide moiety strongly coordinated to a lanthanide metal ion. A calcium-sensing chelator, that in absence of any Ca^{2+} weakly binds to the lanthanide via an amide carbonyl, was placed at the *trans*-position of the macrocylic ring

[8]. Reversible interaction of this chelator with Ca^{2+} induced an alteration in the Ln^{3+} metal center coordination environment resulting in changes in δ_F of the CF_3 group ($\Delta\delta_F$ = 3 ppm (−69 to −66 ppm)) for the Dy complex. A similar chemical shift of 3.9 ppm (−94.6 to −98.5 ppm) was observed for the Tm complex. Another approach using paramagnetic complexes to measure [Ca^{2+}] made use of an aromatic CF_3 incorporated within the calcium chelator. Upon binding of Ca^{2+} to the chelator, the distance from the CF_3 to the lanthanide center changed, as a result of the reorganization within the molecule [30]. A modulation of the ^{1}H MR signal occurred for the Gd complex. In ^{19}F NMR spectra, line broadening of the fluorine signal was observed upon addition of calcium. Of note, in Tb, Dy, and Tm complexes, addition of Ca^{2+} (1 Eq) caused a complete disappearance of ^{19}F NMR signal.

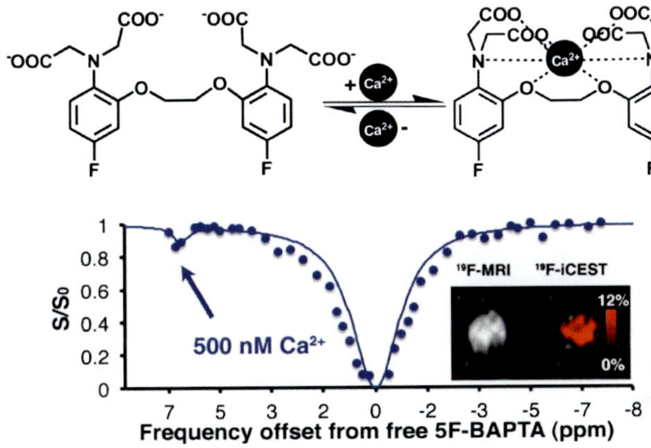

Figure 5.3 Ca^{2+}-binding by 5F-BAPTA. Schematic representation (top panel) of the dynamic exchange process of free and Ca^{2+}-bound 5F-BAPTA. The bottom panel depicts a ^{19}F-iCEST Z-spectrum of a solution containing 0.5 mM 5F-BAPTA and 500 nM Ca^{2+}. The inset depicts a ^{19}F MR image of the sample with an overlaid iCEST image. Adapted with permission from Bar-Shir et al., *J. Am. Chem. Soc.*, 135 (2013), 12164–12167. Copyright (2013) American Chemical Society.

5.3.3 Responsive ^{19}F Probes for Detection of Proteins and Their Function

Proteins are indispensable macromolecules that have numerous and diverse functions in living organisms, such as catalyzing metabolic reactions, molecule transportation, replicating DNA, and are involved in cell signaling as well as signal transduction. Hence, the ability to image specific proteins or their action noninvasively in vivo by ^{19}F MRI is of great importance as this would not only allow insights into the studied process of interest (i.e., neuronal activity, gene expression), but also early detection of abnormalities as indicators of disease states (e.g., overexpression of receptors on cancer cells).

5.3.3.1 Enzyme responsive probes

A number of ^{19}F probes were developed for the detection of enzymes [5]. These essential biocatalysts serve as indicators of disease processes (i.e., overexpression of matrix metalloproteinases on cancer cells) and are important markers of gene expression [31]. Responsiveness of enzyme sensing probes is typically driven by highly specific cleavage or formation of covalent bonds. Advantageously, enzymes regenerate after each reaction cycle and so they can be used repeatedly to convert many single substrate molecules to their products. Therefore, a large ^{19}F signal amplification can potentially be achieved by using such an approach, as enzyme detection is not inherently limited by its concentration. A widespread application is the monitoring of cancer chemotherapeutics metabolism via chemical shifts in ^{19}F MR spectra as reviewed comprehensively by others [32, 33]. For instance, an activation of the prodrug 5-fluorocysteine to the potent cytotoxic 5-fluorouracil (5-FU) by bacterial cytosine deaminase (CD) within tumors was noninvasively detected due to a 1.2 ppm chemical shift difference [34]. This enzyme was introduced specifically to cancer cells by using a monoclonal antibody-CD conjugate [34] or by gene delivery strategies [35]. Carboxypeptidase G2 (CPG2) is another important prodrug-activating bacterial enzyme utilized in targeted chemotherapy strategies, which was monitored by ^{19}F MRS probes [36].

The expression of specific genes and their regulation is commonly assessed through utilization of various bacterial reporter enzymes not present in the studied biological system and thus are easily detectable. One of the first, and still very popular, reporter enzymes is β-galactosidase (β-gal) encoded by the *LacZ* gene. Various ^{19}F MR chemical shift probes for detection of β-gal activity were demonstrated, which consisted of an enzyme cleavable galactose moiety and a fluorinated aglycone [37–39]. More recently, dual-modality approaches have become favored for enzyme detection as a way of acquiring more complementary and unambiguous information about the studied processes of interest. In this regard, we reported on a β-gal responsive dual-modal $^{1}H/^{19}F$ MRI paramagnetic probe, Gd-DOMF-Gal, in which the fluorine bearing unit and a Gd-complex were kept in close proximity to each other via a self-immolative linker coupled to a galactose moiety [9]. Exploiting the well-known phenomenon of intramolecular shortening of ^{19}F relaxation by gadolinium (PRE effect, see above) [40], the fluorine MR signal of Gd-DOMF-Gal was quenched in the absence of β-gal. Enzymatic conversion of Gd-DOMF-Gal resulted in a simultaneous "turning on" of the ^{19}F NMR/MRI signal and an altered effect of the Gd^{3+} complex on the relaxation properties of surrounding water. As a consequence, a decrease of longitudinal relaxivity by 46% and 27% was observed in HEPES and in PBS buffer, respectively, resulting in a pronounced alteration in ^{1}H MR signal intensity as shown in Fig. 5.4. This change can most likely be ascribed to differences in the rotational dynamics between Gd–DOMF–Gal and the cleaved product.

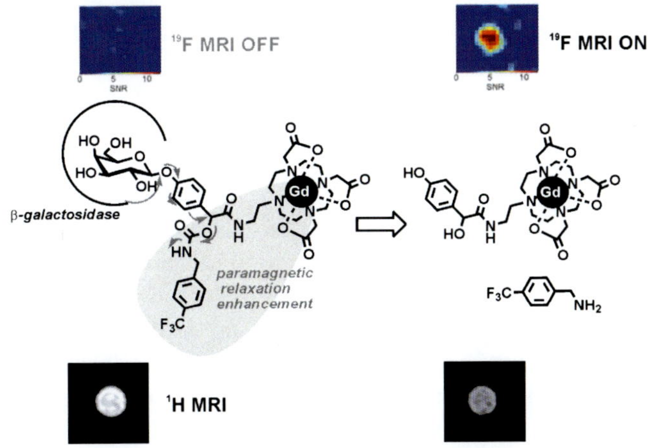

Figure 5.4 Chemical structure of Gd-DOMF-Gal and its activation by β-galactosidase. ^{19}F images of the solution of Gd-DOMF-Gal in the absence of β-gal (top left) and after its enzymatic conversion for 50 min (top right). Representative ^{1}H MR images of the same samples are shown in lower panels.

In another study, Mizukami et al. presented a mono-modal β-gal responsive ^{19}F MR probe, in which a PRE strategy was explored to deliver an off/on response [41]. Mason and co-workers demonstrated another dual ^{1}H/^{19}F approach for the detection of β-gal activity, in which enzymatic hydrolysis of a fluorosalicylaldehyde aroylhydrazone β-D-galactopyranoside induced a ^{19}F chemical shift difference of 7.65 ppm [42]. In the presence of ferric ions, the released aglycone chelator for iron formed a complex that generated a strong ^{1}H MRI contrast in T_2-weighted images and caused the disappearance of the ^{19}F MRI signal. Sensing the activity of intracellular β-gal requires efficient cellular internalization of the probe. In an alternative approach, a cell-surface displayed β-lactamase (β-lac) reporter enzyme and a corresponding ^{19}F NMR/MRI probe (PRE strategy similar to β-gal) have been used to monitor gene expression without the need of probe internalization [43].

Human monoamine oxidase A (hMAO-A) is an important enzyme that degrades monoamine neurotransmitters, such as serotonin, norepinephrine and dopamine. Elevated levels of hMAO-A have been associated with psychiatric disorders, and hMAO-A inhibitors are used for therapy. Therefore, specific detection of hMAO-A activity is diagnostically important. A chemical shift-switching ^{19}F MRI-chromogenic probe was developed [44]. Upon its oxidation by hMAO-A, colored 2-fluoro-4-nitrophenol was produced that showed a ^{19}F chemical shift change of 4.2 ppm compared to the substrate. This difference was sufficient to image each compound selectively by ^{19}F MRI (Fig. 5.5).

Noninvasive monitoring of apoptosis is of great importance for in-depth understanding of metabolic pathways leading to programmed cell death in autoimmune diseases, but also for evaluation of drug efficacy such as those used for induction of apoptosis in tumor cells. Caspase-3 is an important marker of apoptosis associated with initiation of the underlying signal cascade. This enzyme recognizes the peptide sequence Asp-Glu-Val-Asp (DEVD) and selectively cleaves its C-terminal bond. Accordingly, Mizukami et al. reported on a ^{19}F MR probe, Gd-DOTA-DEVD-Tfb (Tfb = para-trifluoromethoxybenzyl) [40], and subsequently developed a dual-modal agent, Gd-DOTA-DEVD-AFC (AFC = 7-amino-4-trifluoromethyl coumarin) for the detection of caspase-3 activity (Fig. 5.5) [45]. Upon enzymatic conversion of

Gd-DOTA-DEVD-AFC, the fluorinated fluorophore, AFC, is cleaved off the peptide. This resulted in augmentation of the initially "quenched" ^{19}F MRI signal (cancelation of intramolecular PRE) and an increase in fluorescence intensity of the released AFC.

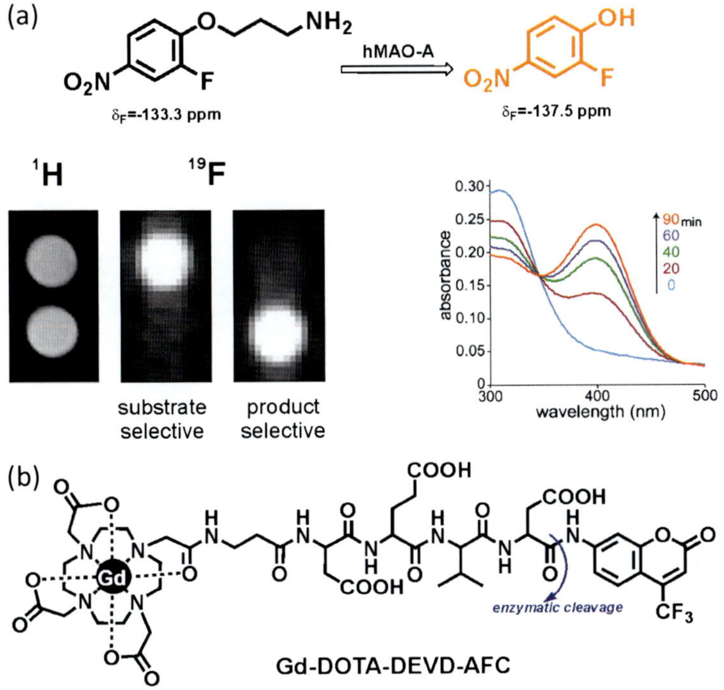

Figure 5.5 Bimodal probes for the detection of enzyme activity. (a) Detection of hMAO-A activity by optical and ^{19}F MR imaging. On top, the conversion of the probe is shown leading to a $\Delta\delta_F = 4.2$ ppm. Below, the ^{1}H and ^{19}F chemical shift selective imaging of substrate and converted product (left), as well as the time-dependent change in absorption spectra during the enzymatic conversion (right) are shown. Adapted with permission from Yamaguchi et al., *J. Am. Chem. Soc.*, 133 (2011) 14208–14211. Copyright (2011) American Chemical Society. (b) Chemical structure of Gd-DOTA-DEVD-AFC, a probe for the visualization of caspase 3 activity by ^{19}F MRI and fluorescence imaging; the arrow indicates the enzymatic cleavage site.

In parallel to enzyme responsive ^{19}F MR probes relying on the PRE effect or chemical shift changes as described above, fluorinated agents were developed, in which an off/on switching of the fluorine signals was achieved through a high restriction of molecular motion leading to very short ^{19}F relaxation times. In this regard, Tanaka et al. demonstrated two bimodal nanoparticle-based probes for the detection of alkaline phosphatase (AP) activity [46, 47].

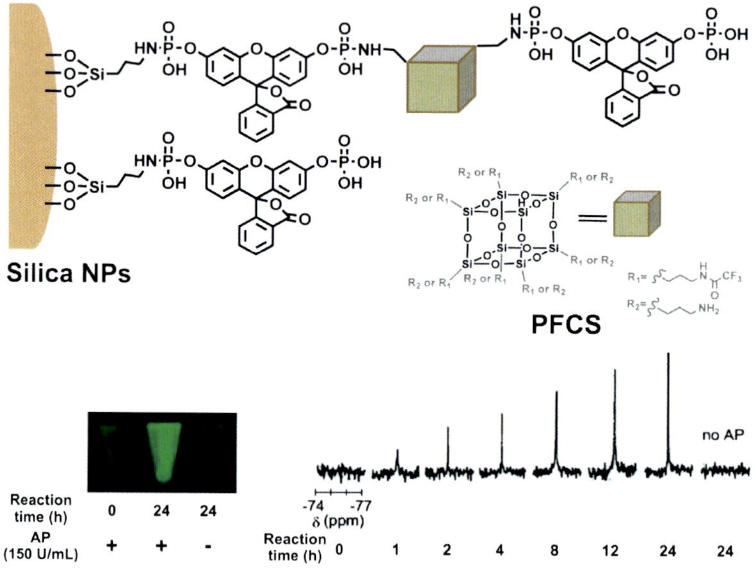

Figure 5.6 Bimodal silica nanoparticle-based probe for the detection of alkaline phosphatase (AP) activity by ^{19}F NMR and fluorescence. Schematic structure of the probe is shown on top (PFCS, perfluorinated cubic silsesquioxanes). The bottom left shows fluorescent images of samples before and after incubation in the presence or absence of AP for 24 h. The bottom right shows the time-dependent conversion of the probe monitored by ^{19}F NMR. Adapted with permission from Tanaka et al., *Bioconjug. Chem.*, 22 (2011) 1484–1490. Copyright (2011) American Chemical Society.

In the first approach, perfluorinated cubic silsesquioxanes (PFCSs) were anchored, via an AP cleavable linker, onto the surface of fluorescein-labeled silica nanoparticles (NPs). As long as the silsesquioxanes were coupled to the NPs, their molecular rotation dynamics were restricted, and thereby the ^{19}F NMR signal remained suppressed. Upon enzymatic activation, a time-dependent emergence of signal was observed by ^{19}F spectroscopy, while fluorescence emission remained constant over a period of 24 h [46]. In a follow-up study, PFCSs were similarly attached to the silica NPs, but via a phosphate-caged fluorescein as linker. In this way, a bimodal probe was created with both ^{19}F MR and fluorescence signal being "switched off" in the absence of AP (Fig. 5.6) [47]. Of note, the detection of glutathione reductase activity was demonstrated using a similar NP based strategy [48]. Here, PFCSs were attached onto the surface of the silica NPs via an enzyme cleavable disulfide linker.

Recently, Hamachi and co-workers presented a very promising alternative approach, in which self-assembling ^{19}F MR probes remained "switched off" in the absence of target enzyme. This occurred due to the effective restriction of molecular motions of the fluorinated moieties. In these nitroreductase (NTR) and matrix metalloproteinase (MMP-2) activity sensing probes, a hydrophobic alkyl chain was linked via an enzyme recognizable sequence (para-nitrobenzene for NTR and the GPLG-VRG peptide for MMP-2) to a ^{19}F-containing hydrophilic moiety [49]. Upon enzymatic cleavage of the recognition motif, the aggregates disassembled and produced a strong fluorine signal. Notably, the activity of endogenously secreted MMP-2 in the supernatant of tumor cells could be assessed by means of ^{19}F MRI using this strategy.

5.3.3.2 Sensing non-enzymatic proteins and nucleic acids

Apart from monitoring enzyme activity, Hamachi´s group applied a self-assembly strategy that relied on the specific binding to the target protein (i.e., human carbonic anhydrase I, avidin, dihydrofolate reductase) and not on the catalytic activity of enzymes [50–52] (Fig. 5.7). These "turn off-on" ^{19}F probes consisted of a fluorinated hydrophobic moiety (tail group) coupled via linker to a hydrophilic protein ligand (head group, cf. Fig. 5.7b) that is specifically recognized by the protein of interest. Upon binding of the ligand to the protein, a disassembly of these "silent" nanoprobes takes place yielding a readily detectable ^{19}F MR signal. Systematic structural studies explored the relationship between the response behavior of such self-assembling probes and the stability of the formed aggregates. Three different types of behavior were observed in the presence of target protein: off/on, always-on and always-off ^{19}F signal. Only probes with a moderate stability of the aggregates created an efficient off/on response [52].

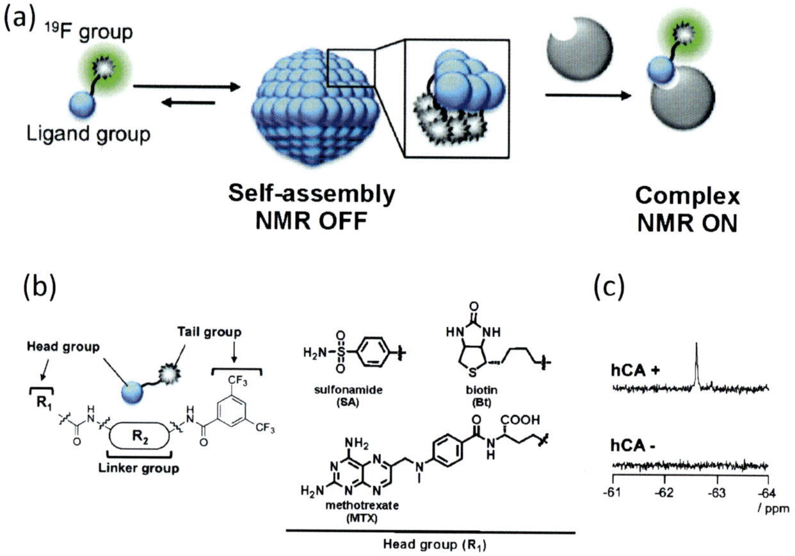

Figure 5.7 Self-assembling nanoprobe for protein detection by ^{19}F MR. (a) Schematic representation of the protein-driven probe activation. (b) General chemical structure of probes and the specific protein-recognition moieties (head group is sulfonamide for human carbonic anhydrase I (hCA); biotin for avidin; methotrexate for dihydrofolate reductase). (c) ^{19}F NMR spectra in the presence and absence of hCA. Adapted with permission from Takaoka et al., *J. Am. Chem. Soc.*, 133 (2011) 11725–11731. Copyright (2011) American Chemical Society.

Furthermore, a similar but "inverse" approach for selective detection of glycoproteins was demonstrated by using a ^{19}F-labeled lectin biosensor [53]. The formation of a complex between probe and target molecule that restricts its molecular mobility due to an increase in apparent molecular mass (M_r), significantly shortened the transverse relaxation time T_2 of fluorine. Thus, the binding of glycoproteins to the lectin biosensor induced a large peak broadening in ^{19}F NMR spectra and "turning off" in ^{19}F MR images. In contrast, upon binding of small sugar molecules like lactose, the ^{19}F NMR peaks remained sharp due to the negligible increase in M_r. This method enabled a specific detection of glycoproteins by means of ^{19}F MR [53].

Recently, an "on-off-on" ^{19}F probe was reported to show a two-step modulation of the fluorine signal [54]. The small hydrophobic molecule (T_2 = 873.6 ms, δ_F = −74.0 ppm) initially adhered to human serum albumin (HSA). This resulted in the formation of a lower mobility complex and a dramatic shortening of ^{19}F T_2 values (14.7 ms), thereby switching off the ^{19}F NMR signal. In the presence of hypochlorous acid, formed during inflammatory processes by myeloperoxidases, a conversion of the adhered ^{19}F probe took place with the release of a hydrophilic fluoroalcohol and subsequent recovery of the ^{19}F NMR signal. This product had a much longer transverse relaxation time (293.8 ms) and showed a sharp peak in the ^{19}F NMR spectrum shifted 2.6 ppm upfield from the substrate and could be specifically detected by chemical shift selective ^{19}F MRI.

In addition to the visualization of enzyme activity and proteins, the specific detection of K-ras mRNA, which is encoded by a gene being frequently mutated in cancer, was recently achieved. The stem-loop structured oligonucleotide ^{19}F NMR probe consisted of a paramagnetic Gd-DOTA chelate and a bis(trifluoro methyl)benzene moiety at its 5′ and 3′ termini, respectively [55]. In the absence of the target mRNA both imaging moieties were placed in close proximity and thus the fluorine signal "quenched" due to a strong PRE effect. Complementary binding of probe to the target molecule increased the distance between both moieties, thus PRE was substantially weakened, and the fluorine signal was restored.

The reported ^{19}F MR techniques to sense (non-enzymatic) proteins and mRNA are promising. However, the main challenge for the visualization of such macromolecules is the need for a direct stable binding of the probe on the target. Hence, the detection limits will depend directly on the total concentration of the target macromolecule, and the signal cannot be amplified by a continuous enzymatic conversion.

5.3.4 ^{19}F Probes Responsive to pO_2

Hypoxia is a pathological condition caused by oxygen deprivation of the tissue as seen in myocardial infarction, stroke, or in solid tumors [56]. Molecular imaging of hypoxia by ^{19}F NMR/MRI has been attempted in vitro and in vivo by several groups and reviewed previously [57]. In brief, fluoronitroimidazoles were mainly used as ^{19}F MR reporters to image hypoxic tissue. After cellular uptake, these molecules undergo intracellular metabolic reduction forming various reactive intermediates. Such reactive species cannot be reoxidized in low oxygen environments and consequently become trapped within various intracellular compartments upon binding to cellular macromolecules and non-protein sulfhydryl compounds (e.g., glutathione) [58]. There are several limitations for using such ^{19}F markers of hypoxia. This includes the production of diverse adducts with varying chemical shifts and at very low concentrations, unfavorable clearance kinetics of non-trapped agent and changes of ^{19}F T_2 values leading to line broadening upon binding to intracellular macromolecules [59]. It is still unclear whether the same species of molecular adducts are responsible for the ^{19}F MR signal in vivo and the histochemical detection of hypoxia. Moreover, there is an ongoing discussion as to whether the observed ^{19}F MR signal in tumor tissue originates from the unreacted reporter or alterations in blood flow in tumor tissue [58]. In addition, this type of ^{19}F probes allows only a qualitative estimation of hypoxia, rather than a quantitative measurement of pO_2 levels in tissue [4].

Another advanced application of responsive ^{19}F MRI probes is the quantitative mapping of the partial pressure of dissolved O_2 (pO_2) by perfluorocarbons (oximetry). The pO_2 can be measured using oxygen-sensitive reporters, particularly perfluorocarbons (e.g., 15-crown-5-ether, perfluorooctyl bromide, hexafluorobenzene (HFB)). PFCs have the ability to dissolve and carry a high payload of oxygen, which is in fast exchange with the surrounding tissue via free diffusion. Given the paramagnetic effect of O_2, oximetry by ^{19}F MRI is based on the linear relationship between ^{19}F longitudinal relaxation rates (R_1) of the sensor and pO_2 levels in tissue. In contrast to the above mentioned hypoxia probes, this linear relationship is valid over the whole range of pO_2 values. Thus, quantitative pO_2 mapping in tissue is also possible under normoxic or hyperoxic conditions, not only in hypoxia. However, since R_1 depends on the magnetic field as well as the temperature, exact calibration curves have to be measured for each probe and field strength. Once calibrated, with proper temperature control, examination of tissue (e.g., liver, spleen, tumors) or vascular oxygenation can be measured, depending on the administration route, e.g., via direct injection into the tissue of interest or systemic injection of PFC emulsions [5]. For instance, ^{19}F MR echo planar imaging of HFB has been shown to have high temporal resolution, and the observed tumor pO_2 values corresponded to those obtained with polarographic oxygen electrodes or fiber optic probes [60]. Nonetheless, distinctly higher absolute pO_2 values were obtained in hypoxic regions of tumors by ^{19}F MRI oximetry than by measurements with oxygen electrodes in several studies [58]. The use of oximetry in vitro, in vivo, and in *human* settings as well as the problems linked to this methodology are summarized in comprehensive reviews [3–5, 58, 61].

5.4 Sensitivity and Detection Levels for ^{19}F MRI/MRS

^{1}H MRI/MRS commonly requires tissue concentrations of a drug or contrast agent at least in the micromolar range. To achieve a similar image quality as in ^{1}H MRI, a high local tissue concentration of fluorine is required. However, a low signal-to-noise-ratio (SNR) is sufficient to detect fluorinated probes in ^{19}F MRI because of the lack of background signal from endogenous fluorine [7].

Despite few reports demonstrating detection of fluorine-labeled compounds in vivo at low micromolar concentrations [2, 62, 63], ^{19}F MRI typically requires millimolar concentrations in the tissue of interest to obtain images with sufficient SNR in a reasonable acquisition time. As an estimate of the required concentration per voxel for responsive ^{19}F probes in vivo, 10^{15}–10^{16} ^{19}F spins per voxel have been reported for PFC labeled cells in vivo for imaging in a reasonable scan time at ≥ 7T [64–67].

One option to circumvent the sensitivity issue is the use of compounds containing multiple chemically equivalent fluorine atoms (e.g., trifluoromethyl or *tris*-trifluoromethyl (*tert*-butyl) groups), or highly fluorinated compounds such as PFCs. Most of the responsive ^{19}F MR probes contain, however, only a limited number of fluorine per molecule, with the exception being PFCs used for oximetry. An increase in the number of fluorine per molecule critically alters several physicochemical properties of the compound (e.g., pK_a, lipophilicity, hydrophobicity, etc.), its stability, and bioavailability [68]. High lipophilicity/hydrophobicity effects probe-macromolecule interactions and the ability to bind to or cross membranes such as the blood brain barrier. Thus, the extent of fluorination not only strongly affects MR visibility, but also the tissue distribution of the probe. The impact of fluorination on the properties of compounds and their disposition and metabolism is reviewed in detail, for instance, in Park et al. [69].

Furthermore, nanostructured or self-assembling probes, accumulating a large number of smaller fluorinated moieties in a single spot, can be used to increase the local concentration of fluorine. However, the resultant shorter relaxation times might have an adverse effect on the detectability. Additionally, dose limitations might exist in vivo due to the potential toxicity of the high concentration of the probe required for obtaining a sufficient SNR in ^{19}F MRI/MRS; this might be one of the reasons why only a few of the responsive ^{19}F MR probes presented in this chapter have been used in vivo to date.

Nevertheless, the detection limit in ^{19}F MRI depends not only on properties of the ^{19}F probe itself (relaxation rates, number of fluorine atoms, etc.), but is also strongly influenced by other key factors such as the ^{19}F MRI pulse sequence, the chosen spatial resolution (voxel size) and the hardware (magnetic field strength, RF coils, etc.) used for detection.

An important goal for ^{19}F MRI is to maximize the SNR per unit time. Both the longitudinal (T_1) and the transverse (T_2) relaxation times of fluorine probes and pharmaceuticals are generally long, with typical T_1 times on the order of seconds and T_2 on the order of hundreds of milliseconds. The long T_2 time allows the use of efficient spatial encoding through long spin echo trains in fast spin echo sequences. However, it impedes rapid signal averaging based on gradient echo sequences. Furthermore, most existing fluorine compounds have an unfavorable T_2/T_1

ratio that ideally should equal 1 in order to maximize SNR. In this line, it was shown that shortening of relaxation times due to PRE can be helpful to improve the detection limit of such probes. For instance, Chalmers et al. reported that concentrations as low as 20 μM were detectable in phantoms by using gradient echo acquisition techniques and PRE [70].

New pulse sequences (i.e., ultrashort echo time, UTE, or zero echo time, ZTE, imaging) developed recently allow the detection of very fast relaxing spins and a considerable increase in SNR by exploiting PRE of ^{19}F probes containing paramagnetic metal ions [71]. However, such sequences, in particular ZTE, primarily enable the acquisition of spin density weighted images, independent of T_2, and thus may not be useful for detecting responsive effects.

In general, the SNR increases with magnetic field strength. Thus, the use of modern (ultra) high field MR scanners may help to lower the detection limit currently required for ^{19}F MRI. In addition, development of dedicated RF coils (e.g., cryo-coils) and optimization of other hardware components can further enhance the detectability of responsive ^{19}F MR probes.

Importantly, spatial resolution has also a considerable impact on the detectability. With increasing voxel size, a sufficient SNR can be achieved much easier. Nevertheless, any spatial information (e.g., about the local distribution of an enzyme) is diminished at large voxel sizes. Since specific anatomical information is of particular importance in studies with responsive probes, it is critical to find a reasonable compromise between detectability and spatial resolution.

5.5 Conclusions

Even though several promising ^{19}F probes have been reported in the past years that are responsive to microenvironmental changes, the field is still in its infancy. The majority of studies were performed in vitro and occasionally in cellular systems. Only a couple of approaches have demonstrated results in vivo, such as hypoxia imaging, oximetry, detection of β-gal activity, and prodrug-activated bacterial enzymes (i.e., CD, CG-2). Thus, there is a need for further improvements in fluorinated responsive probes.

The intrinsically low ^{19}F concentrations and the resulting low SNR remain the main challenge for in vivo applications of ^{19}F responsive probes. However, recent developments in probe design, the use of higher magnetic fields, improvements in RF coils and imaging sequences, as well as reconstruction techniques, will open up new opportunities for ^{19}F MRI and MRS in cellular and molecular imaging.

References

1. Knight, J. C., Edwards, P. G., and Paisey, S. J. (2011). Fluorinated contrast agents for magnetic resonance imaging; a review of recent developments, *RSC Adv.*, **1**, 1415–1425.
2. Reid, D. G., and Murphy, P. S. (2008). Fluorine magnetic resonance in vivo: A powerful tool in the study of drug distribution and metabolism, *Drug Discov. Today*, **13**, pp. 473–480.
3. Ruiz-Cabello, J., Barnett, B. P., Bottomley, P. A., and Bulte, J. W. M. (2010). Fluorine (^{19}F) MRS and MRI in biomedicine, *NMR Biomed.*, **24**, pp. 114–129.
4. Yu, J., Kodibagkar, V. D., Cui, W., and Mason, R. P. (2005). ^{19}F: A versatile reporter for non-invasive physiology and pharmacology using magnetic resonance, *Curr. Med. Chem.*, **12**, pp. 819–848.
5. Yu, J.-X., Hallac, R. R., Chiguru, S., and Mason, R. P. (2013). New frontiers and developing applications in ^{19}F NMR, *Prog. Nucl. Magn. Reson. Spectrosc.*, **70**, pp. 25–49.
6. McFarland, E., Koutcher, J. A., Rosen, B. R., Teicher, B., and Brady, T. J. (1985). In vivo ^{19}F NMR imaging, *J. Comput. Assist. Tomogr.*, **9**, pp. 8–15.
7. Ahrens, E. T., and Zhong, J. (2013). In vivo MRI cell tracking using perfluorocarbon probes and fluorine-19 detection, *NMR Biomed.*, **26**, pp. 860–871.
8. Harvey, P., Chalmers, K. H., De Luca, E., Mishra, A., and Parker, D. (2012). Paramagnetic ^{19}F chemical shift probes that respond selectively to calcium or citrate levels and signal ester hydrolysis, *Chem. Eur. J.*, **18**, pp. 8748–8757.
9. Keliris, A., Mamedov, I., Hagberg, G. E., Logothetis, N. K., Scheffler, K., and Engelmann, J. (2012). A smart ^{19}F and ^{1}H MRI probe with self-immolative linker as a versatile tool for detection of enzymes, *Contrast Media Mol. Imaging*, **7**, pp. 478–483.

10. Senanayake, P. K., Kenwright, A. M., Parker, D., and Hoorn, S. K. van der. (2007). Responsive fluorinated lanthanide probes for ^{19}F magnetic resonance spectroscopy, *Chem. Commun.*, pp. 2923–2925.

11. Bar-Shir, A., Gilad, A. A., Chan, K. W. Y., Liu, G., van Zijl, P. C. M., Bulte, J. W. M., and McMahon, M. T. (2013). Metal ion sensing using ion chemical exchange saturation transfer ^{19}F magnetic resonance imaging, *J. Am. Chem. Soc.*, **135**, pp. 12164–12167.

12. Taylor, J. S., and Deutsch, C. (1983). Fluorinated alpha-methylamino acids as ^{19}F NMR indicators of intracellular pH, *Biophys. J.*, **43**, pp. 261–267.

13. Deutsch, C. J., and Taylor, J. S. (1989). New class of ^{19}F pH indicators: Fluoroanilines, *Biophys. J.*, **55**, pp. 799–804.

14. Mehta, V. D., Kulkarni, P. V., Mason, R. P., Constantinescu, A., Aravind, S., Goomer, N., and Antich, P. P. (1994). 6-Fluoropyridoxol: A novel probe of cellular pH using ^{19}F NMR spectroscopy, *FEBS Lett.*, **349**, pp. 234–238.

15. Chalmers, K. H., De Luca, E., Hogg, N. H. M., Kenwright, A. M., Kuprov, I., Parker, D., Botta, M., Wilson, J. I., and Blamire, A. M. (2010). Design principles and theory of paramagnetic fluorine-labelled lanthanide complexes as probes for ^{19}F magnetic resonance: A proof-of-concept study. *Chem. Eur. J.*, **16**, pp. 134–148.

16. Kenwright, A. M., Kuprov, I., Luca, E. D., Parker, D., Pandya, S. U., Senanayake, P. K., and Smith, D. G. (2008). ^{19}F NMR based pH probes: Lanthanide(III) complexes with pH-sensitive chemical shifts, *Chem. Commun.*, pp. 2514–2516.

17. Chen, S., Yang, Y., Li, H., Zhou, X., and Liu, M. (2014). pH-Triggered Au-fluorescent mesoporous silica nanoparticles for ^{19}F MR/fluorescent multimodal cancer cellular imaging, *Chem. Commun.*, **50**, pp. 283–285.

18. Huang, X., Huang, G., Zhang, S., Sagiyama, K., Togao, O., Ma, X., Wang, Y., Li, Y., Soesbe, T. C., Sumer, B. D., Takahashi, M., Sherry, A. D., and Gao, J. (2013). Multi-chromatic pH-activatable ^{19}F-MRI nanoprobes with binary ON/OFF pH transitions and chemical-shift barcodes, *Angew. Chem. Int. Ed.*, **52**, pp. 8074–8078.

19. Oishi, M., Sumitani, S., Bronich, T. K., Kabanov, A. V., Boska, M. D., and Nagasaki, Y. (2009). Novel ^{19}F MRS/I nanoprobe based on pH-responsive PEGylated nanogel: pH-dependent ^{19}F magnetic resonance studies, *Chem. Lett.*, **38**, pp. 128–129.

20. Gianolio, E., Napolitano, R., Fedeli, F., Arena, F., and Aime, S. (2009). Poly-beta-cyclodextrin based platform for pH mapping via a ratiometric $^{19}F/^{1}H$ MRI method, *Chem. Commun.*, pp. 6044–6046.
21. Levy, L. A., Murphy, E., Raju, B., and London, R. E. (1988). Measurement of cytosolic free magnesium ion concentration by ^{19}F NMR, *Biochemistry*, **27**, pp. 4041–4048.
22. Smith, G. A., Morris, P. G., Hesketh, T. R., and Metcalfe, J. C. (1986). Design of an indicator of intracellular free Na^{+} concentration using ^{19}F-NMR, *Biochim. Biophys. Acta Mol. Cell Res.*, **889**, pp. 72–83.
23. Plenio, H., and Diodone, R. (1996). Covalently bonded fluorine as a σ-donor for groups I and II metal ions in partially fluorinated macrocycles, *J. Am. Chem. Soc.*, **118**, pp. 356–367.
24. Smith, G. A., Hesketh, R. T., Metcalfe, J. C., Feeney, J., and Morris, P. G. (1983). Intracellular calcium measurements by ^{19}F NMR of fluorine-labeled chelators, *Proc. Natl. Acad. Sci.*, **80**, pp. 7178–7182.
25. Badar-Goffer, R. S., Ben-Yoseph, O., Dolin, S. J., Morris, P. G., Smith, G. A., and Bachelard, H. S. (1990). Use of 1,2-bis(2-amino-5-fluorophenoxy)ethane-*N,N,N′,N′*-tetraacetic acid (5FBAPTA) in the measurement of free intracellular calcium in the brain by ^{19}F-nuclear magnetic resonance spectroscopy, *J. Neurochem.*, **55**, pp. 878–884.
26. Marban, E., Kitakaze, M., Kusuoka, H., Porterfield, J. K., Yue, D. T., and Chacko, V. P. (1987). Intracellular free calcium concentration measured with ^{19}F NMR spectroscopy in intact ferret hearts, *Proc. Natl. Acad. Sci.*, **84**, pp. 6005–6009.
27. Metcalfe, J. C., Hesketh, T. R., and Smith, G. A. (1985). Free cytosolic Ca^{2+} measurements with fluorine labelled indicators using ^{19}F NMR, *Cell Calcium*, **6**, pp. 183–195.
28. Song, S. K., Hotchkiss, R. S., Neil, J., Morris, P. E., Hsu, C. Y., and Ackerman, J. J. (1995). Determination of intracellular calcium in vivo via fluorine-19 nuclear magnetic resonance spectroscopy, *Am. J. Physiol.*, **269**, pp. C318–C322.
29. Benters, J., Flögel, U., Schäfer, T., Leibfritz, D., Hechtenberg, S., and Beyersmann, D. (1997). Study of the interactions of cadmium and zinc ions with cellular calcium homoeostasis using ^{19}F-NMR spectroscopy, *Biochem. J.*, **322**, pp. 793–799.
30. Kadjane, P., Platas-Iglesias, C., Boehm-Sturm, P., Truffault, V., Hagberg, G. E., Hoehn, M., Logothetis, N. K., and Angelovski, G. (2014). Dual-frequency calcium-responsive MRI agents, *Chem. Eur. J.*, **20**, pp. 7351–7362.

31. Razgulin, A., Ma, N., and Rao, J. (2011). Strategies for in vivo imaging of enzyme activity: An overview and recent advances, *Chem. Soc. Rev.*, **40**, pp. 4186–4216.

32. Van Laarhoven, H. W. M., Punt, C. J. A., Kamm, Y. J. L., and Heerschap, A. (2005). Monitoring fluoropyrimidine metabolism in solid tumors with in vivo ^{19}F magnetic resonance spectroscopy, *Crit. Rev. Oncol. Hematol.*, **56**, pp. 321–343.

33. Wolf, W., Presant, C. A., and Waluch, V. (2000). ^{19}F-MRS studies of fluorinated drugs in humans, *Adv. Drug Deliv. Rev.*, **41**, pp. 55–74.

34. Aboagye, E. O., Artemov, D., Senter, P. D., and Bhujwalla, Z. M. (1998). Intratumoral conversion of 5-fluorocytosine to 5-fluorouracil by monoclonal antibody-cytosine deaminase conjugates: noninvasive detection of prodrug activation by magnetic resonance spectroscopy and spectroscopic imaging, *Cancer Res.*, **58**, pp. 4075–4078.

35. Huber, B. E., Austin, E. A., Richards, C. A., Davis, S. T., and Good, S. S. (1994). Metabolism of 5-fluorocytosine to 5-fluorouracil in human colorectal tumor cells transduced with the cytosine deaminase gene: Significant antitumor effects when only a small percentage of tumor cells express cytosine deaminase, *Proc. Natl. Acad. Sci.*, **91**, pp. 8302–8306.

36. Jamin, Y., Smyth, L., Robinson, S. P., Poon, E. S. C., Eykyn, T. R., Springer, C. J., Leach, M. O., and Payne, G. S. (2011). Noninvasive detection of carboxypeptidase G2 activity in vivo, *NMR Biomed.*, **24**, pp. 343–350.

37. Liu, L., Kodibagkar, V. D., Yu, J.-X., and Mason, R. P. (2007). ^{19}F-NMR detection of lacZ gene expression via the enzymic hydrolysis of 2-fluoro-4-nitrophenyl β-d-galactopyranoside in vivo in PC3 prostate tumor xenografts in the mouse, *FASEB J.*, **21**, pp. 2014–2019.

38. Yu, J., and Mason, R. P. (2006). Synthesis and characterization of novel lacZ gene reporter molecules: Detection of beta-galactosidase activity by ^{19}F nuclear magnetic resonance of polyglycosylated fluorinated vitamin B6, *J. Med. Chem.*, **49**, pp. 1991–1999.

39. Yu, J.-X., Kodibagkar, V. D., Liu, L., and Mason, R. P. (2008). A ^{19}F-NMR approach using reporter molecule pairs to assess beta-galactosidase in human xenograft tumors in vivo, *NMR Biomed.*, **21**, pp. 704–712.

40. Mizukami, S., Takikawa, R., Sugihara, F., Hori, Y., Tochio, H., Wälchli, M., Shirakawa, M., and Kikuchi, K. (2008). Paramagnetic relaxation-based ^{19}F MRI probe to detect protease activity, *J. Am. Chem. Soc.*, **130**, pp. 794–795.

41. Mizukami, S., Matsushita, H., Takikawa, R., Sugihara, F., Shirakawa, M., and Kikuchi, K. (2011). ^{19}F MRI detection of β-galactosidase activity for imaging of gene expression, *Chem. Sci.*, **2**, pp. 1151–1155.
42. Yu, J.-X., Kodibagkar, V. D., Hallac, R. R., Liu, L., and Mason, R. P. (2012). Dual ^{19}F/^{1}H MR gene reporter molecules for in vivo detection of β-galactosidase, *Bioconjug. Chem.*, **23**, pp. 596–603.
43. Matsushita, H., Mizukami, S., Mori, Y., Sugihara, F., Shirakawa, M., Yoshioka, Y., and Kikuchi, K. (2012). ^{19}F MRI monitoring of gene expression in living cells through cell-surface β-lactamase activity, *Chem Bio Chem*, **13**, pp. 1579–1583.
44. Yamaguchi, K., Ueki, R., Nonaka, H., Sugihara, F., Matsuda, T., and Sando, S. (2011). Design of chemical shift-switching ^{19}F magnetic resonance imaging probe for specific detection of human monoamine oxidase A, *J. Am. Chem. Soc.*, **133**, pp. 14208–14211.
45. Mizukami, S., Takikawa, R., Sugihara, F., Shirakawa, M., and Kikuchi, K. (2009). Dual-function probe to detect protease activity for fluorescence measurement and ^{19}F MRI, *Angew. Chem. Int. Ed.*, **48**, pp. 3641–3643.
46. Tanaka, K., Kitamura, N., Naka, K., and Chujo, Y. (2008). Multi-modal ^{19}F NMR probe using perfluorinated cubic silsesquioxane-coated silica nanoparticles for monitoring enzymatic activity, *Chem. Commun.*, pp. 6176–6178.
47. Tanaka, K., Kitamura, N., and Chujo, Y. (2011). Bimodal quantitative monitoring for enzymatic activity with simultaneous signal increases in ^{19}F NMR and fluorescence using silica nanoparticle-based molecular probes, *Bioconjug. Chem.*, **22**, pp. 1484–1490.
48. Tanaka, K., Kitamura, N., and Chujo, Y. (2012). Heavy metal-free ^{19}F NMR probes for quantitative measurements of glutathione reductase activity using silica nanoparticles as a signal quencher, *Bioorg. Med. Chem.*, **20**, pp. 96–100.
49. Matsuo, K., Kamada, R., Mizusawa, K., Imai, H., Takayama, Y., Narazaki, M., Matsuda, T., Takaoka, Y., and Hamachi, I. (2013). Specific detection and imaging of enzyme activity by signal-amplifiable self-assembling ^{19}F MRI probes, *Chem. Eur. J.*, **19**, pp. 12875–12883.
50. Takaoka, Y., Sakamoto, T., Tsukiji, S., Narazaki, M., Matsuda, T., Tochio, H., Shirakawa, M., and Hamachi, I. (2009). Self-assembling nanoprobes that display off/on ^{19}F nuclear magnetic resonance signals for protein detection and imaging, *Nat. Chem.*, **1**, pp. 557–561.
51. Takaoka, Y., Sun, Y., Tsukiji, S., and Hamachi, I. (2011). Mechanisms of chemical protein ^{19}F-labeling and NMR-based biosensor

construction in vitro and in cells using self-assembling ligand-directed tosylate compounds, *Chem. Sci.*, **2**, pp. 511–520.

52. Takaoka, Y., Kiminami, K., Mizusawa, K., Matsuo, K., Narazaki, M., Matsuda, T., and Hamachi, I. (2011). Systematic study of protein detection mechanism of self-assembling ^{19}F NMR/MRI nanoprobes toward rational design and improved sensitivity, *J. Am. Chem. Soc.*, **133**, pp. 11725–11731.
53. Sun, Y., Takaoka, Y., Tsukiji, S., Narazaki, M., Matsuda, T., and Hamachi, I. (2011). Construction of a ^{19}F-lectin biosensor for glycoprotein imaging by using affinity-guided DMAP chemistry, *Bioorg. Med. Chem. Lett.*, **21**, pp. 4393–4396.
54. Doura, T., Hata, R., Nonaka, H., Sugihara, F., Yoshioka, Y., and Sando, S. (2013). An adhesive ^{19}F MRI chemical probe allows signal off-to-on-type molecular sensing in a biological environment, *Chem. Commun.*, **49**, pp. 11421–11423.
55. Sakamoto, T., Shimizu, Y., Sasaki, J., Hayakawa, H., and Fujimoto, K. (2011). Signal turn-on probe for nucleic acid detection based on ^{19}F nuclear magnetic resonance, *Bioorg. Med. Chem. Lett.*, **21**, pp. 303–306.
56. Gilany, K., and Vafakhah, M. (2010). Hypoxia: A Review, *J. Paramed. Sci.*, **1**, pp. 43–60.
57. Krohn, K. A., Link, J. M., and Mason, R. P. (2008). Molecular Imaging of Hypoxia, *J. Nucl. Med.*, **49**, pp. 129S–148S.
58. Robinson, S. P., and Griffiths, J. R. (2004). Current issues in the utility of ^{19}F nuclear magnetic resonance methodologies for the assessment of tumour hypoxia, *Philos. Trans. R. Soc. Lond. B Biol. Sci.*, **359**, pp. 987–996.
59. Salmon, H. W., and Siemann, D. W. (2004). Utility of ^{19}F MRS detection of the hypoxic cell marker EF5 to assess cellular hypoxia in solid tumors, *Radiother. Oncol.*, **73**, pp. 359–366.
60. Mason, R. P., Rodbumrung, W., and Antich, P. P. (1996). Hexafluorobenzene: A sensitive ^{19}F NMR indicator of tumor oxygenation, *NMR Biomed.*, **9**, pp. 125–134.
61. Chen, J., Lanza, G. M., Wickline, S. A. (2010). Quantitative magnetic resonance fluorine imaging: Today and tomorrow, *Wiley Interdiscip. Rev. Nanomed. Nanobiotechnol.*, **2**, pp. 431–440.
62. Henry, M. E., Moore, C. M., Kaufman, M. J., Michelson, D., Schmidt, M. E., Stoddard, E., Vuckevic, A. J., Berreira, P. J., Cohen, B. M., and Renshaw, P. F. (2000). Brain kinetics of paroxetine and fluoxetine

on the third day of placebo substitution: A fluorine MRS study. *Am. J. Psychiatry*, **157**, pp. 1506–1508.

63. Henry, M. E., Bolo, N. R., Zuo, C. S., Villafuerte, R. A., Cayetano, K., Glue, P., Damle, B. D., Andrews, E., Lauriat, T. L., Rana, N. S., Churchill, J. H., and Renshaw, P. F. (2013). Quantification of brain voriconazole levels in healthy adults using fluorine magnetic resonance spectroscopy, *Antimicrob. Agents Chemother.*, **57**, pp. 5271–5276.

64. Flögel, U., Ding, Z., Hardung, H., Jander, S., Reichmann, G., Jacoby, C., Schubert, R., and Schrader, J. (2008). In vivo monitoring of inflammation after cardiac and cerebral ischemia by fluorine magnetic resonance imaging, *Circulation*, **118**, pp. 140–148.

65. Partlow, K. C., Chen, J., Brant, J. A., Neubauer, A. M., Meyerrose, T. E., Creer, M. H., Nolta, J. A., Caruthers, S. D., Lanza, G. M., and Wickline, S. A. (2007). ^{19}F magnetic resonance imaging for stem/progenitor cell tracking with multiple unique perfluorocarbon nanobeacons, *FASEB J.*, **21**, pp. 1647–1654.

66. Srinivas, M., Heerschap, A., Ahrens, E. T., Figdor, C. G., and de Vries, I. J. M. (2010). ^{19}F MRI for quantitative in vivo cell tracking, *Trends Biotechnol.*, **28**, pp. 363–370.

67. Stoll, G., Basse-Lüsebrink, T., Weise, G., and Jakob, P. (2012). Visualization of inflammation using ^{19}F-magnetic resonance imaging and perfluorocarbons, *Wiley Interdiscip. Rev. Nanomed. Nanobiotechnol.*, **4**, pp. 438–447.

68. Smart, B. E. (2001). Fluorine substituent effects (on bioactivity), *J. Fluor. Chem.*, **109**, pp. 3–11.

69. Park, B. K., Kitteringham, N. R., and O'Neill, P. M. (2001). Metabolism of fluorine-containing drugs, *Annu. Rev. Pharmacol. Toxicol.*, **41**, pp. 443–470.

70. Chalmers, K. H., Kenwright, A. M., Parker, D., and Blamire, A. M. (2011). ^{19}F-lanthanide complexes with increased sensitivity for ^{19}F-MRI: Optimization of the MR acquisition, *Magn. Reson. Med.*, **66**, pp. 931–936.

71. Schmid, F., Höltke, C., Parker, D., and Faber, C. (2013). Boosting ^{19}F MRI—SNR efficient detection of paramagnetic contrast agents using ultrafast sequences, *Magn. Reson. Med.*, **69**, pp. 1056–1062.

Part 3
Inflammation Imaging

Chapter 6

Imaging Acute Organ Transplant Rejection with ^{19}F MRI

T. Kevin Hitchens, Lesley M. Foley, and Qing Ye

Animal Imaging Center,
University of Pittsburgh, Pittsburgh, Pennsylvania 15203, USA

tkh17@pitt.edu

Despite advances in immunosuppressive therapies following organ transplantation, acute rejection remains a leading cause of patient morbidity and mortality. Clinical surveillance of acute rejection relies on serial tissue biopsy, which is an invasive procedure and prone to sampling errors. Research in animal models of acute rejection has demonstrated that cellular MRI of graft-infiltrating macrophages may provide an alternative method for acute organ transplant rejection surveillance. In this chapter, we will review cellular MRI studies used to detect kidney and heart allograft acute rejection, including two studies that have utilized PFC labeling and ^{19}F MRI [1, 2].

6.1 Organ Transplantation

Organ transplantation has become the preferred treatment option for patients facing end-stage organ failure. Solid organs

Fluorine Magnetic Resonance Imaging
Edited by Ulrich Flögel and Eric Ahrens

ISBN 978-981-4745-31-4 (Hardcover), 978-981-4745-32-1 (eBook)
www.panstanford.com

may be transplanted from a deceased donor, or in some cases from a living donor; for example, individuals can donate a single kidney or part of their liver without losing essential organ function. Although organ transplantation has met with great success, two major problems remain: a shortage of donors and rejection of the transplanted organs. There is an ever-widening gap between the number patients on transplantation waiting lists and the number of donor organs available each year, with some patients not surviving long enough for a suitable donor organ to become available. At the same time, unless the organ is transplanted from a genetically identical donor (e.g., an identical twin), recipients face a lifetime threat of their immune systems destroying the foreign, transplanted organ.

While successful transplantation surgery was first reported in the 1950s, outcomes overall have vastly improved over the years. In 1954, Merrill performed the first successful human kidney transplantation surgery between identical twins [3]. At that time, transplantation from non-identical donors remained challenging, mainly due to ineffective methods for controlling the recipients' immune response, with patients rarely surviving beyond 30 days [4]. Clinical transplantation of other organs also proved more difficult, taking more than a decade later to demonstrate therapeutic potential. In 1967, Starzl performed the first successful human liver transplantation [5]. In the same year, Barnard [6] achieved some success in heart transplantation; however, early results were poor with few patients surviving past their hospital stay. For a recent review and historical perspective, see ref. [7]. With advances in modern immunosuppressive therapy, organ preservation, and patient monitoring, the survival rates for allograft transplant patients have improved dramatically over the last several decades. As a result, organ transplantation has now become a widespread clinical treatment for organ failure, with thousands of procedures performed each year. According to Scientific Registry of Transplant Recipients in the United States (http://www.srtr.org), current five-year graft survival is nearly 75% following heart transplantation, and nearly 85% for kidney transplantation from a living donor. Nevertheless, allograft rejection, both acute and chronic, are still the leading causes of morbidity and mortality after organ transplantation.

6.2 Organ Rejection

There are three forms of organ rejection: hyperacute rejection, acute rejection and chronic rejection. Hyperacute rejection occurs within a few minutes to hours after transplantation and the graft must be removed or the patient will die. Chronic rejection can take place over many years as the body's constant immune response to the new tissue slowly damages the organ. Acute allograft rejection generally occurs in the days to months after organ transplantation due to histocompatibility mismatch between the donor organ and the recipient. Even with modern immunosuppressive drugs, during the first 6 months acute rejection episodes still occur in 40% to 70% of heart transplants [8], 30% to 40% of kidney transplants [9], and 20% to 40% of liver transplants [10]. The number and severity of episodes of acute rejection also correlate with the development of chronic rejection [11, 12]. Thus, accurate detection of rejection and evaluation of therapeutic intervention is critical for long-term graft survival.

Clinical surveillance for acute rejection relies on tissue biopsy. The frequency of scheduled biopsies varies between transplantation centers, but a cardiac transplant recipient may have 10 or more endomyocardial biopsy procedures in the first year. Scheduled biopsies for renal transplants remain controversial; however, following clinical indication of renal allograft dysfunction, core biopsy remains the gold standard for the diagnosis of renal transplant abnormality [13]. Scheduled biopsies lead to the detection of silent acute rejection in 13% of renal transplant patients [14], and it has been suggested that the presence of subclinical inflammation may contribute to the development of chronic graft sclerosis [15]. Tissue biopsy is an invasive procedure with associated risks, and it is also prone to sampling error because of the limited sizes and locations of tissue available. Thus, having a non-invasive technique for the routine surveillance and diagnosis of organ rejection would greatly improve patient care.

The hallmark of acute rejection is immune cell infiltration followed by cellular injury and structural tissue deterioration of the transplanted graft. Multiple cell types contribute to allograft rejection, however, T cells and macrophages represent the most abundant infiltrating immune cell types found in acute allograft rejection [16]. The numbers of both macrophages and T cells have been shown to increase as rejection becomes more severe [17–19]. Macrophages participate in both innate and adaptive immune responses [20]. They are increasingly recognized as key inflammatory amplifiers in T cell–driven organ rejection and play various roles in the alloimmune reaction [21]. A number of studies suggest that macrophages play an injurious role in acute cellular allograft rejection, as well as in chronic rejection [22, 23]. Allograft infiltration by macrophages correlates with a decline in graft function and poor prognosis [21, 24, 25]. Macrophages, constituting 38% to 60% of infiltrating cells in acute renal allograft rejection, are key effectors of tissue damage with rejection [19]. Not surprisingly, systemic depletion of macrophages using clodronate-liposomes provides significant protection against acute allograft rejection in rodent models [26–28].

6.3 In vivo Macrophage Labeling and MRI Cell Tracking

Macrophages play a significant role in acute allograft rejection and provide a convenient labeling target for cellular MRI. Since monocytes/macrophages are phagocytes, they can be readily labeled with a wide variety of reagents by endocytosis, either in culture or in circulation by direct intravenous injection without affecting the cell function [29]. Reagents are taken up by monocytes and macrophages that home to sites of inflammation (in this case rejection). In animal models of organ rejection, MRI cell tracking of macrophages have used ultra-small superparamagnetic iron-oxide (USPIO) nanoparticles [28, 30–32], micrometer-sized superparamagnetic iron-oxide (MPIO) particles [23, 33], or perfluorcarbon (PFC) nanoemulations containing perfluoro-15-crown-5 ether [1, 2]. The use of USPIO/MPIO cell labeling to generate image contrast or PFC nanoemulsion as cell tracer agents offer two distinct strategies for cellular MRI and examples of both are shown below for detecting acute rejection in rodent models of kidney and heart transplantation.

Superparamagnetic iron-oxide particles have served as MRI contrast agents for a wide variety of preclinical and clinical applications. Their use as contrast agents for MRI cell tracking is reviewed elsewhere [34] and only a brief description is provided here. USPIO particles consist of a magnetite core surrounded by a biological compatible coating such as dextran or polyethylene glycol, having a hydrodynamic diameter on the order of 30 nm. MPIO particles consist of an inert polymer matrix sphere (~1 μm diameter) with a high magnetite content that is non-biodegradable, thus can be used for long-term macrophage tracking studies, as in the case of longitudinal tracking of macrophages in models of chronic rejection [23]. Intermediate-size particles are generally referred to as SPIO. All these particles cause enhanced nuclear spin relaxation of the surrounding water protons (^{1}H). Although dipolar interactions can be exploited to generate positive (T_1-based) image contrast, the qualitative effect on water relaxation is generally $T_2^* > T_2 >> T_1$; thus, the accumulation of iron-oxide labeled cells is often detected with dark (or negative) T_2^* image contrast. The large magnetic field gradients generated by (U)SPIO/MPIO extend far beyond the boundaries of the cell, and because of the "blooming effect" and high intracellular iron content that can be achieved with MPIO labeling, single cells can be detected in vivo with T_2^*-weighted images collected at much lower than cellular resolution [33, 35]. High sensitivity is one of the advantages of MRI cell tracking using superparamagnetic iron-oxide; however, there are other mechanisms that can cause image hypointensity in T_2 or T_2^*-weighted images, sometimes making unambiguous detection of labeled cells difficult.

The PFC label is directly detected with ^{19}F MR spectroscopy or imaging. Fluorine-19 (^{19}F) is the only abundant and stable isotope of fluorine. It is an NMR-active, spin-1/2 nucleus with a gyromagnetic ratio similar to ^{1}H. In biological systems, fluorine has a very low natural abundance, mainly in the form of fluoride; thus, ^{19}F MRI is highly specific for the PFC label. For cell labeling, PFC molecules with a large number of chemically equivalent fluorine atoms are often formulated into nanoemulsions. Several PFC molecules are suitable for ^{19}F MRI cell tracking [36, 37]. The studies described below used perfluoro-15-crown-5-ether, which has 20 equivalent fluorine atoms and a simple NMR spectrum with a single chemical shift.

Most preclinical MRI scanners are capable of ^{19}F detection with the addition of a suitable radio-frequency resonator, either dual-tuned for $^{19}F/^{1}H$, or a single channel resonator with a tuning range spanning the two nuclei. Since ^{19}F MRI generally does not yield sufficient anatomical information, ^{19}F MRI cell tracking requires two images, a ^{1}H image for anatomy and the ^{19}F image for the labeled cells. If the two images are collected with the same field-of-view (FOV) with the carrier frequencies set precisely to water ^{1}H and the ^{19}F PFC resonance peak, respectively, then the two images can be merged to place the signal from PFC-label in anatomical context without additional image post-processing. The fluorine image is generally collected with a reduced resolution, employing signal averaging to overcome the challenges of imaging dilute ^{19}F nuclear spins. An external concentration standard can be used to compare signals between studies. Of note, PFC T_1 generally decreases with higher magnetic field strength due to increased chemical shift anisotropy (CSA) relaxation [38, 39], thus going to a higher field scanner can provide a significant boost in sensitivity due to the potential for more rapid data acquisition times, as well as increased spin polarization.

6.4 Detection of Acute Kidney Transplant Rejection Using MRI Cell Tracking

Cellular MRI has been used to detect macrophage infiltration in response to renal rejection using an orthotropic kidney transplant model in rats [2, 18, 30]. In these studies, the allograft transplantation pair consisted of a Dark Agouti (DA) rat as the kidney donor and a Brown Norway (BN) rat as the organ recipient. For orthotopic kidney transplantation, the recipient is prepared with a left nephrectomy. The graft renal artery and vein are anastomosed end-to-side with the recipient's abdominal aorta and inferior vena cava, followed by end-to-end anastomosis of the ureter. For the transplantation surgery, the ischemic time for the kidney is on the order of 25 min [18, 30, 40]. The right kidney is left intact and thus provides an internal control and undisturbed renal function. Major histocompatibility complex (MHC) mismatch between strains, DA (RT1a) and BN (RT1n), results in acute kidney rejection that is evident on post-operative day (POD) 3. Without immunosuppressive intervention, the recipient's immune cells continue to infiltrate the graft, and severe rejection is evident by POD 5, corresponding to a peak of infiltrating ED1+ macrophages, as well as CD4+ and CD8+ T cells [30]. With a BN-to-BN transplantation pair, the isograft kidney will not experience rejection, and thus can serve as a control for surgical procedures and ischemic reperfusion injury of the transplanted kidney.

Figure 6.1 shows examples of using both ^{19}F MRI and T_2^*-weighted ^{1}H MRI to detect the accumulation of PFC-labeled macrophages or USPIO-labeled macrophages, respectively, in response to acute renal allograft rejection [2]. In both cases, macrophages were labeled in vivo by direct intravenous (i.v.) injection of the respective cell labeling reagent. For PFC-labeling, a commercially available PFC reagent (VS-580H, Celsense, Inc., Pittsburgh, PA) was used. VS-580H is a 30% w/v formulation of perfluoro-15-crown-5-ether.

Figure 6.1A shows a ^{19}F image displayed in red pseudo-color and overlayed on the ^{1}H anatomical image, immediately following i.v. injection of PFC. Using a ^{19}F gradient-echo sequence, PFC signal is found in circulation and in both the transplanted and native kidneys with approximately equal intensity. The PFC is taken up by circulating monocytes and macrophages, which home to the site of inflammation, in this case rejection. In the follow-up scan 24 h post-injection, PFC signal is found in the transplanted kidney experiencing acute rejection, and only at the level of the noise in the native kidney (Fig. 6.1B). PFC signal is also found in organs of the reticuloendothelial system (RES), including the spleen and liver, which is consistent with the clearance pathway.

Earlier studies demonstrating the utility of cellular MRI to detect acute kidney allograft rejection used T_2*-weighted imaging and in vivo macrophage labeling with USPIO [18, 30]. An example using USPIO for macrophage labeling is shown in Figs. 6.1D,E. Similar to Fig. 6.1A, circulating USPIO is observed in both the native and transplanted kidneys immediately following i.v. injection. Rather than positive signal, T_2*-weighted image hypointensity (Fig. 6.1D) is observed as compared with a transplant recipient receiving saline (Fig. 6.1F). Twenty-four

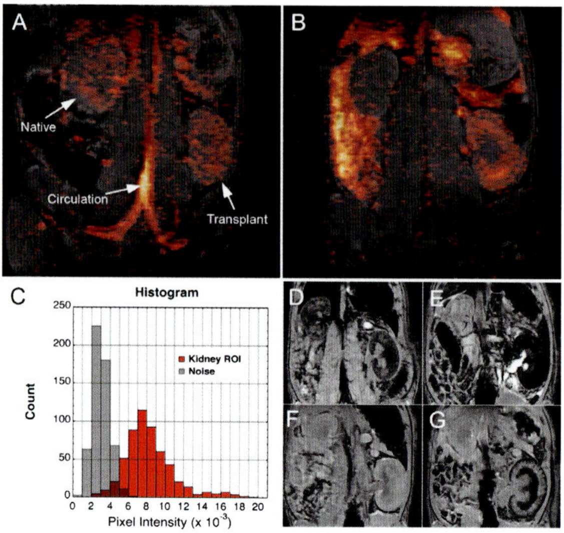

Figure 6.1 PFC and USPIO cell labeling to detect acute allograft kidney rejection by MRI cell tracking. Panels A and B show Composite $^{1}H/^{19}F$ images of an orthotopic kidney transplant recipient. ^{1}H images were collected using a respiratory-gated spin-echo with a 2 mm slice thickness whereas the ^{19}F images were collected with an ungated gradient-echo using a 4 mm slice thickness for increased ^{19}F sensitivity (TR/TE = 300/4 ms, flip angle = 30°, FOV = 7 × 7 cm, 128 × 64 matrix, number of averages (NA) = 192, acquisition time ~1 h). The images are overlaid and ^{19}F is rendered in pseudo-color. Panel A shows images collected approximately 2 h post PFC injection on POD 3; the ^{19}F signal is found in the vasculature and equally in both the native (upper) and graft (lower) kidneys. Panel B shows images collected 24 h post PFC injection. Signal intensity is found in the allograft kidney (SNR = 5.3), as well as other organs, and is virtually undetectable above the image noise in the native kidney. Panel C shows a histogram comparing the signal found in the transplant kidney and an equal area of noise. For comparison with T_2^*-weighted imaging, two allograft recipients are shown in Panels D and E and Panels F and G, respectively. Panel D shows an allograft kidney on POD 4 just following an injection of 6 mg/kg USPIO; 24 h later (Panel E) image darkening can be seen in the cortex and medulla of the allograft kidney. Panel F shows an allotransplant recipient on POD 4 just following an injection of saline (control). Panel G shows the same rat imaged on POD 5; severe rejection results in hemorrhaging, with image hypointensity observed the medulla. Taken from Fig. 6 of ref. [2] with permission.

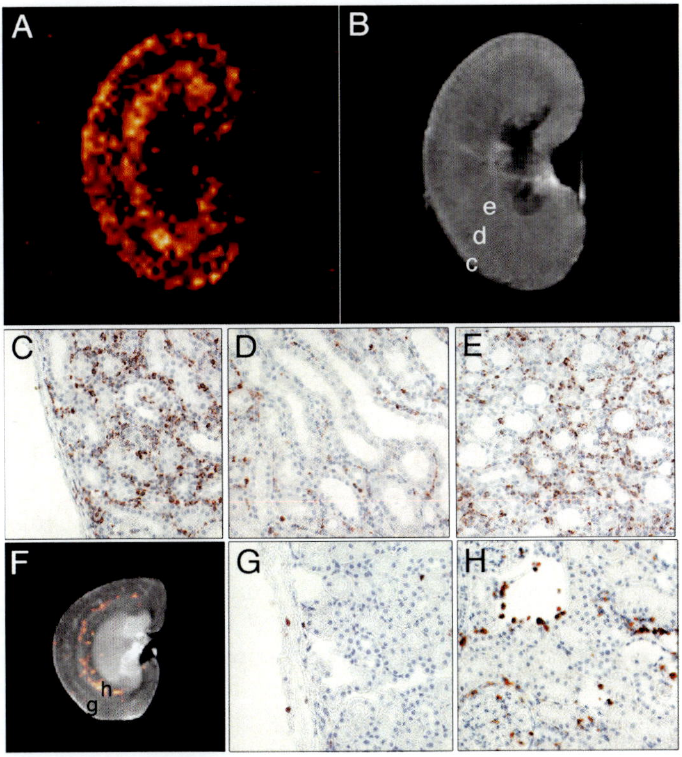

Figure 6.2 Ex vivo magnetic resonance microscopy of kidney allograft correlates with immunohistochemical staining for ED1+ cells. Panels A and B show ^{19}F and ^{1}H images of the allograft kidney, respectively, harvested on POD 4, 24 h following PFC injection. ^{19}F MRI was collected with a spin-echo sequence (2000/7 ms TR/TE, 3 × 3 cm FOV, 64 × 64 matrix, NA = 16 and 1.5 mm slice thickness). The SNR was measured to be 4.65 over an ROI containing the cortex and medulla. On Panel B, the labels c, d, and e indicate the approximate locations for images C (cortex), D (base of medullary pyramids), and E (medulla), respectively. For comparison, the native kidney is shown in Panel H. The ^{19}F imaging parameters were the same as for the allograft, except NA = 512 (32-times greater). SNR was 2.8 in the crescent area containing the ^{19}F signal density. The labels g and h indicate the approximate locations for Panels G and H, respectively. Minimal ED1+ cells are observed in the cortex (Panel G), and the majority of the ^{19}F signal intensity seen in the native kidney appears to be due to macrophages associated with the membranes of the minor calices, shown in H. Taken from Fig. 7 of ref. [2] with permission.

hours post injection (on POD 5), significant MRI signal reduction is observed in the transplanted kidney (Fig. 6.1E). A complication of using USPIO for cell tracking is that T_2^*-weighted image hypointensity is non-specific for labeled cell accumulation. In the case of kidney transplantation, severe acute rejection leads to extensive interstitial hemorrhaging in the medulla and diffuse interstitial hemorrhaging in the cortex [18]. Therefore, negative image contrast is clearly observed in T_2^*-weighted image when no USPIO is given (Fig. 6.1G). Using the DA-to-BN transplantation model, Ye et al. [18] reported an 80% reduction in T_2^*-weighted signal intensity in the medulla as a result of hemorrhaging on POD 5. In the cortex of the kidney, hemorrhaging resulted in a 48% reduction in T_2^*-weighted signal intensity on POD 5, compared with a significant signal loss of 85% when 6 mg Fe/kg USPIO was given 24 h prior.

PFC signal in the kidney correlates well with histological staining for ED1+ macrophages. Magnetic resonance microscopy (MRM) of an allograft kidney harvested on POD 4 showed distinct bands of PFC signal in the cortex and the medulla with little signal seen in the boundary zone, shown in Fig. 6.2A. Tissue sections taken from these same regions display a similar pattern when stained for ED1+ cells (Figs. 6.2C–E). Once PFC clears from circulation, no fluorine signal is observed in the native kidney in vivo; however, ex vivo MRM with significant signal averaging revealed some PFC remains in the native kidney (Fig. 6.2F). The PFC accumulation also correlates with histology for ED1+ macrophages (Figs. 6.2G,H), likely due to macrophages normally found in the vessel endothelium of the kidney [41]. These results demonstrate that ^{19}F MRI combined with in vivo PFC-labeling of macrophage can be used to detect acute allograft kidney rejection.

6.5 Detection of Acute Allograft Rejection in the Heart with MRI Cell Tracking

Several heterotopic heart (or heart-lung) transplantation models in rats and mice have been used to study rejection [42–45]. In these models, the recipient receives a second heart or heart-lung graft that is transplanted outside the thoracic cavity. The native heart then supports life and normal physiology, so that the rejection process can be studied in the graft tissue from the early stages of rejection to the very severe stages of rejection and loss of graft function. Animal models have been developed where a heart is transplanted to cervical [46], inguinal [31], and abdominal [47, 48] regions. The surgical engraftment varies depending on the model, but spontaneous beating generally resumes once circulation to the graft heart has been restored. A "non-working" heterotopic heart transplantation refers to a model where en bloc ligation of the graft pulmonary artery and vein does not allow for normal circulation and cardiac physiology. These models are suitable for immunological studies, but not for functional investigations since all four chambers of the heart are not working, and insufficient loading can result in left ventricular hypertrophy. A schematic of a heterotopic working-heart transplantation model with intact pulmonary circulation is shown in Fig. 6.3 [47]. In this model, the en bloc donor heart and lung are transplanted into the recipient abdomen. Graft aorta and superior vena cava (SVC) are connected with recipient abdominal aorta and inferior vena cava (IVC) using end-to-side anastomosis, respectively. Partial ligation of the Proximal IVC allows proper preloading of the left ventricle.

Using this abdominal working-heart transplantation model, Wu et al. [33, 49] used cellular and functional MRI to demonstrate that rejection in the heart is highly heterogeneous for both macrophage infiltration and myocardial function. With the DA to BN transplantation pair, Grade I acute rejection is evident beginning on POD 3, without immunosuppressive treatment, and progresses to severe, Grade IV rejection and graft failure by POD 7. Using T_2^*-weighted MRI to detect macrophages labeled in vivo with MPIO, a temporal progression pattern for macrophage infiltration was revealed, with macrophage infiltration beginning at the pericardium and increasingly penetrating towards the endocardium as rejection progressed [33]. This pattern had not been previously observed since biopsy samples are very limited in location and size. It is also interesting to note that clinical biopsy samples in the heart are generally taken from the septal wall in the right-ventricle [50] and not from the pericardium. Areas of USPIO-labeled macrophage infiltration also correlate with regional loss of contractile function measured by myocardial tagging and strain analysis, suggesting that a two-pronged approach of combined cellular and functional MRI may provide a safe and sensitive alternative to biopsy for routine surveillance following cardiac transplantation [49].

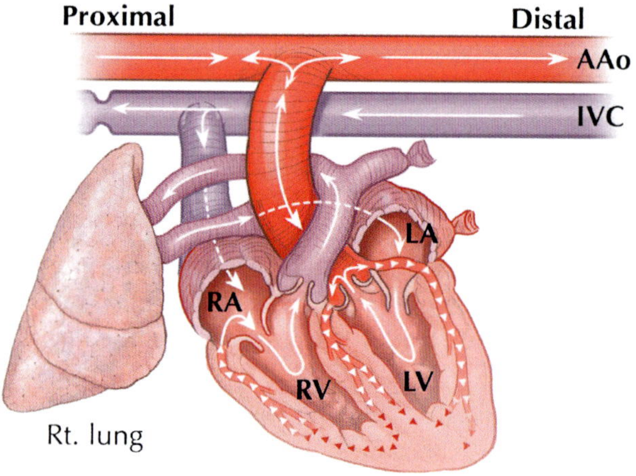

Figure 6.3 Schematic representation of the abdominal heterotopic biventricular working-heart transplantation model. Arrows indicate the direction of blood flow. Models with a single anastomosis of the graft aorta have insufficient loading, via regurgitation only, whereas the double-anastomosis model shown has intact pulmonary circulation and receives proper loading via partial ligation of the proximal the inferior vena cava (IVC). AAo = abdominal aorta; LA = left atrium; LV = left ventricle; RA = right atrium; RV = right ventricle; Rt. lung = donor right lung. Adapted from Fig. 1 of ref. [28] with permission.

^{19}F MRI cell tracking in this model is consistent with the prior studies using superparamagnetic iron-oxide particles for macrophage imaging. Figure 6.4 shows examples of both ^{19}F MRI and T_2^*-weighted ^1H MRI to detect the cardiac rejection in the DA-to-BN rat heterotopic working-heart transplantation model. As with the kidney model described above, this study used a commercially available PFC [2]. In vivo MRI of the heart normally exhibit clear contrast boundaries between the bright myocardial tissue and dark, air-filled, lungs. In this model, however, the transplanted heart is among the abdominal organs.

Twenty-four hours following i.v. administration of PFC, labeled macrophages were found accumulating in the heart, primarily in the pericardium of the allograft (Figs. 6.4A,C), which was confirmed by ex vivo MRM (Figs. 6.4B,D). Only trace amount PFC signal was found in isograft hearts (Fig. 6.4E).

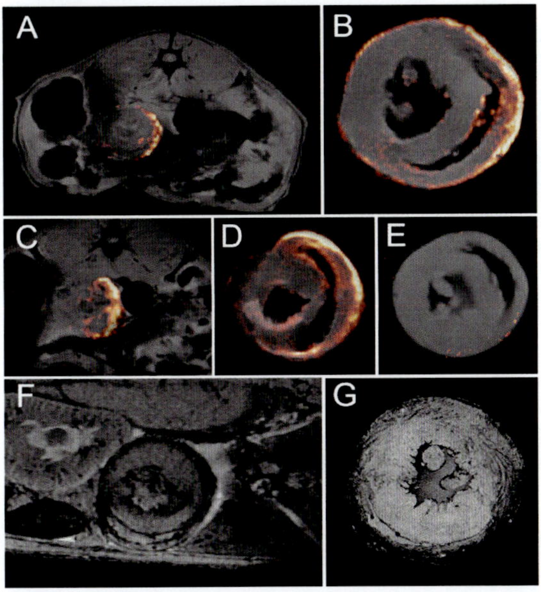

Figure 6.4 Non-invasive detection of cardiac transplant rejection with MRI cell tracking of macrophages. Panels A–E show merged $^{19}F/^{1}H$ images with ^{19}F MRI shown in red pseudo color. Panels A and C show PFC-labeled macrophage accumulation in the allograft heart transplanted in the abdomen at two different slice positions. Other features include the spine, back muscles, and intestines. The ^{19}F image was collected using a spin-echo sequence with respiratory/EKG gating (8 × 8 cm FOV, 128 × 128 matrix, 2.5 mm slice thickness, NA = 4). Panels B and D shows the same heart imaged ex vivo at higher resolution (2.6 cm FOV, 128 × 128 matrix, NA = 8, 1.5 mm slice thickness). Panel E shows an ex vivo $^{19}F/^{1}H$ composite image of an isograft heart harvested on POD 6 using the same parameters as B, except NA = 32; only a trace amount of PFC signal is found in the heart. For comparison, Panel F shows a T_2^*-weighted image of an allograft on POD 6, 24 h following in vivo macrophage labeling with USPIO (TE = 10 ms, 4 cm FOV, 256 × 256 matrix, 1 mm slice thickness, NA = 6). Ex vivo T_2^*-weighted MRM of the heart is shown in Panel G. Taken from Fig. 2 of ref. [2] with permission.

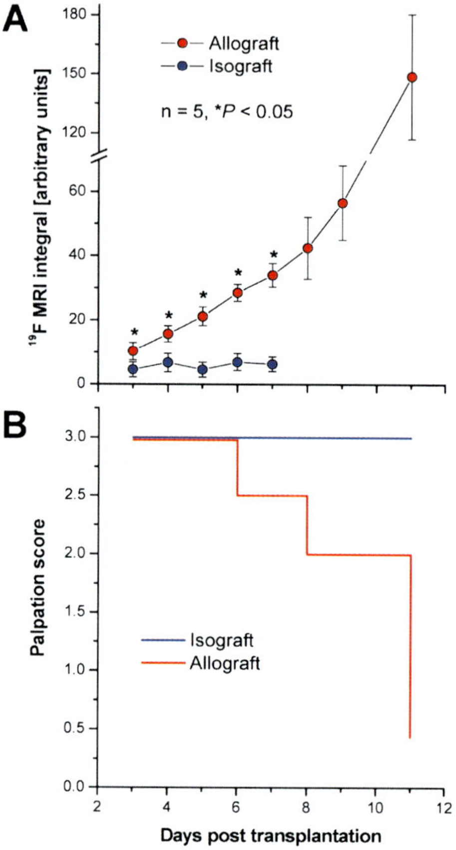

Figure 6.5 PFC detection of cardiac rejection precedes conventional parameters. (a) Quantification of ^{19}F MRI signal and (b) palpatation scores in allografts (red) and isografts (blue) over time. Values are mean ± SD, $n = 5$, ^{19}F signals were significantly different between allo- and isografts at the first MRI examination on POD 3 (*$p < 0.05$). Deterioration of allograft function, by palpatation, was not observed until POD 6. Taken from Fig. 4 of ref. [1] with permission.

An increase in labeled-macrophage infiltration has been shown to correlate with rejection in several imaging studies. Wu et al. [49] showed that the integrated volume of T_2^*-weighted image hypointensity found for USPIO-labeled macrophages increased up to a moderate rejection grade, with no further increase observed for hearts experiencing severe rejection. One reason for the lack of sensitivity at the later stages of rejection could be that the recruitment of additional USPIO-labeled cells to the same area of hypointensity does not increase the measured index. Using ^{19}F NMR spectroscopy of excised tissues, the normalized PFC signal in allograft hearts was found to be greater with severe rejection than moderate rejection, with both having greater signal than isograft or native hearts [2]. Using a heterotopic non-working heart transplantation model in mice, Flögel et al. [1] reported a temporal increase in PFC signal in allograft hearts as rejection progressed. In this study, transplantation between C57B10 and C57BL/6 mice leads to acute rejection and graft failure within 2 weeks [48]. When PFC was give 24 h prior to imaging, the integrated PFC signal in the allograft hearts was significantly higher than isograft hearts at an early stage, POD 3, with a steady increase up to POD 11, Fig. 6.5. Deterioration of allograft function, determined by palpatation, was not observed until POD 6 showing that that ^{19}F MRI cell tracking is a sensitive measure of early allograft rejection. In addition, immunosuppressive treatment with Rapamycin led to a significant decrease in PFC signal, correlating with a decrease in CD11b-positive cells infiltrating allograft hearts [1].

6.6 Conclusions

Since its clinical development in the 1950s, organ transplantation has prolonged the lives of numerous patients facing end-stage organ failure, with survival rates and patient quality of life improving significantly over the last several decades. Despite considerable advancements in the field, invasive tissue biopsy has remained the gold standard for graft surveillance of acute rejection. Acute organ rejection is heterogeneous, making tissue biopsy prone to sampling errors and false-negative results. Thus, patient care can be improved with the development of more accurate and less invasive surveillance methods. Pioneering studies by Ho and coworkers demonstrated the feasibility of using MRI cell tracking to detect organ rejection in rodent models with USPIO-labeled macrophages [18, 30–32]. The more recent studies utilizing ^{19}F MRI cell tracking confirm the utility of macrophage infiltration as a biomarker for acute rejection, providing a unique signal to track. With appropriate clinical-grade reagents, ^{19}F MRI cell tracking could provide a sensitive, non-invasive, technique for surveillance of rejection over the whole organ, potentially reducing or eliminating the need for serial tissue biopsy and improving patient quality of life.

References

1. Flögel, U., Su, S., Kreideweiss, I., Ding, Z., Galbarz, L., Fu, J., Jacoby, C., Witzke, O., and Schrader, J. (2011). Noninvasive detection of graft rejection by in vivo ^{19}F MRI in the early stage. *Am. J. Transplant.*, 11, pp. 235–244.
2. Hitchens, T. K., Ye, Q., Eytan, D. F., Janjic, J. M., Ahrens, E. T., and Ho, C. (2011). 19F MRI detection of acute allograft rejection with in vivo perfluorocarbon labeling of immune cells. *Magn. Reson. Med.*, 65, pp. 1144–1153.
3. Merrill, J. P., Murray, J. E., Harrison, J. H., and Guild, W. R. (1956). Successful homotransplantation of the human kidney between identical twins. *JAMA*, 160, pp. 277–282.
4. Murray, J. E., Merrill, J. P., Dammin, G. J., Dealy, J. B., Jr., Alexandre, G. W., and Harrison, J. H. (1962). Kidney transplantation in modified recipients. *Ann. Surg.*, 156, pp. 337–355.
5. Starzl, T. E., Groth, C. G., Brettschneider, L., Penn, I., Fulginiti, V. A., Moon, J. B., Blanchard, H., Martin, A. J., Jr., and Porter, K. A. (1968). Orthotopic homotransplantation of the human liver. *Ann. Surg.*, 168, pp. 392–415.
6. Barnard, C. N. (1968). Human cardiac transplantation: An evaluation of first 2 operations performed at Groote Schuur Hospital Cape Town. *Am. J. Cardiol.*, 22, pp. 584–596.
7. Watson, C. J. E., and Dark, J. H. (2012). Organ transplantation: Historical perspective and current practice. *Brit. J. Anaesth*, 108, pp. I29–I42.
8. Lindenfeld, J., Miller, G. G., Shakar, S. F., Zolty, R., Lowes, B. D., Wolfel, E. E., Mestroni, L., Page, R. L., and Kobashigawa, J. (2004). Drug therapy in the heart transplant recipient–Part II: Immunosuppressive drugs. *Circulation*, 110, pp. 3858–3865.
9. Roberts, D. M., Jiang, S. H., and Chadban, S. J. (2012). The treatment of acute antibody-mediated rejection in kidney transplant recipients-a systematic review. *Transplantation*, 94, pp. 775–783.
10. Shaked, A., Ghobrial, R. M., Merion, R. M., Shearon, T. H., Emond, J. C., Fair, J. H., Fisher, R. A., Kulik, L. M., Pruett, T. L., and Terrault, N. A. (2009). Incidence and severity of acute cellular rejection in

recipients undergoing adult living donor or deceased donor liver transplantation. *Am. J. Transplant.*, 9, pp. 301–308.

11. Joosten, S. A., Sijpkens, Y. W. J., van Kooten, C., and Paul, L. C. (2005). Chronic renal allograft rejection: Pathophysiologic considerations. *Kidney Int.*, 68, pp. 1–13.
12. Yamani, M. H., Yousufuddin, M., Starling, R. C., Tuzcu, M., Ratliff, N. B., Cook, D. J., Abdo, A., Crowe, T., Hobbs, R., Rincon, G., Bott-Silverman, C., McCarthy, P. M., and Young, J. B. (2004). Does acute cellular rejection correlate with cardiac allograft vasculopathy? *J. Heart Lung Transplant.*, 23, pp. 272–276.
13. Ahmad, I. (2004). Biopsy of the transplanted kidney. *Semin. Intervent Radiol.*, 21, pp. 275–281.
14. Schwarz, A., Mengel, M., Gwinner, W., Eisenberger, U., Hiss, M., Radermacher, J., Fiebeler, A., Abou-Rebyeh, F., and Haller, H. (2002). Protocol biopsy program after renal transplantation: Structure and first results. *Transplant. Proc.*, 34, pp. 2238–2239.
15. Rush, D. N., Jeffery, J. R., and Gough, J. (1995). Sequential protocol biopsies in renal transplant patients. Clinico-pathological correlations using the Banff schema. *Transplantation*, 59, pp. 511–514.
16. Wood, K. J., and Goto, R. (2012). Mechanisms of rejection: Current perspectives. *Transplantation*, 93, pp. 1–10.
17. Grau, V., Herbst, B., and Steiniger, B. (1998). Dynamics of monocytes/macrophages and T lymphocytes in acutely rejecting rat renal allografts. *Cell Tissue Res.*, 291, pp. 117–126.
18. Ye, Q., Yang, D., Williams, M., Williams, D. S., Pluempitiwiriyawej, C., Moura, J. M., and Ho, C. (2002). In vivo detection of acute rat renal allograft rejection by MRI with USPIO particles. *Kidney Int.*, 61, pp. 1124–1135.
19. Jose, M. D., Ikezumi, Y., van Rooijen, N., Atkins, R. C., and Chadban, S. J. (2003). Macrophages act as effectors of tissue damage in acute renal allograft rejection. *Transplantation*, 76, pp. 1015–1022.
20. He, H., Stone, J. R., and Perkins, D. L. (2002). Analysis of robust innate immune response after transplantation in the absence of adaptive immunity. *Transplantation*, 73, pp. 853–861.
21. Wyburn, K. R., Jose, M. D., Wu, H., Atkins, R. C., and Chadban, S. J. (2005). The role of macrophages in allograft rejection. *Transplantation*, 80, pp. 1641–1647.
22. Azuma, H., and Tilney, N. L. (1995). Immune and nonimmune mechanisms of chronic rejection of kidney allografts. *J. Heart Lung Transplant.*, 14, pp. S136–S142.

23. Ye, Q., Wu, Y. L., Foley, L. M., Hitchens, T. K., Eytan, D. F., Shirwan, H., and Ho, C. (2008). Longitudinal tracking of recipient macrophages in a rat chronic cardiac allograft rejection model with noninvasive magnetic resonance imaging using micrometer-sized paramagnetic iron oxide particles. *Circulation*, 118, pp. 149–156.

24. Chadban, S. J., Wu, H., and Hughes, J. (2010). Macrophages and kidney transplantation. *Semin. Nephrol.*, 30, pp. 278–289.

25. Rowshani, A. T., and Vereyken, E. J. (2012). The role of macrophage lineage cells in kidney graft rejection and survival. *Transplantation*, 94, pp. 309–318.

26. Kim, M. G., Boo, C. S., Ko, Y. S., Lee, H. Y., Cho, W. Y., Kim, H. K., and Jo, S. K. (2010). Depletion of kidney CD11c+ F4/80+ cells impairs the recovery process in ischaemia/reperfusion-induced acute kidney injury. *Nephrol. Dial. Transplant.*, 25, pp. 2908–2921.

27. Frantz, S., Hofmann, U., Fraccarollo, D., Schafer, A., Kranepuhl, S., Hagedorn, I., Nieswandt, B., Nahrendorf, M., Wagner, H., Bayer, B., Pachel, C., Schon, M. P., Kneitz, S., Bobinger, T., Weidemann, F., Ertl, G., and Bauersachs, J. (2013). Monocytes/macrophages prevent healing defects and left ventricular thrombus formation after myocardial infarction. *FASEB J.*, 27, pp. 871–881.

28. Wu, Y. L., Ye, Q., Eytan, D. F., Liu, L., Rosario, B. L., Hitchens, T. K., Yeh, F. C., Rooijen van, N., and Ho, C. (2013). Magnetic resonance imaging investigation of macrophages in acute cardiac allograft rejection after heart transplantation. *Circ. Cardiovasc. Imaging*, 6, pp. 965–973.

29. Weissleder, R., Elizondo, G., Wittenberg, J., Rabito, C. A., Bengele, H. H., and Josephson, L. (1990). Ultrasmall superparamagnetic iron oxide: Characterization of a new class of contrast agents for MR imaging. *Radiology*, 175, pp. 489–493.

30. Zhang, Y., Dodd, S. J., Hendrich, K. S., Williams, M., and Ho, C. (2000). Magnetic resonance imaging detection of rat renal transplant rejection by monitoring macrophage infiltration. *Kidney Int.*, 58, pp. 1300–1310.

31. Kanno, S., Lee, P. C., Dodd, S. J., Williams, M., Griffith, B. P., and Ho, C. (2000). A novel approach with magnetic resonance imaging used for the detection of lung allograft rejection. *J. Thorac. Cardiovasc. Surg.*, 120, pp. 923–934.

32. Kanno, S., Wu, Y. J., Lee, P. C., Dodd, S. J., Williams, M., Griffith, B. P., and Ho, C. (2001). Macrophage accumulation associated with rat

cardiac allograft rejection detected by magnetic resonance imaging with ultrasmall superparamagnetic iron oxide particles. *Circulation*, 104, pp. 934–938.

33. Wu, Y. L., Ye, Q., Foley, L. M., Hitchens, T. K., Sato, K., Williams, J. B., and Ho, C. (2006). In situ labeling of immune cells with iron oxide particles: An approach to detect organ rejection by cellular MRI. *Proc. Natl. Acad. Sci. U. S. A.*, 103, pp. 1852–1857.

34. Ho, C., and Hitchens, T. K. (2004). A non-invasive approach to detecting organ rejection by MRI: Monitoring the accumulation of immune cells at the transplanted organ. *Curr. Pharm. Biotechnol.*, 5, pp. 551–566.

35. Shapiro, E. M., Gonzalez-Perez, O., Manuel Garcia-Verdugo, J., Alvarez-Buylla, A., and Koretsky, A. P. (2006). Magnetic resonance imaging of the migration of neuronal precursors generated in the adult rodent brain. *Neuroimage*, 32, pp. 1150–1157.

36. Janjic, J. M., and Ahrens, E. T. (2009). Fluorine-containing nanoemulsions for MRI cell tracking. *Wiley Interdisciplinary Rev. Nanomed. Nanobiotechnol.*, 1, pp. 492–501.

37. Jacoby, C., Temme, S., Mayenfels, F., Benoit, N., Krafft, M. P., Schubert, R., Schrader, J., and Flögel, U. (2014). Probing different perfluorocarbons for in vivo inflammation imaging by ^{19}F MRI: Image reconstruction, biological half-lives and sensitivity. *NMR Biomed.*, 27, pp. 261–271.

38. Kadayakkara, D. K., Damodaran, K., Hitchens, T. K., Bulte, J. W., and Ahrens, E. T. (2014). (19)F spin-lattice relaxation of perfluoropolyethers: Dependence on temperature and magnetic field strength (7.0–14.1T). *J. Magn. Reson.*, 242, pp. 18–22.

39. Shukla, H. P., Mason, R. P., Woessner, D. E., and Antich, P. P. (1995). Comparison of 3 commercial perfluorocarbon emulsions as high-field F-19 NMR probes of oxygen-tension and temperature. *J. Magn. Reson. Series B*, 106, pp. 131–141.

40. Lee, S. (1967). An improved technique of renal transplantation in the rat. *Surgery*, 61, pp. 771–773.

41. Paul, L. C., Grothman, G. T., Benediktsson, H., Davidoff, A., and Rozing, J. (1992). Macrophage subpopulations in normal and transplanted heart and kidney tissues in the rat. *Transplantation*, 53, pp. 157–162.

42. Steinbruchel, D. A., Nielsen, B., Salomon, S., and Kemp, E. (1994). A new model for heterotopic heart transplantation in rodents: Graft atrial septectomy. *Transplant. Proc.*, 26, pp. 1298–1299.

43. Martins, P. N. (2008). Assessment of graft function in rodent models of heart transplantation. *Microsurgery*, 28, pp. 565–570.
44. Asfour, B., Hare, J. M., Kohl, T., Baba, H. A., Kass, D. A., Chen, K., Tjan, T. D., Hammel, D., Weyand, M., Hruban, R. H., Scheld, H. H., and Byrne, B. J. (1999). A simple new model of physiologically working heterotopic rat heart transplantation provides hemodynamic performance equivalent to that of an orthotopic heart. *J. Heart Lung Transplant.*, 18, pp. 927–936.
45. Hasegawa, T., Visovatti, S. H., Hyman, M. C., Hayasaki, T., and Pinsky, D. J. (2007). Heterotopic vascularized murine cardiac transplantation to study graft arteriopathy. *Nat. Protoc.*, 2, pp. 471–480.
46. Gu, X., Xiang, J., Zhou, Y., Qian, S., and Chen, Z. (2007). Improved cuff technique for cervical heart transplantation in mice. *Microsurgery*, 27, pp. 317–319.
47. Wu, Y. J., Sato, K., Ye, Q., and Ho, C. (2004). MRI investigation of graft rejection following organ transplantation using rodent models. *Method Enzymol.*, 386, pp. 73–105.
48. Wu, K., Zhang, J., Fu, J., Wu, S., Philipp, T., Uwe, H., Kribben, A., and Witzke, O. (2006). Novel technique for blood circuit reconstruction in mouse heart transplantation model. *Microsurgery*, 26, pp. 594–598.
49. Wu, Y. L., Ye, Q., Sato, K., Foley, L. M., Hitchens, T. K., and Ho, C. (2009). Noninvasive evaluation of cardiac allograft rejection by cellular and functional cardiac magnetic resonance. *JACC Cardiovasc. Imaging*, 2, pp. 731–741.
50. Baughman, K. L., and Baim, D. S. (2006). Endomyocardial Biopsy, In Baim D. S. (Ed.), *Grossman's Cardiac Catheterization, Angiography and Intervention, Seventh Edition*. (pp. 395–411) Lippincott, Williams, & Wilkins: Philadelphia.

Chapter 7

Cardiac Disease

Ruud B. van Heeswijk,[a,b,c] **Christine Gonzales,**[d,e] **and Juerg Schwitter**[d,e]

[a]*Department of Radiology, University Hospital of Lausanne (CHUV), Lausanne, Switzerland*
[b]*Department of Biology and Medicine, University of Lausanne (UNIL), Lausanne, Switzerland*
[c]*CardioVascular Magnetic Resonance Research Center (CVMR), Center for Biomedical Imaging (CIBM), Lausanne, Switzerland*
[d]*Cardiology Service, University Hospital of Lausanne (CHUV), Lausanne, Switzerland*
[e]*Cardiac Magnetic Resonance Center (CRMC), University Hospital of Lausanne (CHUV), Lausanne, Switzerland*

ruud.van-heeswijk@chuv.ch

7.1 Introduction

While inflammatory processes in the cardiovascular system are common, their implications are manifold, and often these processes are incompletely understood. Inflammation is certainly the main mechanism causing tissue damage in the setting of myocarditis. Myocarditis represents a large and heterogeneous group of diseases with various etiologies including viral as well as parasitic infections (such as Chagas' disease), autoimmune reactions, toxins, electric shock, and others (Sagar et al., 2012).

Fluorine Magnetic Resonance Imaging
Edited by Ulrich Flögel and Eric Ahrens

ISBN 978-981-4745-31-4 (Hardcover), 978-981-4745-32-1 (eBook)
www.panstanford.com

The method of choice to establish the diagnosis and type of myocarditis is based on immunohistology, which is performed on endomyocardial biopsies (EMB). There is thus general agreement that an invasive approach is needed to work up patients with known or suspected myocarditis and to establish the final diagnosis (Cooper et al., 2007). However, selecting patients to undergo EMB is not an easy task, and the patients' clinical symptoms, physical findings, ECG-changes, and cardiac imaging information are typically required for this decision.

In recent years, cardiac magnetic resonance (CMR) has emerged as one of the most valuable imaging methods to detect myocarditis, and it is now recommended in European Guidelines to be applied as a first-line method in heart failure patients with suspected myocarditis (McMurray et al., 2012). In the setting of myocarditis, CMR detects inflammatory lesions by the late-gadolinium enhancement (LGE) technique, which is able to identify very small lesions of cell necrosis. If such lesions are located in the sub-epicardial layer, typically in the infero-lateral region of the left ventricle, the diagnosis of myocarditis is highly likely (Mahrholdt et al., 2004). This LGE CMR approach detects acute myocarditis proven by EMB with a sensitivity and specificity of 70% and 90%, respectively (Schwitter, 2012). The relatively low sensitivity of approximately 70% is most likely due to the fact that the LGE approach detects myocarditis only if it is causing cell necrosis.

In addition to necrosis detection, there is an interest in using T_2-weighted CMR imaging to detect myocardial edema, which can also be a marker of myocarditis. This approach yields sensitivities and specificities to detect acute myocarditis of 75% and 75%, respectively (Schwitter, 2012). While the concept of detecting myocardial edema is attractive, the current technique of T_2-weighted CMR is probably not sufficiently stable to yield sensitivities and specificities high enough to justify clinical decision making. Several groups are thus actively developing quantitative T_2-mapping approaches to overcome this limitation (Giri et al., 2009; van Heeswijk et al., 2015b) and preliminary results are encouraging (Thavendiranathan et al., 2012). Nevertheless, while novel CMR techniques allow detecting and quantifying necrosis and edema in the setting of inflammation, with these techniques one cannot obtain information on the cellular composition of inflammatory lesions, and the description of the inflammatory process by necrosis and edema alone remains incomplete.

It is also well established that in some situations acute myocarditis can transform into a chronic disease with activation of autoimmune inflammatory processes (Eriksson et al., 2003b). In such a group of patients, a placebo-controlled randomized clinical trial was performed, where the phase of myocarditis was determined in EMBs and an immunosuppressive treatment was successful when started in "autoimmune phase" of the myocarditis (Frustaci et al., 2009). Conversely, an immunosuppressive treatment could even accelerate tissue damage if applied during the "pathogen-driven" phase of the disease. Thus, a better understanding of the pathophysiology of myocarditis and a non-invasive technique to characterize its phase could help to individually tailor the treatment of patients and to monitor its efficacy. Myocarditis can also develop as peri-myocarditis. While LGE CMR is highly sensitive to detect acute pericarditis (Feng et al., 2011), little is known about the pathophysiological interrelationship between the two entities. As current CMR techniques detect myocarditis and pericarditis by tissue characterization (necrosis and edema), adding ^{19}F CMR could extend inflammation characterization to the next level since inflammatory cell composition can report on the type and phase of an inflammatory process. Accordingly, the possibility to identify specific cell types involved in the inflammatory process by means of ^{19}F CMR could open new ways to assess patients with myocarditis and to select the most appropriate treatment by considering the phase and the cell-composition of myocarditis and/or pericarditis at different stages of the disease.

An autoimmune-driven form of myocarditis can also be found in the context of acute or chronic rejection in patients after orthotopic heart transplantation. Its diagnosis in patients requires repetitive EMBs, thus a non-invasive approach based on macrophage ^{19}F MRI would be highly desirable, and positive proof-of-concept studies will be discussed in this chapter.

Another cardiac pathology that involves inflammatory mechanisms is myocardial infarction. Acute myocardial infarction (AMI) results in the highest death toll for a single disease in most industrialized countries (Nichols et al., 2012). Progression of coronary artery disease (CAD) is driven by repetitive plaque ruptures (Burke et al., 2001) that undergo repair mechanisms that involve inflammation. Typically, ruptures of low-grade stenotic plaques do not result in total vessel occlusions and AMI, but the repair of such ruptures commonly results in an increase in stenosis degree (Burke et al., 2001). Therefore, repetitive ruptures promote the degree of coronary artery stenosis, which later on reach hemodynamic significance and can thus cause ischemia during situations of increased oxygen demand. At a later stage of CAD, when highly stenotic lesions undergo rupture, the probability of vessel occlusions, and thus of developing AMI, increases. For example, lesions of 30–40% stenosis when rupturing result in vessel occlusions in approximately 1%/year, while high-degree stenosis of 70–80% diameter reduction result in total vessel occlusions in up to 10%/year (Schwitter, 2006). A technique that can detect inflammatory process in instable plaques would enable monitoring of the progression of CAD; when combined with stenosis degree, this technique could allow the detection of plaques at risk for acute occlusions, i.e., to identify patients at immediate risk for AMI.

If a plaque rupture is accompanied by a total occlusion (typically due to intra-luminal thrombus formation) of a vessel with insufficient collateral blood supply, AMI develops unless the coronary artery is re-opened by an invasive percutaneous intervention (PCI) in due time. With increasing ischemia time, necrosis expands within the area at risk from the endocardial layer towards the epicardium (wave front phenomenon). Accordingly, ESC Guidelines (Steg et al., 2012) recommend a primary PCI in AMI, to be performed within 90 min after first medical contact. While primary PCI in AMI improved survival, during the phase of reperfusion complex reactions are initiated in the reperfused tissue that involve endothelial damage and leakage, intracellular Ca^{2+}-overload, mitochondrial damage, but also inflammatory responses that may become self-perpetuating and deleterious (Hausenloy et al., 2013). The protective and reparative role of post-reperfusion inflammation and its potentially negative effects are poorly understood. In a randomized placebo-controlled clinical trial, STEMI patients were treated with a fibrin-derived peptide to block monocyte infiltration during the reperfusion phase (Atar et al., 2009). The LGE CMR technique could document a reduction of necrosis mass versus placebo in this trial, supporting the concept of post-reperfusion inflammation being involved in extending post-reperfusion tissue damage. These and other data illustrate the need for a better understanding of the role of inflammation in the ischemia-reperfusion situation. As primary PCI of AMI is increasingly used, the inflammatory response might represent a new therapeutic target for this growing patient population. A technique that allows following the inflammatory response in AMI non-invasively and in longitudinal studies could substantially promote research in this area and might also allow for clinical decision making once available on clinical MR scanners.

In the following sections, technical aspects specific for cardiac ^{19}F applications will be discussed followed by presentations of ^{19}F CMR studies in the field of myocarditis and ischemic heart disease.

7.2 Motion Compensation and Pulse Sequences

From a technical point of view, cardiac MRI is particularly challenging as it aims to image a fast moving organ. Since MRI exploits relatively low signals to generate an image, it typically samples signals over many heartbeats, which requires correction of the complex motion of the heart that results from (i) cardiac contraction and relaxation and (ii) respiratory motion. Several strategies are currently in use to deal with this motion problem (van Heeswijk et al., 2012a). An additional challenge for ^{19}F MRI is obtaining a sufficiently high signal-to-noise ratio (SNR). This means that SNR-efficient pulse sequences with multiple repetitions are preferred, which increases the likelihood of motion artifacts. Motion causes *k*-space data to be misaligned within an image and between repetitions, which in turn causes significant blurring and ghosting artifacts.

7.2.1 Cardiac Motion

The primary source of motion when studying cardiovascular pathologies is of course the heart itself. The contraction of the heart is a complex motion that consists of a shortening of both ventricles in the direction of the long axis combined with a circumferential shortening (Rademakers et al., 1994) and thickening in the short axis followed by relaxation and "untwisting" (Stuber et al., 1999c). Resting heart rates in humans are usually on the order of 40–80 bpm, while those in small animals such as mice can easily reach 600 bpm, or 10 heartbeats per second. Since ^{19}F CMR is typically acquired as an average of several repetitions, cardiac motion can be compensated with the use of electrocardiogram (ECG) triggering. Here, the QRS-complex of the ECG is used to synchronize the acquisitions so that they occur in the same phase of the heartbeat. This avoids motion blurring and misalignment of the data, which is especially important in the case of a low SNR or when studying very small structures, since after motion correction all signal adds up in the correct location in the image (with a smaller area compared to that in a motion-blurred image), resulting in sharper delineation and higher SNR. For cardiac motion correction, the ECG is ideally suited, and it is also needed to monitor the patient during the CMR examination, particularly during examinations where stress procedures are applied. ECG-triggering is especially useful in human subjects and large animals, since they generally have a long diastolic rest period of 100–400 ms. However, in small animals such as mice, this resting period lasts only on the order of ~20 ms, which means that many pulse sequences will be inefficient. However, given the short duration of a mouse heartbeat (as low as 100 ms) compared to the repetition time of a turbo-SE pulse sequence (500–5000 ms), only a relatively short time is lost waiting for the next exploitable ECG trigger.

A straightforward alternative to ECG-triggering is to simply ignore cardiac motion. This requires a relatively high number of repetitions as well as a relatively high SNR. When a high number of repeated acquisitions are used, the average ^{19}F MR signal will add up in the location of the ^{19}F signal source, while the motion artifacts will have a much lower signal that could fall within the noise or can be removed by simple thresholding. The main disadvantage of the technique is that the signal source can be blurred, compromising an accurate identification and quantification of the ^{19}F signal within an organ.

7.2.2 Respiratory Motion

The second type of motion is respiration, with a cycle length of typically 2–5 s in humans and 0.3–1 s in small animals, which affects the entire thorax and the upper abdomen. Two main strategies are used to deal with respiratory motion. One is based on data acquisition during a breath-hold (Atkinson et al., 1991), while the other approach tries to track respiratory motion during free breathing. In clinical routine cardiac MRI, most often data are acquired over 10–15 heart beats using a *k*-space segmented approach (Pearlman et al., 1994) which allows the suppression of respiratory motion through breath-holding. This approach is limited by the fact that very long breath-holds are not physiologically feasible, and therefore, compromises may be needed in temporal and spatial resolution as well as in SNR to acquire the data in a breath-hold. Alternatively, an elegant way to extend acquisition duration beyond a breath-hold is to monitor the position of the heart during free-breathing and to correct the data in *k*-space (or during image reconstruction) accordingly. Conventionally, in the free breathing approach, data were acquired only when the heart resides within an "acceptance window" (Stuber et al., 1999b). Typically, the position of the right hemi-diaphragm (i.e., the lung-liver interface) is monitored with a "pencil-beam" MR technique which is used as a surrogate for the heart position (Stuber et al., 1999a). This "diaphragmatic" navigator technique is susceptible to irregular breathing patterns, and its scan efficiency is limited, as it acquires data during a short period of the respiratory cycle. Nowadays the heart position itself can be derived from the *k*-space data (so-called self-navigation technique) (Larson et al., 2004; Piccini et al., 2012; Stehning et al., 2005). As this technique uses position information of the heart itself, and not of the right hemi-diaphragm, it is expected to yield more robust quality and recent studies are encouraging (Monney et al., 2015; Piccini et al., 2014). It is expected that such techniques for respiratory motion correction will be applied for ^{19}F CMR as well. Recently, a highly accelerated MRI technique emerged: the compressed sensing approach. Compressed sensing requires sparsity of data in at least one dimension so that significant undersampling can be performed. Since cardiac cine images essentially consist of a series of nearly identical images, this sparsity condition is met for cardiac cine datasets, and this technique is thus feasible in clinical CMR (Vincenti et al., 2014), Compressed sensing is also of interest for combination with ^{19}F

CMR, as the ^{19}F signal features sparsity as well, as has already been successfully demonstrated in a non-cardiac ^{19}F MRI study (Zhong et al., 2013a).

Of note, large animals under anesthesia are commonly ventilated, which means that the respiratory motion is regular and well-known, and that an end-expiration or end-inspiration rest period for imaging can be easily established. In small animals, breathing can be accurately monitored through respiratory bellows or an air cushion underneath the animal. Under anesthesia, these animals also tend to have a relatively long end-expiratory rest period that can be used for imaging. Alternatively, like for cardiac motion, respiratory motion can be ignored at the cost of motion blurring and aliasing in the resulting images.

7.2.3 Bulk Motion

The last type of motion is bulk motion, which occurs when a subject voluntarily or involuntarily moves their entire body. It differs from cardiac and respiratory motion in that it is not periodic and its time of occurrence can usually not be predicted. However, it can normally be avoided through external and indirect measures. In humans, it can be largely countered by instructing them not to move before a scan, optionally assisted by physical constraints such as straps, bands, and cushions. In the case of animals, the cancellation of bulk motion is normally achieved through some form of anesthesia (Fish et al., 2008). Here, care should be taken so that the physiological conditions such as the temperature and respiration rate are kept constant, since hypothermia can lead to significantly lower heart rates, while a too low respiration rate can result in gasping, which can substantially decrease the efficiency of cardiac and respiratory motion compensation, respectively.

When imaging vessels in and near the neck, such as the carotid arteries, swallowing motion can also cause motion artifacts. Like other non-repetitive bulk motion, it can be avoided through patient instruction or animal anesthesia, while several types of navigator gating can be used to counter involuntary swallowing (Crowe et al., 2005).

7.3 Animal Models of Cardiovascular Diseases

In the following subsections, the advantages of several cardiovascular applications of ^{19}F MRI in animals, as well as their main limitations will be discussed.

7.3.1 Angiography

One of the first cardiovascular applications of ^{19}F MRI was angiography. Joseph et al. demonstrated fluorine signal detection directly after injection in the rat in 1985 (Joseph et al., 1985), while high-resolution ^{19}F MRI of isolated perfused organs was demonstrated by Börnert et al. in 1989 (Bornert et al., 1989). The advantage of ^{19}F for angiography is that in most regions, the PFC emulsion strictly remains within the blood vessel lumen, thus becoming a highly specific intravascular contrast agent with high vessel sharpness. However, this advantage is limited by the relatively low concentration (and thus SNR) of the PFC emulsion, which requires choosing a balance between spatial resolution and scan time. With recent advances in hardware development and emulsion design, first pass angiography is now possible (Neubauer et al., 2007), while lower concentrations of injected PFC can be visualized as well (van Heeswijk et al., 2012b). It should be noted that the immune system can absorb the PFC emulsion, which results in additional ^{19}F signals slowly appearing in the lymphatic system, spleen and liver (through Kupffer cells) after approximately one hour post-contrast agent injection for ^{19}F MR angiography.

7.3.2 Myocarditis

While PFCs are not (yet) approved for diagnostic applications in humans, several animal models of myocarditis have been used to assess the potential of ^{19}F CMR to detect and characterize myocarditis. For this application, PFC emulsion is typically administered intravenously. The emulsion droplets (nanoparticles) are then taken up predominantly by monocytes and macrophages which accumulate in the inflammatory foci. ^{19}F CMR can then selectively image the ^{19}F signal and thus yields an image on monocyte/macrophage distribution in tissue. In several well-characterized animal models of myocarditis the ^{19}F CMR technique was successfully applied. It has been demonstrated in both autoimmune (van Heeswijk et al., 2013) and viral (Jacoby et al., 2014) myocarditis models that ^{19}F MRI can be used to visualize and quantify the inflammation (Fig. 7.1).

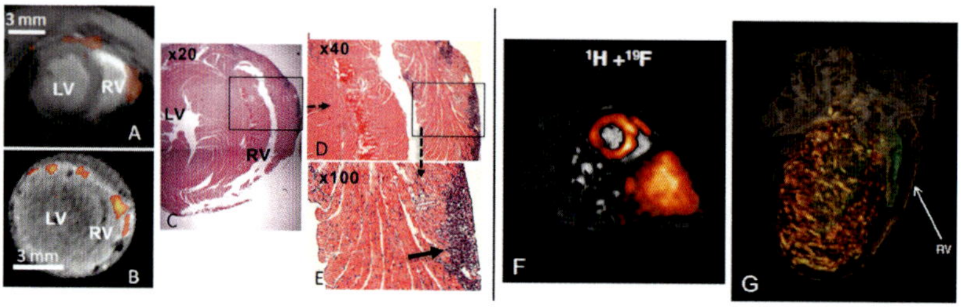

Figure 7.1 Non-invasive inflammation imaging in autoimmune (A-E) and viral (F-G) myocarditis. (A) Short-axis in vivo fused ^{19}F and ^{1}H MR image of the left (LV) and right (RV) ventricles of a mouse with autoimmune myocarditis. Colored patches in both the anterior LV and the lateral RV indicate inflammation. (B) The equivalent slice from a 3D ex vivo ^{19}F MR image. The ventricular cavities have collapsed to a smaller volume, but ^{19}F signal can still be observed at similar locations as in the in vivo image. (C) Overview of a hematoxylin-eosin (HE) stained heart section at the same anatomic level, where the right and left ventricles (RV and LV) can be clearly discerned. (D, E) Magnifications of the previous HE-stained heart section. Infiltrating immune cells can be observed in a thin subepicardial layer of the RV myocardium (solid arrow). (F) Short-axis in vivo fused ^{19}F and ^{1}H MRI of viral myocarditis induced by coxsackievirus B3. A large signal can be observed throughout the LV. (G) High-resolution 3D-rendered ^{1}H and ^{19}F MR data sets acquired ex vivo. Semitransparent surface contours of the heart and ventricles (green) were reconstructed from ^{1}H MRI and overlaid with the volume-rendered anatomically registered ^{19}F data (red). Images republished with permission from van Heeswijk et al. (2013) and Jacoby et al. (2014).

The experimental autoimmune myocarditis (EAM) model is created by injecting wild-type mice with an emulsion of alpha myosin heavy chain (a cardiac muscle protein) and Freund's adjuvant (an antigen emulsion that activates the immune system); this results in auto-inflammation of the myocardium that has a clear peak 21 days after immunization (Eriksson et al., 2003a). This immunological time course has been previously thoroughly characterized (Blyszczuk et al., 2009; Kania et al., 2009) including quantification of inflammation via histology. Therefore, this EAM model was used to demonstrate the ability of ^{19}F CMR to non-invasively detect and quantify the inflammatory response in the heart. On the two days prior to the peak of inflammation, mice were injected with an emulsion of perfluoro-polyether (PFPE)—pure or coupled with the fluorophore FITC (fluorescein isothiocyanate). ECG- and respiratory triggered GRE ^{1}H MRI and non-triggered TSE ^{19}F MRI were then performed at the peak of inflammation at day 21. ^{19}F signal was consistently observed throughout the heart, including small foci in the thin right ventricle in vivo (Fig. 7.1A), which was confirmed ex vivo (Fig. 7.1B). Post-mortem histology demonstrated infiltrated immune cells at the location of the ^{19}F signal (Fig. 7.1C,E). Immunohistochemistry showed co-localization of FITC with inflammatory cells, demonstrating that the PFC emulsion was absorbed by the inflammatory cells. Flow cytometry was furthermore used to demonstrate that the PFC was taken up by macrophages, dendritic cells, and granulocytes, but not by lymphocytes, which confirmed that ^{19}F signals were originating mainly from infiltrating myeloid inflammatory cells.

A viral form of myocarditis can be induced by an intra-peritoneal injection of an enterovirus (coxsackievirus B3) in immunodeficient (ABY/SnJ) mice. An emulsion of perfluoro-15-crown ether was injected 12 days after application of this virus, which was then followed by ECG- and respiration-triggered GRE ^{1}H MRI and non-triggered TSE ^{19}F MRI 48 h later (Jacoby et al., 2014). Subsequently, post-mortem MRI showed a patchy ^{19}F pattern that was homogeneously distributed over the entire left ventricle, while only minor PFC depositions were observed in the right ventricle both in vivo (Fig. 7.1F) and ex vivo (Fig. 7.1G). Flow cytometry demonstrated that mainly macrophages and dendritic cells had absorbed PFC, and that the ^{19}F signal integral precisely reflected the total number of infiltrated immune cells. T lymphocytes had also absorbed a small amount of PFC, which differs from the observations in the autoimmune myocarditis model.

7.3.3 Heart Transplantation

In clinical practice, one treatment of choice of end-stage heart failure is orthotopic heart transplantation. The most common and recurring complications after successful transplantation are episodes of autoimmune rejection of the graft, i.e., the autoimmune mediated inflammation of the myocardium. Episodes of rejection are detected by serial EMBs, which are then treated by intensified immunosuppressive therapy (Stehlik et al., 2011). The invasive nature of the EMBs together with the need for guidance with X-ray radiation renders a non-invasive alternative highly desirable. Animal models of transplantation in both mice (Flögel et al., 2011) and rats (Hitchens et al., 2011) consist of the grafting of a heart from one strain of rodent onto the vena cava of a second strain (sketched in Fig. 7.2A). This heterotopic second heart will then develop rejection, while an iso-graft (i.e., transplantation within the same strain) functions as a control. In such a rat model, Hitchens et al. (2011) have shown that immune cells can be labeled in vivo by direct injection of a ^{19}F cellular tracer agent, and immune cell accumulation at the site of organ rejection can be detected in vivo (Fig. 7.2B) and ex vivo (Fig. 7.2C) by ^{19}F MRI. These results are consistent with previous studies in these models using iron-oxide cell labeling and T_2^*-weighted image contrast for detecting macrophage accumulation due to rejection (Kanno et al., 2001).

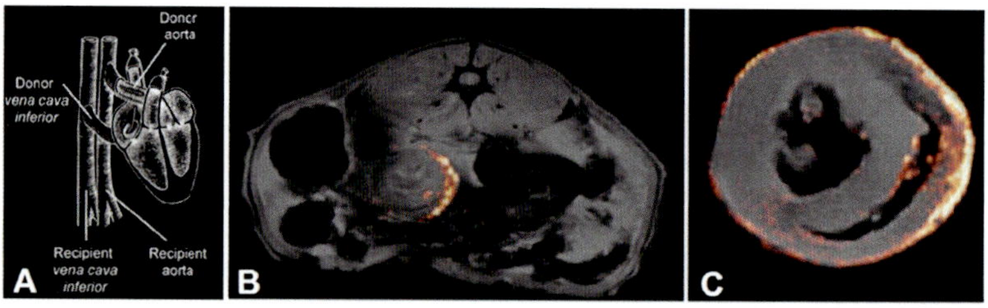

Figure 7.2 ^{19}F MRI to monitor rejection after cardiac transplantation. (A) Schematic drawing of the heterotopic heart transplantation model. (B) Fused in vivo ^{19}F and ^{1}H MR images show the detection of PFC-labeled immune cells in the myocardium of the transplanted heart graft 24 h following injection. (C) An ex vivo fused image confirms the detection in the heart in panel B. Images republished with permission from Flögel et al. (2011) and Hitchens et al. (2011).

Similarly, Flögel et al. (2011) demonstrated in a mouse model that ^{19}F MRI enabled the detection of an initial immune response not later than 3 days after surgery, when conventional parameters did yet not reveal any signs of rejection. In separate experiments, the immunosuppressant rapamycin was furthermore used to demonstrate the ability of ^{19}F MRI to monitor immunosuppressive therapy. Since PFC deposition is directly related to amount of immune cells in a tissue, the highly specific ^{19}F signal should also directly reflect the degree of rejection. Thus, with this ^{19}F technique, immunosuppressive therapy could potentially be monitored and investigated quantitatively (Hitchens et al., 2011).

7.3.4 Myocardial Infarction

Most experimental and clinical research currently focuses on AMI and post-infarct remodeling. Rodent models of infarction consist of wild-type mice or rats, in which a coronary artery (commonly the left anterior-descending coronary artery) is ligated at a set time before the MR examination, ensuring similarly affected coronary territories and disease progress.

The first ^{19}F CMR study of MI by Flögel et al. (Flögel et al., 2008) consisted of an LAD ligation in mice, followed by tail-vein injection of the PFC perfluoro-15-crown-5 ether 2 h later. Serial ^{1}H and ^{19}F MRI examinations were performed 5 times within 7 days after surgery, and demonstrated a progressive appearance of ^{19}F signal over time in the infarcted region of the left ventricle, while no signal appeared in the (non-affected) septal region. Centrifuged blood samples demonstrated that the PFC preferably accumulated in the mononuclear immune cells. The blood and hearts of mice that underwent similar MR experiments with a PFC emulsion labeled with the fluorophore rhodamine demonstrated that the PFC emulsion is mainly taken up by monocytes in the blood, and that the emulsion was exclusively taken up by the injured region of the myocardium. ^{19}F MRI can thus be used to study the involved cell populations and to characterize the spread and intensity of inflammation in AMI.

In an alternative rodent model, the ligated LAD can be re-opened after a predetermined time, and the involvement of inflammation in the reperfusion process can be studied. It is well established that MI with microvascular obstruction (MVO) goes together with an impaired patient prognosis and adverse LV remodeling (Nijveldt et al., 2009). Therefore, ^{19}F CMR has also been used together with LGE and T_2^*-weighted imaging to explore the role of monocyte recruitment in different territories of reperfused, acutely infarcted myocardium, e.g., in a rat model of reperfused AMI (Ye et al., 2013). Here, the LAD was ligated for a duration of 120 min, after which the snare was removed. After IV injection of an emulsion of perfluoro-15-crown-5 ether one day later, in vivo ^{19}F/^{1}H MRI examinations were performed longitudinally 3, 7, and 14 days after reperfusion (Figs. 7.3B,G,H, respectively). At the same time points, intramyocardial hemorrhage was also detected through ^{1}H-T_2^*-weighted imaging (Fig. 7.3C), while early gadolinium enhancement was used to detected MVO (Fig. 7.3D), and LGE was used to delineate the infarcted area (Fig. 7.3E). These indices were then combined into a parametric image that related the spatial distribution of hemorrhage, infarction, and inflammation (Fig. 7.3F). Rats were also scanned with the same protocol on day 28 after reperfusion to study the long-term outcomes. The resulting images demonstrated that in the absence of MVO, the monocyte infiltration was more intense in infarcted regions associated with serious reperfusion injuries, as indicated by severe intramyocardial hemorrhage. However, monocyte recruitment was significantly impaired in MVO areas, even when accompanied by severe intramyocardial hemorrhage. These results were confirmed by post-mortem histology and flow cytometry in this study. Compared with AMI with isolated intramyocardial hemorrhage, AMI with MVO furthermore resulted in significantly worse pump function of the left ventricle 28 days after the infarction. The study thus elegantly demonstrates that the impaired monocyte infiltration in MVO regions could be related to delayed healing and consequently to a worse functional outcome in the long term.

A major future goal of ^{19}F MRI of inflammation is the translation of the technique to the clinical setting, which includes using clinical scanners and animal models that approximate human dimensions. One important translational step was made by Zhong and colleagues, who investigated the use of ^{19}F MRI in a porcine model of MI in a clinical 3 T scanner (Zhong et al., 2013b). Their results indicated that ^{19}F imaging on a clinical scanner is feasible for the visualization of recent MI. Here, the use

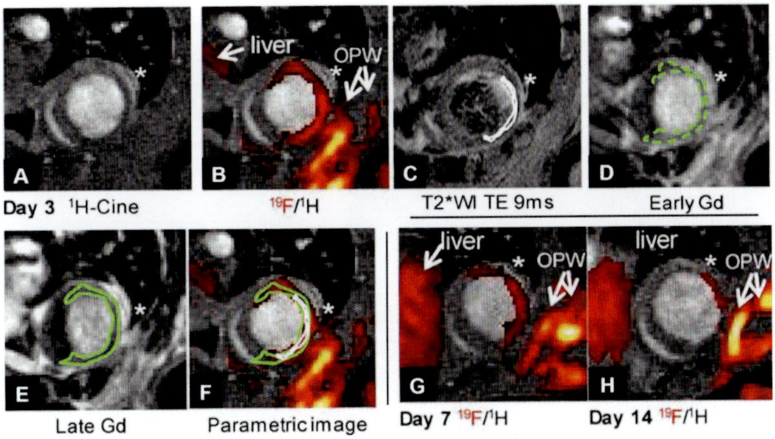

Figure 7.3 Longitudinal observations of the spatiotemporal distribution of ^{19}F signal in reperfused areas of myocardial infarction (MI) with isolated intramyocardial hemorrhage (IMH). (A) ^{1}H cine image 3 days after infarction and reperfusion, with * indicating the site of infarction. (B) ^{19}F image overlaid on the cine image demonstrates inflammation in the reperfused region (OPW = operational wound). (C) IMH region on a T_2^*-weighted image. (D) Early gadolinium enhancement and (E) late gadolinium enhancement (LGE). (F) The parametric image relates the spatial distribution of IMH (solid white line), MI area (solid green line), and ^{19}F signal. (G) and (H) ^{19}F signal is shown to decrease over time. Images republished with permission from Ye et al. (2013).

of the PFC agents had the advantage that they are taken up predominantly by macrophages, which are only present in recent infarcts, thus providing a well-defined signal of known biological origin that can be quantified against a null background. The next important step further in the application of ^{19}F-MRI in acute infarction was presented by Bönner et al. (2015), who successfully used perfluoro-octylbromide (PFOB), a PFC with a short biological half-life of 3–8 days, which can be used for serial studies. They also developed a novel pulse sequence which elegantly handles the multiple spectral peaks of PFOB. Very importantly, this PFOB compound was tested in the past as a plasma expander (due to some oxygen-transport properties of PFC), and phase III clinical trials of the Food and Drug Administration of the United States (FDA) demonstrated its safety when administered in large quantities to patients with blood loss during surgery (Spahn et al., 2002). Furthermore, in the study of Bönner et al., a close correlation was found between the ^{19}F-signal located in macrophages of the infarct territory and immunohistology (Fig. 7.4).

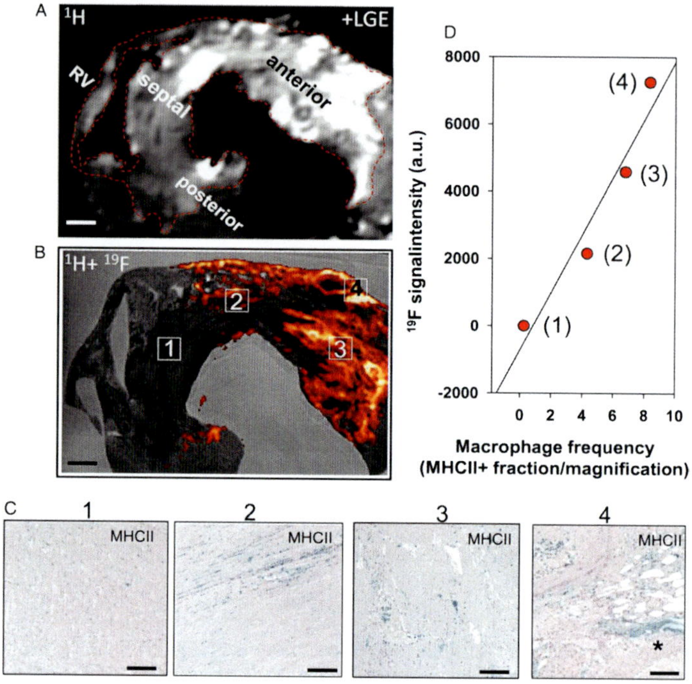

Figure 7.4 ^{19}F signal is distributed in a "patchy" fashion within LGE delineated area and correlates with macrophage density. (A) ^{1}H short axis image with transmural LGE (2 SDs above background delineated by red dashed line) of the anterio-septal myocardium and the right ventricle (RV) at 3 T. (B) Merged high-resolution ^{1}H and ^{19}F image at 9.4 T of the LGE positive myocardium (A). ^{19}F signal (red) is distributed in a "patchy" fashion across the infarcted myocardium (areas 1–4). (C) Subsequently, areas 1–4, as depicted in (B), were analyzed histological for regional analysis of macrophage frequency (MHCII+ cells = turquoise spots, magnification × 100). Note, area 1 without any fluorine signal and area 4 with the highest fluorine signal in the vicinity of a vessel (star). (D) ^{19}F signal intensity of the four regions correlates with macrophage frequency within myocardial infarction ($p < 0.05$, $R^2 = 0.94$). Images republished with permission from Bönner et al. (2015).

7.3.5 Atherosclerosis

The infiltration of various types of immune cells plays a pivotal role throughout the development and progression of atherosclerotic plaque. This inflammation is furthermore a major determinant of the vulnerability of plaques, which indicates their likelihood to rupture and lead to myocardial infarction or stroke (Stone et al., 2011). The detection of immune cells in atherosclerotic plaque could therefore help to better characterize their vulnerability. To evaluate the potential of ^{19}F MRI to assess the inflammation burden, a PFC emulsion was serially injected in a mouse model of atherosclerosis (apolipoprotein-E-knockout or ApoE$^{-/-}$ mice subjected to 12 weeks of high fat diet), with the aim to subsequently detected the ^{19}F signal of immune cells in vivo in plaques in the aorta and its branches (van Heeswijk et al., 2015a).

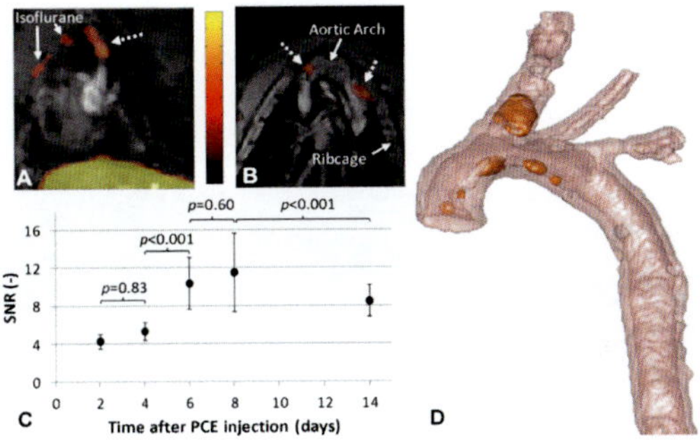

Figure 7.5 ^{19}F and ^{1}H MRI of atherosclerotic plaque in mice. (A) Sagittal ^{19}F/^{1}H fusion visualization of the liver, the left side of the heart and the brachiocephalic artery. An atherosclerotic plaque (dotted arrow) is confirmed by histology. ^{19}F signals are also detected in liver (PFC uptake by Kupffer cells) and in the subcutaneous fat (uptake of the anesthetic isoflurane). (B) Two plaques in the aortic arch are shown in another animal in a ^{19}F/^{1}H fusion image. (C) The ^{19}F SNR in the plaques does not significantly change 6 days after last PFC administration. (D) Ex vivo high-resolution 3D rendering of an aorta and its ^{19}F signal. Both the inner and outer surface of the aortic wall have been visualized in transparent pink, while the ^{19}F signal in orange is observed on the inner surface of the aortic arch and the brachiocephalic artery. Images republished with permission from van Heeswijk et al. (2015a).

Indeed, at day 2 after the last PFC administration, the presence of PFC in plaque macrophages was confirmed by ^{19}F MRI (Fig. 7.5A,B), and the ^{19}F signal increased up to day 6 after PFC administration (Fig. 7.5C). These findings were confirmed in excised vessels through high-resolution ex vivo ^{19}F MRI (Fig. 7.5D). Finally, bright-field and immunofluorescent histology were used to demonstrate that the ^{19}F MR signal indeed comes from the atherosclerotic plaque itself. After localization of a plaque (Fig. 7.6A,B), the PFC was detected through the fluorophore in the emulsion (Fig. 7.6C), while macrophages were detected via the Mac2 antibody (Fig. 7.6D). The exact overlap of the signals of these highly specific images in the merged image then confirmed the presence of PFC within the macrophages in the plaques.

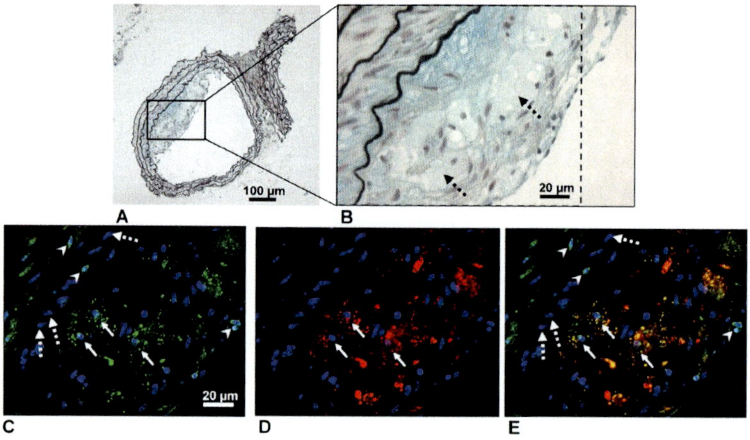

Figure 7.6 Histology of an atherosclerotic plaque in a brachiocephalic artery. (A,B) Movat's pentachrome staining shows that the plaque is characterized by the presence of macrophage foam cells (black dotted arrows), a necrotic core (in blue), and a thin fibrous cap. (C,D) Immunofluorescent staining of the same region of the plaque (the dashed region in (B) demonstrates the presence of PFC (FITC in green) and macrophages (Mac-2 in red) in the plaque. Blue indicates cellular nuclei (from DAPI). (E) The merged image of C and D demonstrates that most of the PFC is contained within the macrophages (solid arrows). Cells that are not macrophages that absorbed PFC (arrowheads) and cells that are not macrophages and did not absorb PFC (dotted arrows) can also be observed. Images republished with permission from van Heeswijk et al. (2015a).

7.4 In vitro ^{19}F-Labeling of Inflammatory Cells

As discussed above, direct IV injection of PFC emulsions has proven effective to track macrophages in vivo in animal models of viral or autoimmune myocarditis as well as in AMI, i.e., in ischemia/reperfusion. However, two potential limitations of this IV administration need to be mentioned. First, additional examinations like immunohistology or flow cytometry might be needed to verify that in a particular model or clinical situation the injected PFC is preferentially incorporated into monocytes/macrophages and not other cell types. Secondly, rather large quantities of PFC are commonly administered in such studies, since a high percentage will be retained by resident immune cells in the liver and spleen. Therefore, in vitro labeling of cells would be advantageous, as an exact characterization of the labeled cells is possible, and the cell population tracked non-invasively by ^{19}F CMR is thus precisely defined. In addition, the ^{19}F content per cell can be measured, e.g., in a portion of the ^{19}F-labeled injected cells (Fig. 7.7), which then allows for absolute quantification of the number of ^{19}F-labeled cells in the target. And finally, by injecting ^{19}F-labeled cells, it is expected that the total amount of injected PFC is low, which would be desirable from a safety point of view.

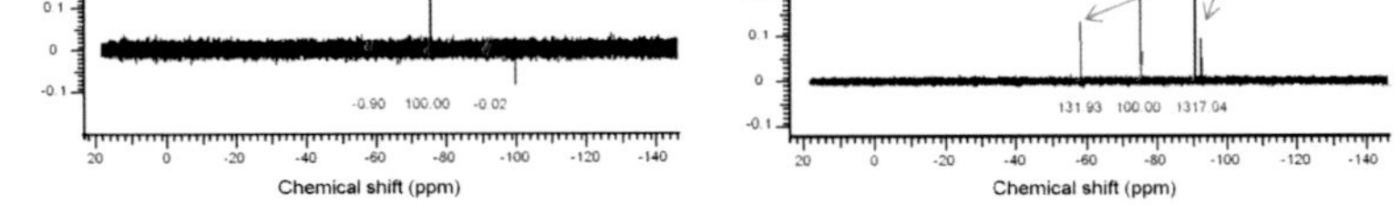

Figure 7.7 Fluorine incorporation in wild-type BMM. MR spectroscopy profiles of unlabeled cells (A) and after 18 h incubation with PFC (B). TFA (trifluoroacetic acid) was used as an internal control for ^{19}F quantification.

With this idea, bone marrow macrophages (BMM) were extracted from Balb/c mice and in vitro labeled with PFC before being injected into mice with autoimmune myocarditis (Eriksson et al., 2003a), in order to study their fate by means of ^{19}F CMR. After in vitro labeling with a perfluoro-polyether (PFPE) based PFC (from Celsense, Inc., Pittsburgh, PA), BMM exhibited a high level of ^{19}F content ($2 \times 10^{15} \pm 10\%$ ^{19}F spins /cell, see MR spectroscopic profiles in Fig. 7.7). ^{19}F-labeled BMM were injected IV at day 9 post immunization and ^{19}F CMR was performed 24 h (day 10) and/or 72 h (day 12) following cell injection. In the myocarditis animals, ^{19}F CMR consistently detected ^{19}F-labeled BMM in the liver and lungs at day 10–12 after immunization (Fig. 7.8A), but occasionally small ^{19}F signals were also detectable in the LV myocardium (Fig. 7.8B). Thus, such preliminary results indicate that the migration of ^{19}F-labeled macrophages can also be detected in vivo by ^{19}F CMR after in vitro ^{19}F labeling.

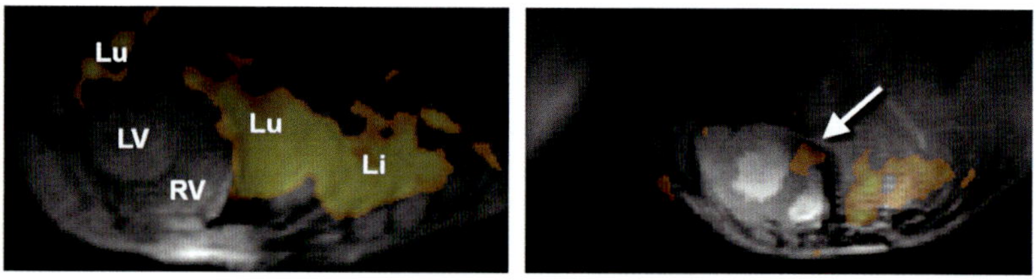

Figure 7.8 In vivo ^{19}F CMR performed at day 10 after immunization and 24 h after injection of PFC-labeled BMM. The left panel shows fusion of ^{1}H CMR (grey) and ^{19}F CMR (orange) images at the same anatomic location in short axis orientation. ^{1}H CMR depicts the anatomy including RV and LV, lungs (Lu) and liver (Li). ^{19}F signal can be observed in the lungs and the liver. The right panel shows, in addition to a ^{19}F signal in the liver, a small region with a faint ^{19}F signal is detected in the inferior wall of the LV (arrow) after IV injection of ex vivo PFC-labeled BMMs.

7.5 Conclusions and Perspectives

In the field of myocarditis and rejection-associated inflammation, several positive studies demonstrated that the ^{19}F CMR technique is able to detect specifically the infiltration of macrophages into the inflamed myocardium. Even small infiltrates at the surface of the right ventricle are detectable in the beating hearts of mice with myocarditis. There is thus strong evidence that this technique may become a useful tool to investigate the pathophysiology of myocarditis and graft rejection in animal models. Theoretical considerations and preliminary results obtained on 3 T clinical scanners point to the possibility that ^{19}F CMR may be used in the near future in patients for clinical decision making as this technique may help define the various stages of inflammation in order to tailor the patients' treatments.

Similarly, in the field of ischemia-reperfusion research, ^{19}F CMR provided new insights into the role of inflammatory cells during reperfusion. Therefore, ^{19}F CMR is likely to play an important role in future animal studies to investigate inflammation and autoimmune processes involved in the reperfusion injury. As primary PCI is increasingly used to treat AMI patients, the ^{19}F CMR technique will hopefully facilitate the assessment of novel drug treatments aimed at reducing ischemia-reperfusion injury. In the past, most clinical studies on ischemia-reperfusion in patients could not confirm experimental findings. The opportunity to non-invasively track inflammatory cells involved in the ischemia-reperfusion process by ^{19}F CMR is expected to promote a robust transition of experimentally successful treatments into a clinical application by allowing for a better characterization of treatment effects in animal studies before large clinical trials are started.

In the field of atherosclerosis, ^{19}F CMR is in its very first stages, but promising preliminary results are available. This ^{19}F CMR technique could open new opportunities to investigate the pathophysiology of plaque development and progression and to test the effect of anti-atherosclerotic treatments in animal models. Such studies could help to build up evidence to design large clinical studies in patients with hard clinical endpoints.

In vitro ^{19}F labeling of inflammatory cells, and their in vivo tracking by ^{19}F CMR, is also feasible. This approach allows tracking of well-defined cell populations in vivo and thus, may provide more detailed insight into inflammation and autoimmune mechanisms involved in myocarditis, pericarditis, and graft rejection, but also in plaque pathophysiology and ischemia-reperfusion injury.

In summary, this novel ^{19}F MRI technique offers a large spectrum of promising applications in the cardiovascular system, ranging from detection and monitoring of inflammation in the myocardium (myocarditis, infarction, rejection) to characterization of vessel wall plaques in atherosclerosis. While currently available MRI techniques report on general tissue characteristics like necrosis, edema, hemorrhage etc., ^{19}F MRI yields information on the inflammatory cell composition and can thus boost diagnostics to a next level. To get there, however, further research will be needed in the domains of design and modifications of PFC, in pulse sequence design, and developing robust in vitro labeling strategies (Schwitter, 2015). Furthermore, when ^{19}F MRI becomes available in humans, it will be able to substantially facilitate translational research, as the same non-invasive techniques and parameters can be applied in both animals and humans. Finally, as the cellular information provided by the ^{19}F MRI technique is obtained without radiation exposure, the technique is ideally suited for patient monitoring.

References

Atar D, Petzelbauer P, Schwitter J, Huber K, Rensing B, Kasprzak J, Butter C, Grip L, Hansen P, Süselbeck T, Clemmensen P, Marin-Galiano M, Geudelin B, Buser P, Investigators ftF (2009), Effect of intravenous FX06 as an adjunct to primary percutaneous coronary intervention for acute ST-segment elevation myocardial infarction, *J Am Coll Cardiol*, **53**, 720–729.

Atkinson DJ, Edelman RR (1991), Cineangiography of the heart in a single breath hold with a segmented turboFLASH sequence, *Radiology*, **178**(2), 357–360.

Blyszczuk P, Kania G, Dieterle T, Marty RR, Valaperti A, Berthonneche C, Pedrazzini T, Berger CT, Dirnhofer S, Matter CM, Penninger JM, Luscher TF, Eriksson U (2009), Myeloid differentiation factor-88/interleukin-1 signaling controls cardiac fibrosis and heart failure progression in inflammatory dilated cardiomyopathy, *Circ Res*, **105**(9), 912–920.

Bonner F, Merx MW, Klingel K, Begovatz P, Flögel U, Sager M, Temme S, Jacoby C, Salehi Ravesh M, Grapentin C, Schubert R, Bunke J, Roden M, Kelm M, Schrader J (2015), Monocyte imaging after myocardial infarction with 19F MRI at 3 T: A pilot study in explanted porcine hearts, *Eur Heart J Cardiovasc Imaging*, **16**(6), 612–620.

Bornert P, Dreher W, Schneider W (1989), ^{19}F chemical shift imaging in perfluorocarbons, *NMR Biomed*, **2**(5–6), 278–283.

Burke AP, Kolodgie FD, Farb A, Weber DK, Malcom GT, Smialek J, Virmani R (2001), Healed plaque ruptures and sudden coronary death:

Evidence that subclinical rupture has a role in plaque progression, *Circulation*, **103**(7), 934–940.

Cooper LT, Baughman KL, Feldman AM, Frustaci A, Jessup M, Kuhl U, Levine GN, Narula J, Starling RC, Towbin J, Virmani R (2007), The role of endomyocardial biopsy in the management of cardiovascular disease: A scientific statement from the American Heart Association, the American College of Cardiology, and the European Society of Cardiology endorsed by the Heart Failure Society of America and the Heart Failure Association of the European Society of Cardiology, *J Am Coll Cardiol*, **50**(19), 1914–1931.

Crowe LA, Keegan J, Gatehouse PD, Mohiaddin RH, Varghese A, Symmonds K, Cannell TM, Yang GZ, Firmin DN (2005), 3D volume-selective turbo spin echo for carotid artery wall imaging with navigator detection of swallowing, *J Magn Reson Imaging*, **22**(4), 583–588.

Eriksson U, Kurrer MO, Sonderegger I, Iezzi G, Tafuri A, Hunziker L, Suzuki S, Bachmaier K, Bingisser RM, Penninger JM, Kopf M (2003a), Activation of dendritic cells through the interleukin 1 receptor 1 is critical for the induction of autoimmune myocarditis, *J Exp Med*, **197**(3), 323–331.

Eriksson U, Ricci R, Hunziker L, Kurrer MO, Oudit GY, Watts TH, Sonderegger I, Bachmaier K, Kopf M, Penninger JM (2003b), Dendritic cell-induced autoimmune heart failure requires cooperation between adaptive and innate immunity, *Nat Med*, **9**(12), 1484–1490.

Feng D, Glockner J, Kim K, Martinez M, Syed I, Araoz P, Breen J, Espinosa R, Sundt T, Schaff H, Oh J (2011), Cardiac magnetic resonance imaging pericardial late gadolinium enhancement and elevated inflammatory markers can predict the reversibility of constrictive pericarditis after antiinflammatory medical therapy: A pilot study, *Circulation*, **124**, 1830–1837.

Fish RE, Brown MJ, Danneman PJ, Karas AZ (2008), *Anesthesia and Analgesia in Laboratory Animals*. Waltham, Massachusetts: Academic Press.

Flögel U, Ding Z, Hardung H, Jander S, Reichmann G, Jacoby C, Schubert R, Schrader J (2008), In vivo monitoring of inflammation after cardiac and cerebral ischemia by fluorine magnetic resonance imaging, *Circulation*, **118**(2), 140–148.

Flögel U, Su S, Kreideweiss I, Ding Z, Galbarz L, Fu J, Jacoby C, Witzke O, Schrader J (2011), Noninvasive detection of graft rejection by in vivo (19) F MRI in the early stage, *Am J Transplant*, **11**(2), 235–244.

Frustaci A, Russo MA, Chimenti C (2009), Randomized study on the efficacy of immunosuppressive therapy in patients with virus-negative inflammatory cardiomyopathy: The TIMIC study, *Eur Heart J*, **30**(16), 1995–2002.

Giri S, Chung Y-C, Merchant A, Mihai G, Rajagopalan S, Raman S, Simonetti O (2009), T_2 quantification for improved detection of myocardial edema, *J Cardiovasc Magn Reson*, **11**(1), 56.

Hausenloy D, Yellon D (2013), Myocardial ischemia-reperfusion injury: A neglected therapeutic target, *J Clin Invest*, **123**(1), 92–100.

Hitchens TK, Ye Q, Eytan DF, Janjic JM, Ahrens ET, Ho C (2011), ^{19}F MRI detection of acute allograft rejection with in vivo perfluorocarbon labeling of immune cells, *Magn Reson Med*, **65**(4), 1144–1153.

Jacoby C, Borg N, Heusch P, Sauter M, Bonner F, Kandolf R, Klingel K, Schrader J, Flögel U (2014), Visualization of immune cell infiltration in experimental viral myocarditis by (19)F MRI in vivo, *Magma*, **27**(1), 101–106.

Joseph PM, Fishman JE, Mukherji B, Sloviter HA (1985), In vivo 19F NMR imaging of the cardiovascular system, *J Comput Assist Tomogr*, **9**(6), 1012–1019.

Kania G, Blyszczuk P, Stein S, Valaperti A, Germano D, Dirnhofer S, Hunziker L, Matter CM, Eriksson U (2009), Heart-infiltrating prominin-1+/CD133+ progenitor cells represent the cellular source of transforming growth factor beta-mediated cardiac fibrosis in experimental autoimmune myocarditis, *Circ Res*, **105**(5), 462–470.

Kanno S, Wu YJ, Lee PC, Dodd SJ, Williams M, Griffith BP, Ho C (2001), Macrophage accumulation associated with rat cardiac allograft rejection detected by magnetic resonance imaging with ultrasmall superparamagnetic iron oxide particles, *Circulation*, **104**(8), 934–938.

Larson AC, White RD, Laub G, McVeigh ER, Li D, Simonetti OP (2004), Self-gated cardiac cine MRI, *Magn Reson Med*, **51**(1), 93–102.

Mahrholdt H, Goedecke C, Wagner A, Meinhardt G, Athanasiadis A, Vogelsberg H, Fritz P, Klingel K, Kandolf R, Sechtem U (2004), Cardiovascular magnetic resonance assessment of human myocarditis: A comparison to histology and molecular pathology., *Circulation*, **109**, 1250–1258.

McMurray JJV, Adamopoulos S, Anker SD, Auricchio A, Böhm M, Dickstein K, Falk V, Filippatos G, Fonseca C, Sanchez MAG, Jaarsma T, Køber L, Lip GYH, Maggioni AP, Parkhomenko A, Pieske BM, Popescu BA, Rønnevik PK, Rutten FH, Schwitter J, Seferovic P, Stepinska J, Trindade PT, Voors AA, Zannad F, Zeiher A (2012), ESC Guidelines for the diagnosis and treatment of acute and chronic heart failure 2012, *Eur Heart J*, **33**(14), 1787–1847.

Monney P, Piccini D, Rutz T, Vincenti G, Coppo S, Koestner SC, Sekarski N, di Bernardo S, Bouchardy J, Stuber M, Schwitter J (2015), Single centre experience of the application of self navigated 3D whole heart CMR for the assessment of cardiac anatomy in congenital heart disease, *J Cardiovasc Magn Reson*, **9**, 17:55.

Neubauer AM, Caruthers SD, Hockett FD, Cyrus T, Robertson JD, Allen JS, Williams TD, Fuhrhop RW, Lanza GM, Wickline SA (2007), Fluorine cardiovascular magnetic resonance angiography in vivo at 1.5 T with perfluorocarbon nanoparticle contrast agents, *J Cardiovasc Magn Reson*, **9**(3), 565–573.

Nichols M, Townsend N, Luengo-Fernandez R, Leal J, Gray A, Scarborough P, Rayner M. (2012), European Cardiovascular Disease Statistics 2012. European Heart Network, Brussels, European Society of Cardiology, Sophia Antipolis.

Nijveldt R, Hofman MBM, Hirsch A, Beek AM, Umans VAWM, Algra PR, Piek JJ, van Rossum AC (2009), Assessment of microvascular obstruction and prediction of short-term remodeling after acute myocardial infarction: Cardiac MR imaging study, *Radiology*, **250**(2), 363–370.

Pearlman J, Edelman R (1994), Ultrafast magnetic resonance imaging. Segmented turboflash, echo-planar, and real-time nuclear magnetic resonance., *Radiol Clin North Am*, **32**(2), 593–612.

Piccini D, Littmann A, Nielles-Vallespin S, Zenge MO (2012), Respiratory self-navigation for whole-heart bright-blood coronary MRI: Methods for robust isolation and automatic segmentation of the blood pool, *Magn Reson Med*, **68**(2), 571–579.

Piccini D, Monney P, Sierro C, Coppo S, Bonanno G, van Heeswijk R, Chaptinel J, Vincenti G, de Blois J, Koestner S, Rutz T, Littmann A, Zenge M, Schwitter J, Stuber M (2014), Respiratory self-navigated postcontrast whole-heart coronary MR angiography: Initial experience in patients, *Radiology*, **270**, 378–386.

Rademakers FE, Rogers WJ, Guier WH, Hutchins GM, Siu CO, Weisfeldt ML, Weiss JL, Shapiro EP (1994), Relation of regional cross-fiber shortening to wall thickening in the intact heart. Three-dimensional strain analysis by NMR tagging, *Circulation*, **89**(3), 1174–1182.

Sagar S, Liu PP, Cooper LT, Jr (2012), Myocarditis, *Lancet*, **379**(9817), 738–747.

Schwitter J (2006), Future strategies in the management of coronary artery disease, *Future Cardiol*, **2**(5), 555–565.

Schwitter J (2012), *CMR-Update*. 2nd ed. Lausanne, Switzerland: Schwitter, J. www.herz-mri.ch.

Schwitter J (2015), Cellular imaging: A bright future for 19F CMR, *Eur Heart J Cardiovasc Imaging*, **16**(6), 606–608.

Spahn DR, Waschke KF, Standl T, Motsch J, Van Huynegem L, Welte M, Gombotz H, Coriat P, Verkh L, Faithfull S, Keipert P, European Perflubron Emulsion in Non-Cardiac Surgery Study G (2002), Use of perflubron emulsion to decrease allogeneic blood transfusion in high-blood-loss non-cardiac surgery: Results of a European phase 3 study, *Anesthesiology*, **97**(6), 1338–1349.

Steg PG, James SK, Atar D, Badano LP, Blomstrom-Lundqvist C, Borger MA, Di Mario C, Dickstein K, Ducrocq G, Fernandez-Aviles F, Gershlick AH, Giannuzzi P, Halvorsen S, Huber K, Juni P, Kastrati A, Knuuti J, Lenzen MJ, Mahaffey KW, Valgimigli M, van 't Hof A, Widimsky P, Zahger D (2012), ESC Guidelines for the management of acute myocardial infarction in patients presenting with ST-segment elevation, *Eur Heart J*, **33**(20), 2569–2619.

Stehlik J, Edwards LB, Kucheryavaya AY, Benden C, Christie JD, Dobbels F, Kirk R, Rahmel AO, Hertz MI (2011), The Registry of the International Society for Heart and Lung Transplantation: Twenty-eighth Adult Heart Transplant Report—2011, *J Heart Lung Transplant*, **30**(10), 1078–1094.

Stehning C, Bornert P, Nehrke K, Eggers H, Stuber M (2005), Free-breathing whole-heart coronary MRA with 3D radial SSFP and self-navigated image reconstruction, *Magn Reson Med*, **54**(2), 476–480.

Stone GW, Maehara A, Lansky AJ, de Bruyne B, Cristea E, Mintz GS, Mehran R, McPherson J, Farhat N, Marso SP, Parise H, Templin B, White R, Zhang Z, Serruys PW (2011), A prospective natural-history study of coronary atherosclerosis, *New Engl J Med*, **364**(3), 226–235.

Stuber M, Botnar RM, Danias PG, Kissinger KV, Manning WJ (1999a), Submillimeter three-dimensional coronary MR angiography with real-time navigator correction: Comparison of navigator locations, *Radiology*, **212**(2), 579–587.

Stuber M, Botnar RM, Danias PG, Sodickson DK, Kissinger KV, Van Cauteren M, De Becker J, Manning WJ (1999b), Double-oblique free-breathing high resolution three-dimensional coronary magnetic resonance angiography, *J Am Coll Cardiol*, **34**(2), 524–531.

Stuber M, Scheidegger MB, Fischer SE, Nagel E, Steinemann F, Hess OM, Boesiger P (1999c), Alterations in the local myocardial motion pattern in patients suffering from pressure overload due to aortic stenosis, *Circulation*, **100**(4), 361–368.

Thavendiranathan P, Walls M, Giri S, Verhaert D, Rajagopalan S, Moore S, Simonetti OP, Raman SV (2012), Improved detection of myocardial involvement in acute inflammatory cardiomyopathies using T_2 mapping, *Circ Cardiovasc Imaging*, **5**(1), 102–110.

van Heeswijk RB, Bonanno G, Coppo S, Coristine A, Kober T, Stuber M (2012a), Motion compensation strategies in magnetic resonance imaging, *Crit Rev Biomed Eng*, **40**(2), 99–119.

van Heeswijk RB, De Blois J, Kania G, Gonzales C, Blyszczuk P, Stuber M, Eriksson U, Schwitter J (2013), Selective in vivo visualization of immune-cell infiltration in a mouse model of autoimmune myocarditis by fluorine-19 cardiac magnetic resonance, *Circ Cardiovasc imaging*, **6**(2), 277–284.

van Heeswijk RB, Pellegrin M, Flögel U, Gonzales C, Aubert JF, Mazzolai L, Schwitter J, Stuber M (2015a), Fluorine magnetic resonance imaging of inflammation in atherosclerotic plaque in vivo, *Radiology*, Radiology, **275**(2), 421–429.

van Heeswijk RB, Piccini D, Feliciano H, Hullin R, Schwitter J, Stuber M (2015b), Self-navigated isotropic three-dimensional cardiac T_2 mapping, *Magn Reson Med*, **73**(4), 1549–1554.

van Heeswijk RB, Pilloud Y, Flögel U, Schwitter J, Stuber M (2012b), Fluorine-19 magnetic resonance angiography of the mouse, *PloS One*, **7**(7), e42236.

Vincenti G, Monney P, Chaptinel J, Rutz T, Coppo S, Zenge MO, Schmidt M, Nadar MS, Piccini D, Chevre P, Stuber M, Schwitter J (2014), A novel compressed sensing single-breathhold multi-slice magnetic resonance approach for fast quantification of left ventricular function, volumes, and mass, *J Am Coll Cardiol Cardiovasc Imaging*, **7**(9), 882–892.

Ye YX, Basse-Lusebrink TC, Arias-Loza PA, Kocoski V, Kampf T, Gan Q, Bauer E, Sparka S, Helluy X, Hu K, Hiller KH, Boivin-Jahns V, Jakob PM, Jahns R, Bauer WR (2013), Monitoring of monocyte recruitment in reperfused myocardial infarction with intramyocardial hemorrhage and microvascular obstruction by combined fluorine 19 and proton cardiac magnetic resonance imaging, *Circulation*, **128**(17), 1878–1888.

Zhong J, Mills PH, Hitchens TK, Ahrens ET (2013a), Accelerated fluorine-19 MRI cell tracking using compressed sensing, *Magn Reson Med*, **69**(6), 1683–1690.

Zhong J, Schwartzman D, Schirda C, Balducci A, Helfer B, Wesa A, Ahrens ET (2013b), In vivo fluorine-19 MRI at 3 Tesla to visualize myocardial infarction inflammation in a porcine model. In: *Proc. Intl. Soc. Mag. Reson. Med. 21. Salt Lake City*, Abstract 577.

Part 4

Monitoring of Specific Cell Populations

Chapter 8

Tracking Lymphocytes in vivo

Ghaith Bakdash and Mangala Srinivas

Department of Tumor Immunology,
Radboud Institute for Molecular Life Sciences (RIMLS),
Radboud University Medical Center,
Geert Grooteplein 26-28, Nijmegen, 6500HB, The Netherlands

M.Srinivas@radboudumc.nl

Lymphocytes are often the "effectors" of an immune reaction. Thus, they typically need to traffic to the relevant region or to draining lymph nodes. This trafficking is key to their functionality and can be monitored using MRI. However, research using ^{19}F MRI for tracking these cells in vivo has thus far only been carried out by a few groups, in part because lymphocytes are difficult to label with sufficient ^{19}F to overcome detection limits. Furthermore, these cells can undergo rapid expansion, which further complicates tracking, especially quantification of cell numbers. Regardless, the technique is very promising and the current state of the art is presented here. We also briefly discuss other techniques for tracking lymphocytes in vivo, as well as multimodal imaging strategies.

Fluorine Magnetic Resonance Imaging
Edited by Ulrich Flögel and Eric Ahrens

ISBN 978-981-4745-31-4 (Hardcover), 978-981-4745-32-1 (eBook)
www.panstanford.com

8.1 Introduction

MRI is a very attractive modality to study cell trafficking in vivo. The advantages and disadvantages of ^{19}F MRI for cell tracking are covered in other sections of this book, and elsewhere [1, 2]. Briefly, the ability to detect ^{19}F agents indefinitely is very important, as some subsets of lymphocytes are long-lived and can survive for several days or longer. Thus, other cell tracers, such as radioactive agents, may not have sufficient lifetime for tracking labeled cells. On the other hand, the higher sensitivity of radioligands, particularly PET agents, can allow transient in situ labeling. Hence, there are several factors that must be taken into account when designing a cell tracking study involving cells that are challenging from an imaging perspective.

8.2 Lymphocytes

8.2.1 Function

Protection against pathogenic invaders is provided by white blood cells, also referred to as leukocytes. This heterogeneous cellular compartment consists of multiple populations including granulocytes (~65%), monocytes (~5%) and lymphocytes (~30%). Pathogen-specific immunity is conferred by the adaptive arm of the immune system which, includes lymphocytes. This executive arm consists of two types of cells, mainly T and B lymphocytes. Whereas B cells provide protection through secreting antibodies, T cells play a central role through their cellular effector functions. In this chapter, we will mainly focus on T lymphocytes. Conventional T cells are initially generated in the bone marrow as lymphoid progenitors, which migrate to the thymus where they undergo a process of proliferation and maturation. By the end of this process, which also entails negative selection of self-reactive T cells, standard $CD4^+$ and $CD8^+$ T cells are generated and exported to secondary lymphoid organs through blood [3]. Never encountering a pathogen before, these lymph-node resident T cells are also referred to as naïve T cells.

Mounting an adaptive immune response requires effective activation of naïve T cells, a process elegantly orchestrated by antigen-presenting cells, especially dendritic cells [4]. Upon sensing danger signals, resulting from external invasion or internal stress, dendritic cells migrate to lymph nodes, where naïve T cells reside. Subsequently, dendritic cells license naïve T cells to become effector T cells in a three-signal based paradigm (Fig. 8.1) [5]. The first signal results from the ligation of T cell receptors to pathogen-derived peptide antigens that are presented by major histocompatibility complex (MHC) molecules of dendritic cells. This principal stimulation signal is important to assure antigen specificity of the immune response. The second signal is the co-stimulatory signal, which results from the engagement of the T cell CD28 with the dendritic cell co-stimulatory molecules of the B7 family (CD80 and CD86). Co-stimulation is crucial for the clonal expansion of antigen-specific T cells and initiating protective immunity [6]. The selective development of naïve T cells into pathogen class-specific protective types of effector cells is dictated by a third signal. This T cell polarization signal 3 is mediated by cytokines and surface molecules expressed by dendritic cells [7].

Naturally, optimal defense against the wide variety of invading pathogens requires specialized pathogen-specific T cells. $CD8^+$ cytotoxic T lymphocytes, characterized by their killing capacity, are pivotal for the fight against viral infections and cancerous cells. Furthermore, different subsets of $CD4^+$ T helper cells (Th) provide protection against a wide spectrum of microorganisms. Th1 cells, characterized by interferon-γ (IFN-γ) production, are crucial for protection against intracellular infections like viruses and tuberculosis. Th2 cells, secreting interleukin-4 (IL-4), IL-5 and IL-13, mediate defense against, for example, helminthes [8]. Th17 cells, characterized by IL-17 secretion, contribute to host defense against extracellular bacteria and fungi [9]. Recent years have also witnessed the identification of additional populations of effector $CD4^+$ T cells, namely the IL-9-producing Th9 cells [10] and the IL-22-secreting Th22 cells [11]. In addition to the different classes of effector T cells, crucial for protection against infections, $CD4^+$ T cells also contribute to immune tolerance by developing into regulatory T cells [12].

8.2.2 Migration

In order to be fully functional, effector T cells should home to the site of infection or injury. T cell trafficking is mediated by chemokine receptors that lead lymphocytes to a wide variety of tissues. Factually, the expression pattern of chemokine receptors is distinctive for different effector classes of $CD4^+$ T cells. This is the case for CXCR3, expressed on Th1 cells, CCR4, expressed on Th2 cells, and CCR6, expressed on Th17 cells [13]. Furthermore, accumulating evidence points towards a fourth putative dendritic cell-derived signal that provides T cells with trafficking cues (Fig. 8.1) [14]. For instance, retinoic acid produced by intestinal $CD103^+$ DCs induce the mucosal homing markers integrin

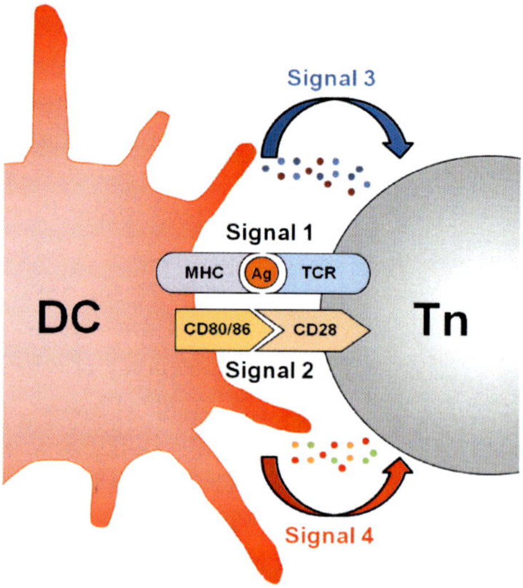

Figure 8.1 Dendritic cells educate naïve T cells into effector T cells. Upon recognizing danger, DCs undergo fundamental changes, referred to as maturation. Mature DCs migrate to draining lymph nodes where they initiate adaptive immune responses by priming naïve T cells (Tn) into fully functional effector T cells. This priming is mediated through three signals. The primary trigger is provided through the antigen presented within the MHC molecule to the T cell receptor (TCR) (Signal 1). Clonal expansion of antigen-specific T cells is mediated through co-stimulatory molecules (CD80/86) activating CD28 on T cells (Signal 2). Furthermore, DCs provide polarizing signals (Signal 3) skewing activated T cells into the appropriate type of helper T cells (Th) or cytotoxic T cells in case of $CD8^+$ T cells. Finally, T cells are provided by homing instructions (putative Signal 4) allowing them to exert their effector functions where they are required. Cited with permission from [17].

heterodimer α4β7 and chemokine receptor 9 (CCR9) on developing effector T cells enabling them to migrate from lymphoid tissues towards intestinal lamina propria [15]. Alternatively, DCs may utilize vitamin D to program T cell expression of the chemokine receptor 10 (CCR10), which directs T cells towards skin epidermis [16].

Once recruited to the site of danger, recently primed T cells exert their effector functions, conferring protection against microbes and cancerous cells.

8.2.3 Autoimmune Disease, Cancer and Transplant Rejection

Thus, T cell trafficking is an integral element of adaptive immunity and failure to home to the right site may lead to inadequate immune response. Therefore, monitoring T cell trafficking is of prime importance to (i) determine the efficiency of T cell-dependent immune responses against microbes or tumors, (ii) to elucidate the pathogenesis of T cell–mediated inflammation, (iii) to reveal the homing sites and its temporal distribution, and (iv) to visualize the mode of action of T cell-based immunotherapies. Applying T cell tracking could assist in the early diagnosis and treatment of a wide variety of diseases ranging from microbial infections to tumors. It will also allow the assessment of emerging novel therapies. For instance, adoptive transfer of autologous tumor-infiltrating T cells is a promising immunotherapeutic approach in the treatment of metastatic melanoma [18] and other cancers. Possessing the ability to track these transferred T cells will allow deeper insight into their homing patterns and their infiltration efficiency of primary and metastatic tumor lesions. Furthermore, T cell monitoring can be extended to cellular therapy with regulatory T cells, which is emerging as a clinical strategy for sustainable immunosuppression in the fields of hematopoietic stem cell transplantation, autoimmune diseases, and solid organ transplantation. Determining the viability and tissue distribution of the administered regulatory T cells is vital for assessing their immunoregulatory effects in vivo. Collectively, T cell tracking has the potential to significantly enhance the quality of currently available therapeutic approaches and will facilitate the development of novel therapies.

8.3 Lymphocyte Tracking with Other Imaging Modalities

8.3.1 Nuclear Imaging Techniques

This section includes scintigraphy, SPECT (single photon emission computed tomography) and PET (positron emission tomography). All of these techniques are used in the clinic. PET is frequently combined with CT these days, for anatomic context. In all cases, the specific cells must be labeled with a radioligand, such as ^{111}In, or with a specific injectable tracer, such as ^{18}F. The half-lives of these nuclei range from about an hour to nearly 3 days. Nuclear imaging techniques can be very sensitive, up to a couple of orders of magnitude higher in sensitivity than ^{19}F MRI. These factors are summarized in Table 8.1.

Table 8.1 Summary of different imaging techniques relevant to cell tracking (adapted from [19])

Modality	Relative sensitivity	Clinical use	Quantification of cell numbers
Scintigraphy	++	Yes	++
SPECT	++	Yes	+++
PET	+++	Yes	+++
MRI (^{1}H)	++	Yes*	+
MRI (^{19}F)	+	Yes*	++
Fluorescence	+	No	+
Bioluminescence	++	No	+

Note: MRI (^{1}H) refers to the use of contrast agents, such as iron oxides. Note that the relative values shown (+, ++ or +++) are only indications of low, medium and high sensitivity or applicability; these should be considered only as rough guides, as these parameters are heavily dependent on the study in question. An asterix (*) in the clinical use column indicates that these uses are only in small trials. It is important to keep in mind that several other relevant factors can play a role, as detailed in the text.

The earliest examples of lymphocyte tracking from the 1960s involve scintigraphy of labeled leukocytes [20], where cell accumulation was used to detect abscesses or sites of inflammation. Labeled leukocyte imaging by scintigraphy is still used to locate foci of infection, differentiate infection from tumor or postoperative changes, and in acute care settings [21, 22]. The cells are typically labeled ex vivo with the tracer, e.g., ^{111}In-oxine or ^{99m}Tc-exametazime. It is important to note that up to ~70% of leukocytes consist of neutrophils, not lymphocytes, thus these techniques are not specific to lymphocytes. An example of a scintigraphy image is shown in Fig. 8.2.

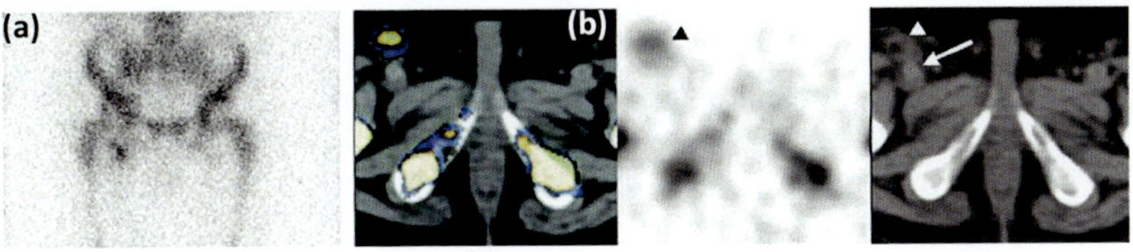

Figure 8.2 Example of a SPECT image of ^{111}In-labeled white blood cells in a patient with suspected infection at a vascular graft at the site of a femoral-popliteal bypass. The planar anterior image shows abnormal signal near this region (arrow). Figure reprinted from open access paper [24].

^{18}F-FDG (fluorodeoxyglucose) PET is gaining in popularity due to the higher resolution and sensitivity of PET, and the fact that the tracer can be directly injected in vivo without the need to purify cells for labeling. FDG is a glucose analogue, and it is taken up in higher amounts by activated inflammatory cells due to their increased expression of cellular glucose transporters [22]. Clinical ^{18}F-FLT (fluorothymidine) PET has also been used to study lymphocyte function [23] after dendritic cell vaccination. FLT is a nucleotide analogue, and thus is taken up in increased amounts by actively dividing cells, such as activated lymphocytes. Since lymphocyte division primarily occurs in lymph nodes, this technique is not suitable for "tracking", but is useful for assessing functionality (i.e., dividing lymphocytes). These techniques are generally well-accepted and are often used for routine clinical practice.

8.3.2 Fluorescence Imaging and Microscopy

Fluorescence imaging has proved to be a valuable tool for high resolution tracking of lymphocytes in animal models. This includes techniques such as multiphoton microscopy, bioluminescence, and intravital imaging. However, due to the nature of light waves, imaging is restricted by penetration depth, thus limiting its use in larger organisms. However, for mice, particularly when imaging superficial lymph nodes, these techniques are powerful enough to detect single cell interactions or even intracellular events such as Ca^{2+} fluctuations [25]. This is very different from nuclear medicine or MRI-based techniques which are intrinsically macroscopic, where only the behavior of a population of cells is studied. Thus, these methods are complementary.

Several reviews have been published on the use of microscopy to study lymphocytes [26–28]; selected highlights are presented here. Two-photon and multiphoton microscopy, particularly in combination with intravital imaging allow for tracking single cell events. The technique is typically limited to the field of view being examined (usually a lymph node or tumor site), although new developments for "discontinuous" tracking are being made [29]. Furthermore, in mice at least, creative strategies, for example, moving tissues of interest to more light-accessible anatomic locations such as the ear can be realized [30]. Photoconversion, which results in a permanent shift in the fluorescence emission spectrum of a suitable dye allows tracking of single cells, and is particularly useful for tracking the progeny of dividing cells, such as activated T cells [31].

A few B cell studies have been carried out, for example, to examine the mechanism of B cell depletion in the liver using intravital microscopy [32]. Importantly, fluorescence agents are almost always combined with ^{19}F agents for ex vivo validation and histology of the labeled cells, as this offers a powerful combination of quantitative, macroscopic imaging with detailed microscopy. Examples are discussed in Section 8.5.

8.4 MRI for Tracking Lymphocytes

MRI is in several ways an ideal technique for tracking lymphocytes: The label lifetime is not restricted by half-life decay, whole body imaging is possible, cell numbers can be quantified from image data, no ionizing radiation is required and repeated imaging sessions are possible. The main disadvantages are sensitivity and dilution or even loss of the agent with cell division. Various strategies have been developed to overcome these issues, and are described in the following sections. A more detailed discussion of the different types of labeling is presented elsewhere [33]. Specific issues involving ^{19}F MRI for quantification of cell numbers are discussed below [2].

8.4.1 Iron-Based Imaging

Iron-based labels, such as super paramagnetic iron oxide (SPIO), are the most commonly used agents for MRI-based cell tracking. Their popularity is due to the relatively high detection sensitivity and the commercial availability of several iron agents. The first published clinical trial involving MRI for cell tracking used iron-labeled dendritic cells in melanoma patients [34]. Unfortunately, these agents are no longer manufactured by the larger companies for clinical use due to poor economic viability.

The main advantage of iron-based contrast agents is the high sensitivity. However, absolute quantification can be difficult as the contrast is concentration-dependent in a nonlinear manner (see Figs. 8.3b and 8.4b). Due to this complex relationship, image contrast alone is generally not sufficient to determine the concentration of the agent or cell number.

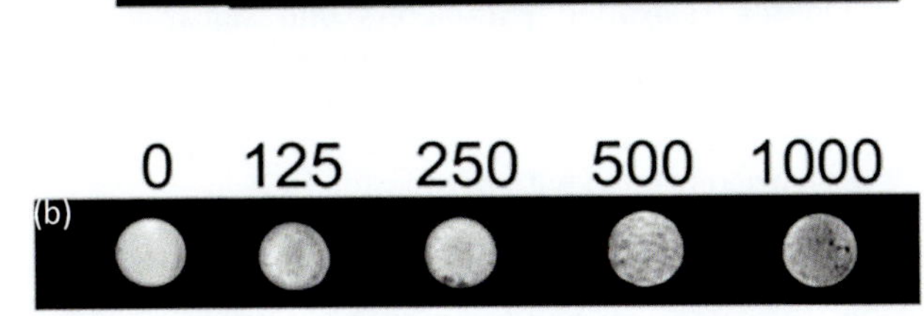

Figure 8.3 MR images of primary human DCs labeled with a (a) ^{19}F agent and a Gd agent and (b) an SPIO agent. Note that panel (a) is in 1000 cells/voxel, while (b) is the actual number of cells per voxel. This is to accommodate the higher sensitivity achieved with iron oxide agents. Figure reprinted with permission [35].

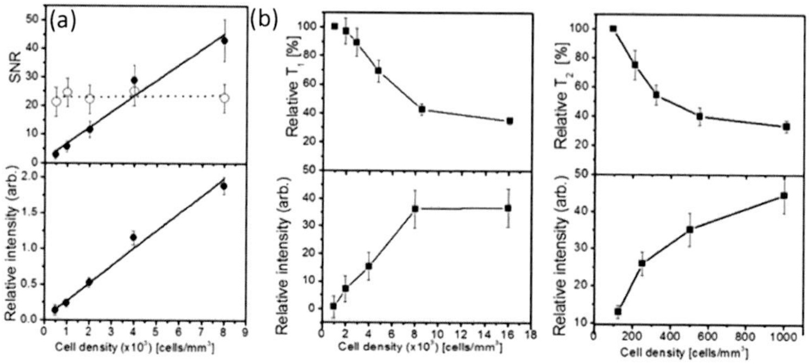

Figure 8.4 Cell number quantification from 7 T MRI data. Cells were labeled, suspended at various densities in gelatin and imaged using the appropriate spin-density, T_1 or T_2^*-weighted sequence. (a) The signal intensity of the labeled cells relative to a calibrated ^{19}F reference with constant ^{19}F content was calculated and plotted, showing the linear relationship between cell density and signal intensity. The plots show the SNR (upper panel) and signal intensity relative to the reference (lower panel) for the ^{19}F-labeled cells and a reference (closed and open circles respectively). Relative intensity here is the ratio between the SNR of the labeled cells (full circles in the upper panel) and SNR of the reference sample (open circles in the upper panel). (b) Similar data were plotted for cells labeled with Gd and SPIO. Cells labeled with Gd (left) were imaged using a T_1-weighted scan, and the relative signal intensity (lower panel) and T_1 plotted with increasing cell density (upper panel). Relative intensity is the change in contrast due to the presence of label. Similar plots for SPIO-labeled cells (right) show the relative T_2^* (upper panel) and intensity (lower panel). All values for the Gd and SPIO cells are normalized to the relaxation parameters and signal intensity over a reference containing nonlabeled cells. The observed departure from linear behavior shows the difficulties of quantification when contrast agents other than ^{19}F are used. Figure reprinted with permission [35].

Finally, one other technique that has been applied to cell imaging is the use of reporter genes, such as iron-binding ferritin [36]. These may be suited for T cell tracking as they are not diluted by cell division, although the challenges with quantification of contrast agent concentration from image data still apply. Thus far, there are no reported studies involving lymphocytes labeled with reporter genes for MRI.

8.4.2 Gadolinium-Based Imaging

T cells have been labeled with Gd agents, but these were used solely for mass spectrometric analysis and not MRI [37]. The authors were able to load purified human $CD4^+$ T cells with up to 10^8 Gd atoms/cell. Thus far, there have been no in vivo studies involving Gd-labeled T cells, due to the issues discussed previously, chiefly ambiguity in localization and low sensitivity relative to SPIO agents.

8.5 ^{19}F MRI for Tracking Lymphocytes

Thus far, there have only been a handful of studies using ^{19}F MRI for in vivo tracking of lymphocytes. Lymphocyte tracking is challenging for several reasons, discussed in a following section. However, successful strategies have been developed and these are discussed here.

8.5.1 Labels and Cell Loading

All the in vivo cell tracking work carried out with lymphocytes has used emulsion-based labels, which consist of a perfluorocarbon droplet coated in surfactant. The cell loading achieved has varied, by up to an order of magnitude, and different strategies have been applied to enhance labeling.

Table 8.2 summarizes the cell labels that have been used with lymphocytes thus far. The labeling achieved is in line with that achieved with dendritic cells, which is rather surprising as lymphocytes are not as phagocytic and are also smaller [43]. However, the cell loading achieved varies nearly 100-fold based on how the label is formulated and on whether or not extra treatment, such as a transfection agent was used, even with similar T cells from the same research group (the first 3 studies listed in the table). One study used an emulsion of PFPE together with a transfection agent, FuGENE 6, to load T cells with up to 10^{13} ^{19}F's/cell [38]. A similar emulsion without the transfection agent, resulted in 10-fold less T cell loading [39]. However, this was still considered advantageous as the transfection agent could lower cell viability and affect functionality. The details of the emulsion formulation process matter as to the achievable cell loading; for example, using a similar PFPE-pluronic which was formulated using microfluidization instead of sonication, resulted in a 10-fold decrease in cell loading [40]. Note that this decrease in cell loading is balanced by the ability of microfluidization to make emulsions in larger quantities than techniques such as sonication, and these emulsions are more homogeneous and stable over longer periods of time.

Table 8.2 A summary of the labels used with murine T cells thus far

Label			Extra	
^{19}F agent	Surfactant	^{19}F/cell	treatment	Reference
In vivo cell tracking				
PFPE	Poloxamer	2.2×10^{13}	Transfection agent	[38]
PFPE	Poloxamer	1.7×10^{12}	None	[39]
PFPE amide	Poloxamer	10^{11}	None	[40]
Ex vivo studies				
PFPE amide	Poloxamer		None	[41]*
PFPE	Poloxamer	10^{11}	Electroporation	[42]

Note: PFPE is perfluoropolyether, which is a linear perfluorocarbon. ^{19}F/cell indicates the number of fluorine nuclei per cell. In these examples, the PFPE had an ester endgroup, unless specified as the amide version. Poloaxomers are often also known as pluronics. Asterisk (*) denotes in vivo study where tissue cell numbers were quantified in excised organs using ^{19}F NMR.

Nevertheless, cell tracking using ^{19}F MRI is forever at the edge of detection and decreases in cell loading, if not balanced by a commensurate increase in cell numbers per voxel, will result in inability to detect the cells in vivo. This was the case with another study, by the same group [41], where the T cells could not be imaged in vivo, but were instead only measured in excised organs ex vivo. However, such ex vivo quantification is also a powerful quantitative technique, as it is rapid and generally more sensitive than in vivo imaging. Furthermore, since NMR does not damage the tissues; the whole, excised organs remain available for further analyses such as histology. To aid this, the organs were excised and imaged in a fixative (paraformaldehyde) which does not affect the ^{19}F signal. These techniques are useful to corroborate in vivo data if the interpretation is difficult. Overall, it can be concluded that it is difficult to obtain sufficient signal for imaging, unless either the ^{19}F cell loading is high or large cell numbers are used.

8.5.2 In vivo Imaging Data

The first study using ^{19}F MRI for quantitative cell tracking involved T cells that were specific to pancreatic antigens, and thus could induce the early stages of diabetes [38]. In this work, T cells were isolated from the spleens of NOD BDC2.5 mice, labeled and activated, and then transferred intraperitoneally to NOD SCID host mice. These mice were imaged 48 h later, and T cell homing to the pancreas was detected (Fig. 8.5). Imaging was carried out using a RARE sequence. The total imaging session was ~2–3 h for all imaging and procedures. Injected T cells ranged from 2–6 × 10^6 in each mouse. After imaging, relevant organs, such as the pancreas and spleen, were harvested for ex vivo NMR and histology. Ex vivo NMR was used to validate the in vivo quantification data, while the histology confirmed that the cells were indeed in the pancreas and that the label was still contained within the relevant T cells (with the aid of an attached fluorescent dye).

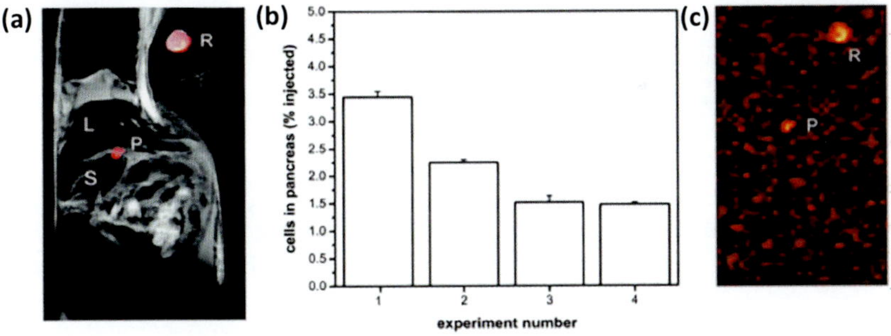

Figure 8.5 MRI data of pancreas-specific T cells labeled with a ^{19}F agent. These cells home to the pancreas, once activated, and can induce early diabetes. (a) Shows homing of the labeled cells to the pancreas (P = pancreas, L = liver, S = spleen, R = reference). An external reference was included in the image for quantification. (b) The numbers of cells homing to the pancreas at 48 h after transfer is shown. The authors use the term "apparent cells" as they do not account for cell division here. (c) Shown is an example of the raw ^{19}F magnitude image with hotspots at the pancreas and reference. Such images were used for quantification of apparent cell numbers.

Control mice injected with cell-free label (with emulsion dose equivalent to 10^7 labeled T cells), or MHC-mismatched labeled, activated T cells showed different localizations in the ^{19}F images, indicating that presence of the label at the pancreas is due to active T cell homing.

A follow-up study from the same group extended the in vivo imaging and cell number quantification to a period of 3 weeks in a different mouse model [39]. As in the previous study, T cells were isolated from the spleens of DO11.10 mice that have all T cells specific for a single antigen, ovalbumin. The cells were labeled ex vivo, activated and then transferred intraperitoneally to a host mouse that also received a subcutaneous injection of the antigen, ovalbumin. These host mice were then imaged five times over a period of 21 days. The ^{19}F signal was detected only in the lymph node draining the ovalbumin, and quantification of this signal was also carried out (see Fig. 8.6). As before, a fluorescent dye was coupled to the ^{19}F probe, which, in this study, was used for both in vivo fluorescence imaging and ex vivo histology.

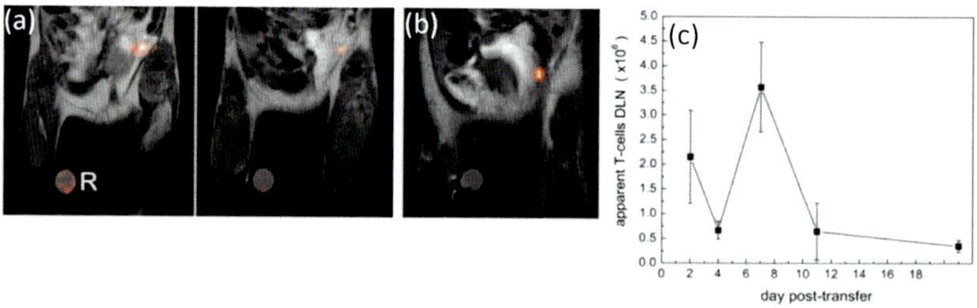

Figure 8.6 Longitudinal in vivo MRI of ^{19}F-labeled T cells in a mouse model of inflammation. (a) Representative images acquired at day 4 from a mouse that received labeled T cells and antigen with adjuvant. Shown are contiguous coronal slices where ^{19}F is rendered in pseudocolor and ^{1}H is in grayscale. T cells are detectable primarily in the draining lymph node (the inguinal node), and no labeled cells were detected in the contralateral lymph node. The external reference capillary used for quantification is labeled *R*. (b) A single slice from the same mouse at day 21. Some label is still detectable in the draining lymph node, although the total integrated signal is ≈ 10-fold weaker than the maximum. (c) Apparent T cell quantification in the draining lymph node from in vivo MRI data. Reprinted with permission from [39].

The authors also carried out more extensive experiments to estimate the in vivo T cell division rate, as this directly affects the cell number calculations. This was done using a conventional fluorescent agent, carboxy-fluorescein diacetate succinimidyl ester (CFSE) (Fig. 8.7) together with flow cytometry (FACS) to identify labeled cells. However, the assay could only be carried out for 7 days, due to dilution of the label. These data showed that extensive cell division occurred within 4 days after transfer. Some of these daughter cells also migrated to the site of antigen transfer (Fig. 8.6B). However, these cells were not detectable in the in vivo MR images. Furthermore, note that the cell numbers calculated using flow cytometry are in close agreement with those determined using in vivo MRI at early timepoints.

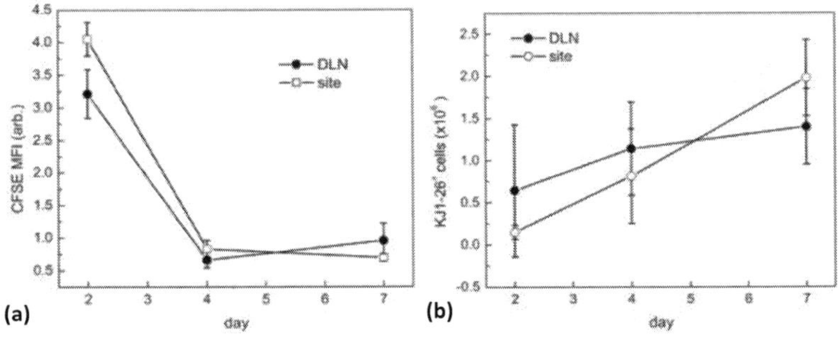

Figure 8.7 FACS-based analysis of T cell division in mouse inflammation model. (a) CFSE mean fluorescence intensity (MFI) for the draining lymph node (DLN) and site of antigen transfer over a 7-day period. The MFI drops significantly between days 2 to 4, indicative of rapid proliferation. (b) The number of antigen-specific T cells in the DLN and site of antigen transfer, estimated from the FACS analyses. Reprinted with permission from [39].

In these ^{19}F studies, one potential complication that can arise is that care must be used with isoflurane, the most commonly-used gaseous anesthetic in imaging studies, as isoflurane can have a diffuse, but detectable ^{19}F signal which overlaps with that of many perfluorocarbons. However, if the imaging is fast enough to avoid significant isoflurane accumulation, the regions imaged do not overlap with isoflurane depots, or if the signal from the isoflurane can be separated from that of the agent, than its use remains possible. A control experiment with the animal receiving isoflurane only, without any additional agent, is suggested. Alternative anesthetics could be used if the in vivo location of isoflurane is a problem.

Finally, the last study involving ^{19}F MRI for in vivo imaging of lymphocytes also used DO11.10 T cells in a murine diabetes model [40]. Here the focus of the work is on the chemical syntheses of the labels, and the lymphocyte tracking image is only shown as a proof of principle. No quantification was carried out. Here, the fluorous component of the dye was covalently linked to a fluorescent agent, ensuring that both components would stay together in vivo and intracellularly.

8.5.3 Ex vivo Studies

Ex vivo studies often face fewer limitations on scanning time. Thus, longer scans with more extensive averaging can be carried out for more "exotic" applications with ^{19}F-labeled lymphocytes. These studies are often carried out on T cell lines (T-lymphoblastoids). For example, cells have been labeled with hexafluorobenzene for oxygen sensing to determine the effect of different tocotrienol antioxidants [44]. Other studies have used ^{19}F agents to gauge cell viability in 3D cultures [45], intracellular pH measurements [46, 47] and Na^{+} concentration measurements in porcine lymphocytes [48]. These are highly specialized applications and are probably limited to ex vivo experiments due to detection limits.

In another study, different subsets of T cells were labeled ex vivo before transfer [41] into a diabetic mouse model. After 48 h, relevant tissues such as the pancreas and certain lymph nodes were removed, and fixed in paraformaldehyde before ^{19}F NMR was carried out. This allowed for quantification of T cells in specific tissues, with complete certainty of localization. Furthermore, as ^{19}F NMR is a nondestructive technique and can be carried out even on fixed tissues, the tissues remained available for further analyses such as histology. In vivo MRI was not feasible here due to low ^{19}F accumulation per voxel, resulting from various factors such as low cell loading and low numbers of cells accumulating per voxel in the region of interest. However, as shown in this chapter, ex vivo ^{19}F NMR of labeled migratory cells is a powerful technique in its own right which can readily be combined with more traditional techniques such as histology and flow cytometry.

8.6 Conclusion

The in vivo ^{19}F MRI studies thus far have used T cells that were labeled ex vivo before transfer to the subject. Furthermore, the cells were activated before transfer. This approach could be useful for following emerging T cell cytotherapies. However, the approach may not be useful for following T cells under physiological conditions where generally a small number of T cells are activated and undergo expansion, when encoutering antigen presented by dendritic cells. In situ labeling has not been attempted with ^{19}F agents for lymphocyte tracking.

Also, there have not yet been any studies on B cell migration, although B cell homing is often critical to effective immune responses, and B cells can be induced to home to regions such as the gut, bone marrow and sites of inflammation [49]. Very limited work has been done to track B cells using MRI, and only using SPIO [50]. This is certainly an interesting area for future research.

In conclusion, cell tracking using ^{19}F MRI is a relatively new technique [51]. Improvements in label design, cell labeling techniques, imaging sequences and imaging hardware have allowed the fledgling technique to move on to more challenging cell and disease types in a relatively short time.

Acknowledgments

GB is supported by a Netherlands Organization for Scientific Research (NWO) grant 951.03.002. MS is supported by the personal grants NWO VENI 700.10.409 and a European Research Council (ERC) Starting Grant ERC-2014-StG-336454-CoNQUeST.

References

1. Lanza, G. M., et al., $^{1}H/^{19}F$ magnetic resonance molecular imaging with perfluorocarbon nanoparticles. *Curr. Top. Dev. Biol.*, 2005. **70**, 57–76.
2. Srinivas, M., et al., ^{19}F MRI for quantitative in vivo cell tracking. *Trends Biotechnol.*, 2010. **28**(7), 363–370.

3. Yates, A. J., Theories and quantification of thymic selection. *Front Immunol.*, 2014. **5**, 13.
4. Banchereau, J., et al., Immunobiology of dendritic cells. *Annu. Rev. Immunol.*, 2000. **18**, 767–811.
5. Bakdash, G., et al., The nature of activatory and tolerogenic dendritic cell-derived signal II. *Front Immunol.*, 2013. **4**, 53.
6. Cunningham, A. J., and K. J. Lafferty, A simple, conservative explanation of the H-2 restriction of interactions between lymphocytes. *Scand. J. Immunol.*, 1977. **6**(1–2), 1–6.
7. Kapsenberg, M. L., Dendritic-cell control of pathogen-driven T-cell polarization. *Nat. Rev. Immunol.*, 2003/12. **3**(12), 984–993.
8. Agnello, D., et al., Cytokines and transcription factors that regulate T helper cell differentiation: New players and new insights. *J. Clin. Immunol.*, 2003/5. **23**(3), 147–161.
9. Harrington, L. E., et al., Interleukin 17-producing CD4+ effector T cells develop via a lineage distinct from the T helper type 1 and 2 lineages. *Nat. Immunol.*, 2005/11. **6**(11), 1123–1132.
10. Chang, H. C., et al., The transcription factor PU.1 is required for the development of IL-9-producing T cells and allergic inflammation. *Nat. Immunol.*, 2010/6. **11**(6), 527–534.
11. Duhen, T., et al., Production of interleukin 22 but not interleukin 17 by a subset of human skin-homing memory T cells. *Nat. Immunol.*, 2009/8. **10**(8), 857–863.
12. Wan, Y. Y., and R. A. Flavell, The roles for cytokines in the generation and maintenance of regulatory T cells. *Immunol. Rev.*, 2006/8. **212**, 114–130.
13. Sallusto, F., and A. Lanzavecchia, Heterogeneity of CD4+ memory T cells: Functional modules for tailored immunity. *Eur. J. Immunol.*, 2009/8. **39**(8), 2076–2082.
14. Campbell, D. J., and E. C. Butcher, Rapid acquisition of tissue-specific homing phenotypes by CD4(+) T cells activated in cutaneous or mucosal lymphoid tissues. *J. Exp. Med.*, 2002/1/7. **195**(1), 135–141.
15. Iwata, M., et al., Retinoic acid imprints gut-homing specificity on T cells. *Immunity*, 2004/10. **21**(4), 527–538.
16. Sigmundsdottir, H., et al., DCs metabolize sunlight-induced vitamin D3 to "program" T cell attraction to the epidermal chemokine CCL27. *Nat. Immunol.*, 2007/3. **8**(3), 285–293.

17. Bakdash, G., Harnessing dendritic cells to promote immune tolerance: Opportunities for allergen-specific immunotherapy, in *Cell Biology & Histology*. 2013, Academic Medical Center, University of Amsterdam: Amsterdam. p. 169.

18. Besser, M. J., et al., Clinical responses in a phase II study using adoptive transfer of short-term cultured tumor infiltration lymphocytes in metastatic melanoma patients. *Clin. Cancer Res.*, 2010/5/1. **16**(9), 2646–2655.

19. Srinivas, M., et al., Imaging of cellular therapies. *Adv. Drug Deliv. Rev.*, 2010. **62**(11), 1080–1093.

20. Winkelman, J., C. J. Collica, and S. G. Sandler, The delineation of abscesses by scintiphotography using Cr51 labeled leukocytes. *Am. J. Roentgenol. Radium. Ther. Nucl. Med.*, 1968. **103**(4), 881–885.

21. Palestro, C. J., C. Love, and K. K. Bhargava, Labeled leukocyte imaging: Current status and future directions. *Q. J. Nucl. Med. Mol. Imaging*, 2009. **53**(1), 105–123.

22. Love, C., and C. J. Palestro, Radionuclide imaging of inflammation and infection in the acute care setting. *Semin. Nucl. Med.*, 2013. **43**(2), 102–113.

23. Aarntzen, E. H., et al., Early identification of antigen-specific immune responses in vivo by [18F]-labeled 3′-fluoro-3′-deoxy-thymidine ([18F]FLT) PET imaging. *Proc. Natl. Acad. Sci. U. S. A.*, 2011. **108**(45), 18396–18399.

24. Bar-Shalom, R., et al., SPECT/CT using 67Ga and 111In-labeled leukocyte scintigraphy for diagnosis of infection. *J. Nucl. Med.*, 2006. **47**(4), 587–594.

25. Randriamampita, C., and A. C. Lellouch, Imaging early signaling events in T lymphocytes with fluorescent biosensors. *Biotechnol. J.*, 2014. **9**(2), 203–212.

26. Germain, R. N., E. A. Robey, and M. D. Cahalan, A decade of imaging cellular motility and interaction dynamics in the immune system. *Science*, 2012. **336**(6089), 1676–1681.

27. Hauser, A. E., S. M. Kerfoot, and A. M. Haberman, Cellular choreography in the germinal center: New visions from in vivo imaging. *Semin. Immunopathol.*, 2010. **32**(3), 239–255.

28. Jacobelli, J., R. S. Lindsay, and R. S. Friedman, Peripheral tolerance and autoimmunity: Lessons from in vivo imaging. *Immunol. Res.*, 2013. **55**(1–3), 146–154.

29. Chtanova, T., et al., Real-time interactive two-photon photoconversion of recirculating lymphocytes for discontinuous cell tracking in live adult mice. *J. Biophotonics*, 2014. **7**(6), 425–433.

30. Chen, B. J., et al., Long-term in vivo imaging of multiple organs at the single cell level. *PLoS One*, 2013. **8**(1), e52087.

31. Carlson, A. L., et al., Tracking single cells in live animals using a photoconvertible near-infrared cell membrane label. *PLoS One*, 2013. **8**(8), e69257.

32. Montalvao, F., et al., The mechanism of anti-CD20-mediated B cell depletion revealed by intravital imaging. *J. Clin. Invest.*, 2013. **123**(12), 5098–5103.

33. Ahrens, E. T., and J. W. Bulte, Tracking immune cells in vivo using magnetic resonance imaging. *Nat. Rev. Immunol.*, 2013. **13**(10), 755–763.

34. de Vries, I. J., et al., Magnetic resonance tracking of dendritic cells in melanoma patients for monitoring of cellular therapy. *Nat. Biotechnol.*, 2005. **23**(11), 1407–1413.

35. Bonetto, F., et al., A novel (19)F agent for detection and quantification of human dendritic cells using magnetic resonance imaging. *Int. J. Cancer*, 2011. **129**(2), 365–373.

36. Gilad, A. A., et al., Developing MR reporter genes: Promises and pitfalls. *NMR Biomed.*, 2007. **20**(3), 275–290.

37. Managh, A. J., et al., Single cell tracking of gadolinium labeled $CD4^+$ T cells by laser ablation inductively coupled plasma mass spectrometry. *Anal. Chem.*, 2013. **85**(22), 10627–10634.

38. Srinivas, M., et al., Fluorine-19 MRI for visualization and quantification of cell migration in a diabetes model. *Magn. Reson. Med.*, 2007. **58**(4), 725–734.

39. Srinivas, M., et al., In vivo cytometry of antigen-specific t cells using 19F MRI. *Magn. Reson. Med.*, 2009. **62**(3), 747–753.

40. Janjic, J. M., et al., Self-delivering nanoemulsions for dual fluorine-19 MRI and fluorescence detection. *J. Am. Chem. Soc.*, 2008. **130**(9), 2832–2841.

41. Morel, P. A., et al., Gene expression analysis of dendritic cells that prevent diabetes in NOD mice: Analysis of chemokines and costimulatory molecules. *J. Leukoc. Biol.*, 2011. **90**(3), 539–550.

42. Srinivas, M., Development and application of a ^{19}F platform for in vivo MRI: Tracking and quantification of T cells in murine disease

models and the effect of immunotherapy on T cell trafficking. 2008 PhD thesis, Carnegie Mellon University: Pittsburgh. 178 pp.

43. Srinivas, M., et al., Labeling cells for in vivo tracking using (19)F MRI. *Biomaterials*, 2012. **33**(34), 8830–8840.
44. Bartusik, D., et al., ^{19}F MRI of 3D CEM cells to study the effects of tocopherols and tocotrienols. *J. Pharm. Biomed. Anal.*, 2010. **53**(3), 599–602.
45. Bartusik, D., et al., The efficacy of new colchicine derivatives and viability of the T-Lymphoblastoid cells in three-dimensional culture using 19F MRI and HPLC-UV ex vivo. *Bioorg. Chem.*, 2009. **37**(6), 193–201.
46. Deutsch, C. J., and J. S. Taylor, Intracellular pH as measured by ^{19}F NMR. *Ann. N. Y. Acad. Sci.*, 1987. **508**, 33–47.
47. Deutsch, C. J., and J. S. Taylor, New class of ^{19}F pH indicators: Fluoroanilines. *Biophys. J.*, 1989. **55**(4), 799–804.
48. Smith, G. A., et al., Design of an indicator of intracellular free Na^+ concentration using ^{19}F-NMR. *Biochim. Biophys. Acta*, 1986. **889**(1), 72–83.
49. Mora, J. R., M. Iwata, and U. H. von Andrian, Vitamin effects on the immune system: Vitamins A and D take centre stage. *Nat. Rev. Immunol.*, 2008. **8**(9), 685–698.
50. Thorek, D. L., et al., In vivo, multimodal imaging of B cell distribution and response to antibody immunotherapy in mice. *PLoS One*, 2010. **5**(5), e10655.
51. Ahrens, E. T., et al., In vivo imaging platform for tracking immunotherapeutic cells. *Nat. Biotechnol.*, 2005. **23**(8), 983–987.

Chapter 9

Tracking of Dendritic Cells

Sonia Waiczies, MinChi Ku, and Thoralf Niendorf

Berlin Ultrahigh Field Facility, Max Delbrueck Centre for Molecular Medicine, Robert Rössle Str. 10, Berlin 13353, Germany

sonia.waiczies@mdc-berlin.de

9.1 Introduction

Dendritic cells (DCs) are professional antigen-presenting cells that mediate both adaptive as well as innate immune responses. Depending on subtype and activation state, DCs interact with naïve T cells by presenting antigen to initiate an immune response or tolerance. In this chapter, we will discuss the crucial need for developing methods to track DCs in vivo. These methods are fundamental not only for understanding the migratory behavior of DCs in health and disease but also for monitoring the behavior of DCs administered as cellular therapeutics for a multitude of diseases. Different methods for tracking DCs will be described as well as initial clinical trials that have been implemented using these methods to monitor DC therapy in cancer patients. Tracking of cells using MRI has gained remarkable attention in the past years. We will also focus on the labeling of DCs with fluorine (^{19}F) rich nanoparticles, their application

Fluorine Magnetic Resonance Imaging
Edited by Ulrich Flögel and Eric Ahrens

ISBN 978-981-4745-31-4 (Hardcover), 978-981-4745-32-1 (eBook)
www.panstanford.com

in vivo, and monitoring the extent of their migration to the draining lymph node with $^{19}F/^{1}H$ MRI and ^{19}F MRS.

9.2 About Dendritic Cells

DCs were originally characterized through their morphological appearance [1]. In their 1973 seminal work, Ralph Steinman and Zanvil Cohn described these cells for the first time. Morphological observations indicated that DCs assume a variety of branching forms and constantly extend and retract many fine cell processes called dendrites [1].

Today these cells are understood to represent a highly heterogeneous group with multiple functions, regulatory circuits, and phenotypes [2, 3]. However, the main function of antigen-loaded DCs is to migrate to T cell zones in the lymph node and present the antigen to naïve T cells [4]. This process, as well as the strength with which DCs interact with T cells, is a crucial behavior for programming the subsequent T cell phenotypes, which determine the priming or tolerance of the immune response [5].

DCs arise from the bone marrow (BM) and seed within peripheral tissues and lymphoid organs (Fig. 9.1) [1]. In BM, the common myeloid progenitor develops into a macrophage-DC progenitor cell [6, 7]. The common DC progenitor expresses

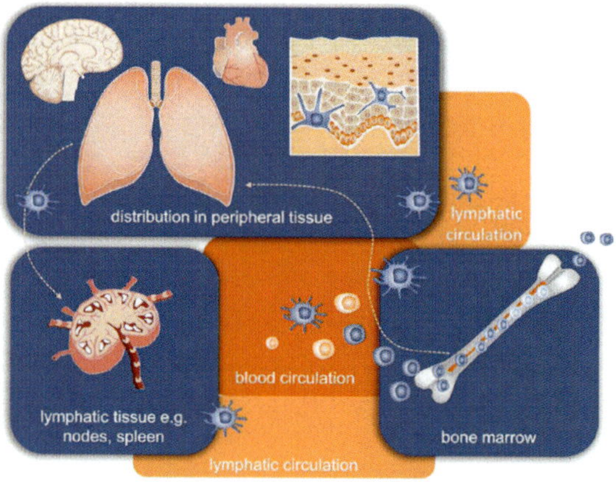

Figure 9.1 DCs arise from the bone marrow (BM) and then migrate to peripheral tissues and lymphoid organs.

macrophage colony-stimulating factor receptor (M-CSFR) and fms-related tyrosine kinase 3 (Flt3), which ultimately leads to plasmacytoid DC (pDC) and conventional DC (cDC) subtypes [8, 9]. Subsequent to lineage commitment, DCs respond to a wide range of stimuli such as Flt3 ligand (Flt3L) and GM-CSF (Csf-2) to perform different functions [10–13].

9.2.1 Dendritic Cell Classification: Challenges Ahead

A mechanistic classification for DC development is not trivial since refinements in the definitions of committed progenitors are still under way [14]. In addition, categorizing cell populations can prove to be challenging considering the phenotypic heterogeneity, even amongst isogenic populations, as a result of stochastic switching in fluctuating environments [15] and overlapping generations [16]. The method currently used to study the constant state of flux in immune cells is rather rigid; it implements preprogrammed cell-type hierarchies and requires a priori definitions of specific cell types [17], which includes precise definition and enumeration using specific cell-surface markers [2].

Emerging single cell–based technologies, such as single-cell mass cytometry, using systems-wide approaches [18] will ultimately shed more light on unpredicted functions of specific cell types, which will be of great use, e.g., during pharmacologic intervention. A study indicated a much more complicated diversity for hematopoietic cell types including DCs. The authors modeled single-cell transcriptional states in DCs to extrapolate variable and dynamic situations and uncovered heterogeneity both under steady state and pathogen activation [19].

Nevertheless, great efforts have been made to sort out specific DC populations within living organisms. Four different lineages of DCs are typically recognized, including (i) classical or conventional DCs (cDCs) [1], (ii) plasmacytoid DCs (pDCs) [20, 21], (iii) monocyte-derived DCs (moDCs) [22–24] and (iv) Langerhans cells (LCs) [25]. Another broad classification for DCs is the subdivision between lymphoid tissue-resident DCs and non-lymphoid tissue DCs, also referred to as migratory or tissue DC, such as LCs or interstitial DCs [5, 26]. The heterogeneity of DCs has also led to a distinction between DCs arising in the steady state versus during an inflammatory response [27, 28].

9.2.2 Dendritic Cells in Health and Disease

Similar to other immune cells, such as T cells and macrophages, the broad DC terminology accommodates for both DCs that play a role in initiating and orchestrating immune responses against harmful insults (such as tumors and infectious agents) and those that act as regulatory safeguards to prevent exaggerated reactions towards normally innocuous stimuli (such as self-tissue or environmental factors) as occurs in autoimmunity or allergy. As with other cellular systems, a fine balance between these "Janus faces" is crucial to provide the necessary immunity against harmful antigens while preventing exaggerated responses, towards trivial or even self-tissue antigens via mechanisms of immunological tolerance. The notion of tolerance induction by DCs started at the turn of the century [29, 30], but first evidence for T cell tolerance towards self-antigens came from studies using superior in vivo antigen targeting and delivery methods under steady-state conditions [31]. Overall, DCs are endowed with a key characteristic—functional plasticity—that results in a prompt and correct response to all presenting stimuli.

9.2.2.1 Dendritic cells in autoimmunity

DCs play an ambivalent role in autoimmunity. They can either induce tolerance or aggravate the autoimmune response. Earlier work reported that adoptive transfer of DCs presenting self-antigens induces and maintains potent autoreactive T cells and severe autoimmune diabetes [32]. However, it later became clear that DCs are also vital for orchestrating central tolerance; both thymic DCs [33] and pDCs [34] may orchestrate tolerance by presenting self-antigens to developing T cells in the thymus and instructing them to become tolerant. In autoimmune rheumatoid arthritis (RA), DCs are found in excessive numbers in serum and synovial fluid [10] and drive the generation of ectopic lymphocytic infiltrates in the inflamed rheumatoid synovium [11]. In vivo application of Flt3L in an animal model of RA increased accumulation of cDCs and pDCs in lymphatic tissue, subsequently increased regulatory T cell numbers and modulated the autoimmune response [35]. Interestingly, the number of circulating pDCs that induce expansion of IL-10 producing regulatory T cells increases in RA patients during disease remission [36]. In multiple sclerosis, monocyte-derived DCs secreted more pro-inflammatory cytokines [37]. Various studies in the experimental autoimmune encephalomyelitis (EAE) model have reported intracerebral recruitment of DCs crucial for the local stimulation and maintenance of autoreactive immune responses during neuroinflammation [38–40]. To support this, inhibition of Flt3L signaling significantly improved the course of established disease in EAE [41]. However, more recent reports showed that a depletion of CD11c+ DCs resulted in more severe disease in EAE [42]. Of note, DCs are also present in the brain in the steady state; these DCs show major differences from microglia and resemble Flt3L-dependent DCs [43], but do not precipitate an autoimmune response [28]. It appears that these locally available DCs act as "instructors" in the CNS during the onset of immune activity in the brain. Under steady state, Langerin-positive migratory DCs induce an in vivo expansion of antigen-specific regulatory T cells and attenuate the disease process in animal models of EAE [44].

9.2.2.2 Dendritic cells in tumor and infectious disease

As key regulators of innate and adaptive immune responses, DCs are critical for the development of immunity against tumors and infectious agents [45]. For this to occur, a migration of functional tumor-specific immune cells into sites of tumor development or infection is crucial. Cerebral spinal fluid (CSF) from patients suffering from bacterial meningitis or meningoencephalitis possesses higher levels of cDCs and pDCs than patients suffering from non-inflammatory neurological diseases [46, 47]. In contrast to cDCs, pDCs respond very early during infection [48], producing copious amounts of type I IFNs, which promote early protective antiviral activity by activating cDCs, which then orchestrate T cell-mediated immunity [49]. Indeed, the depletion of pDCs during viral infection has been found to exacerbate immunopathology [50]. Also during viral infections, activated pDCs may also differentiate into mature DCs to present antigen to T cells. However, in contrast to cDCs, maturing human pDCs retain the ability to induce regulatory T cells and therefore act to induce a state of immunological tolerance towards the infection [51]. Here again, a dual role for DCs is crucial; just as hosts need to arm the necessary immunological responses that identify and eliminate pathogens, damage limitation is also necessary and may involve mechanisms that make the host become more tolerant to the infection [52].

DCs also play a vital role in immunosurveillance against tumor growth. Although in most cases the immune system detects cancerous tissue, sometimes tumor cells manage to evade elimination. One mechanism of tumor evasion is the manipulation of the local environment to keep DCs in an immature state [53]. Tumor cells capable of attenuating the immune response escape the equilibrium that tumors are commonly kept in and progress; this process is known as tumor immunoediting [54]. Particularly for CNS tumors, it will be essential to understand the localization of DCs during disease in order to interpret the implications of in vitro engineered DCs with respect to their area of therapeutic action. DCs present in the meninges and choroid plexus of the healthy brain are in the steady state and show major differences from microglia [43]. When the balance is broken by an inflammatory insult and self-antigen becomes more accessible, these locally available DCs will accelerate the onset of neuroinflammation [39, 40, 55]. It has been proposed that both DCs under steady state and those active in inflammation [39] may detect interstitial antigens delivered through the CSF via a paravenous bulk outflow [56]. This organization of CSF flux was coined the glymphatic system in light of its dependence on glial water flux and its subservience of lymphatic function in interstitial solute clearance [56].

9.3 Why Is Tracking of Dendritic Cells So Important?

The dynamic nature of DCs in antigen recognition, antigen uptake, migration, and antigen presentation is very clear. DC development in the BM, distribution in the peripheral tissue, and exchange between tissue and lymphatic system is an active and dynamic process. The distribution of DCs requires close attention, not only to further understand their physiological function in health, but also from a clinical point of view, to elucidate the mechanisms behind failed functions during disease. One major clinical application for tracking DCs is to monitor these cells when administered as cellular therapies in several disease scenarios. Furthermore, it will be important in the coming years to develop a feasible method for tracking DC vaccines in more complex organs, particularly in the CNS over the therapeutic time window. This would be of particular benefit in patients harboring CNS tumors since DC therapy appears to be an emerging therapy. It will be essential to employ imaging modalities to monitor DCs in translational studies to effectively interpret failed or successful DC therapies in glioma with respect to their localization in the lymphatic or glymphatic system.

9.3.1 Dendritic Cell Immunotherapy

The concept of DC immunotherapy was developed initially for cancer treatment in the mid-nineties [57]. Over the turn of the century, DCs have become the subject of intensive research for use in vaccination protocols to enhance insufficient immune responses to infectious diseases and cancer, or attenuate excessive immune response in allergy and autoimmunity [58].

For tumor therapy, patients are immunized with autologous DCs that are loaded with recognized tumor antigens ex vivo. In contrast to passive immunotherapy via direct administration of tumor-specific T cells, active immunotherapy with tumor antigen or tumor antigen-loaded DCs induces a natural and specific anti-tumor T cell response with a long-term memory necessary for staving off future tumor relapses [59]. Although T cell therapy has shown promise in controlling established tumors [60], the maintenance of long-term immunity with this kind of passive immunotherapy remains elusive [61]. Clinical trials with unmodified tumor antigen-loaded DC vaccines have thus far provided a therapeutic proof-of-principle [62]. Some phase I clinical trials showed partial stabilization of tumor progression following administration of DCs loaded with tumor cells or lysates, respectively [63, 64]. A multi-centered phase III trial comparing DC vaccination to standard chemotherapy failed to demonstrate clinical efficacy for the DC therapy in metastatic melanoma, although limitations typically associated with multi-centered trials, e.g., suboptimal quality and quantity of cells, were reported [65]. A similar placebo-controlled trial showed that vaccination with antigen-loaded DC did have some clinical benefit in metastatic hormone refractory prostate cancer (HRPC) patients [66]. Further clinical trials with DC vaccines, particularly in glioblastoma multiforme, are encouraging [67]. These CNS tumors are generally the most aggressive primary tumors and display poor therapy response [68], especially those derived from non-neuronal tissue [69]. Further phase III trials with DC immunotherapy are currently under way for advanced glioma (ClinicalTrials.gov NCT00045968), renal cell cancer (ClinicalTrials.gov NCT01582672) and melanoma (ClinicalTrials.gov NCT01875653). In these trials overall survival is considered as the primary outcome measure for treatment efficacy, since it is more informative than the short term tumor regression [59]. Predominantly for CNS tumors, it will be essential to understand the localization of DCs following their application in patients in order to interpret the therapeutic implications of in vitro engineered DCs with respect to their site of therapeutic action.

DC immunotherapy has also been explored as a strategy to induce tolerance in transplantation and autoimmunity, although it has not been as intensively studied in the clinic compared to tumor disease [70]. Several animal studies confirm that administration of tolerogenic DCs prevent autoimmunity and graft-versus-host disease (GVHD) [71]. One strategy has been to expose DCs to antigen in the absence of a maturation signal and then transplant them back to induce regulatory T cells in vivo [72, 73]. One possible challenge when dealing with immature DCs (iDCs) as cellular therapies for GVHD or autoimmune diseases is that the inflammatory environment may lead to DC maturation, which would promote an immune reactivation rather than the desired down-modulation of the immune response. However, several efforts worldwide, including the European consortium "One Study," are developing protocols for preparing human tolerogenic DCs and other forms of regulatory innate and adaptive immune cell therapies for clinical testing in organ transplantation [74]. Interestingly, in a study where melanoma subjects were being administered daily subcutaneous (s.c.) injections of Flt3L for 14 days as part of a clinical trial [75], the authors reported on an expansion of regulatory T cells [76]. Important, while DCs need to be steered away from their tolerogenic nature in the case of tumor therapy [31], DCs need to remain stably tolerogenic when preventing GVHD or treating autoimmune diseases, thus for both therapeutic and safety purposes infused cell mixtures should be contaminant-free.

9.3.2 In vitro Generation of Mouse and Human DCs

Two main methods utilizing either GM-CSF (Csf-2) or Flt3L exist for DC generation [5]. These methods result in different DCs that are phenotypically and functionally similar to their in vivo counterparts. The first studies employed Csf-2 to produce large numbers of DCs from BM cells [77]. The DCs generated with Csf-2 resemble inflammatory DCs [78]. Csf-2 transforms both mouse and human hematopoietic progenitors and monocytes into cells that resemble mouse splenic cDCs [77, 79, 80] but does not give rise to pDCs [81]. Csf-2 also appears to be a critical regulator of cDC survival in non-lymphoid but not lymphoid tissue [82]. Similarly, Flt3L promotes the expansion of cDCs, but the resulting subsets phenotypically resemble lymphoid tissue–resident cDCs [13]. In contrast to the inflammatory-type of DCs that appear to be generated by the Csf-2 protocol, DCs generated by the Flt3L protocols represent the steady-state resident DCs [78]. In vitro generated human DCs are commonly prepared by either culturing CD34+ hematopoietic progenitors with Csf-2 and TNF-α to produce cells resembling LCs [83] or culturing monocytes with Csf-2 and IL-4, which produces dermal-like cDCs [79]. The latter methods have been predominantly used to prepare both immunogenic and tolerogenic clinical-grade DCs under strict good manufacturing practice standards [84]. Choice of protocol to culture DCs in vitro should be dictated by the need of the experiment or the clinical situation. Csf-2 reliably produces large numbers of cDCs, which may prove invaluable when designing therapies for human disease. In fact, Csf-2 remains to date a key cytokine for generating DC-based vaccines for clinical use [85]. On the other hand, Flt3L has proven instrumental in recent years in delineating the stages of the pathway by which DCs enter the steady state and on the functional specialization of DC subsets [5].

For clinical purposes, the cost and the level of logistics for standard operational procedures to efficiently receive, process and deliver material back to the patient for each individualized DC immunotherapy is significantly higher than traditional therapies [86]. In the long term, this therapy might nevertheless prove cost-effective, especially if it primes the necessary natural immune response (against tumor or infection) or if it achieves tolerance restoration (for autoimmune disease or GVHD) to prevent disease progression.

9.3.3 How Can We Modulate Dendritic Cells as Therapies

Notwithstanding the challenges behind preparation of cell-based vaccines in clinical scenarios, this therapeutic strategy provides the possibility to engineer, in vitro, a patient's own DCs prior to administration in order to trigger the best possible immune response in vivo.

For GVHD and autoimmune diseases, one clinical approach would be to engineer stable tolerogenic iDC ex vivo via pharmacological manipulation of these cells during or after their generation. One group of drugs that inhibits the maturation status of differentiated iDC are the 3-hydroxy-3-methylglutaryl-CoA (HMG-CoA)-reductase inhibitors (HMGCRI), a family of cholesterol-lowering drugs also known as statins [87, 88]. By depleting isoprenoid stores, we previously reported that HMGCRI inhibit Rho GTPase pathways necessary for actomyosin reorganization in myelin-reactive T cells [89]. We showed that HMGCRI also interferes with cytoskeletal alterations in iDCs during their differentiation from BM precursor cells [90]. Patient-derived DCs modified to be tolerogenic by HMGCRI treatment in vitro could be considered as a future therapeutic strategy for autoimmune diseases. Two studies, one in experimental autoimmune myasthenia gravis [91], and one in experimental autoimmune neuritis [92], revealed the therapeutic benefit of intraperitoneally applying DCs pre-treated with HMGCRI in culture. Both studies also report on the generation of regulatory T cells and downregulation of inflammatory autoimmune responses [91, 92].

For cancer therapy, one particularly promising strategy is to manipulate DC vaccines ex vivo by rewiring their molecular pathways from "pro-tumor" DCs into "anti-tumor" DCs [93]. This can be done by targeting specific molecules that influence the capacity of DCs to combat tumor growth, e.g., by promoting tumor antigen presentation or by promoting their migration to tumor relevant sites. In earlier work we had shown the therapeutic impact of HMGCRI in autoimmune disease [94, 95], but also an increased sustained activation of the extracellular related kinase (ERK) MAPK3 in vitro following incubation with HMGCRI [96]. Later, we also observed a more severe autoimmune response when MAPK3 was absent from the hematopoietic system [97]. It is becoming increasingly clear that ERKs are required for induction of tolerogenic DC [98] as well as induction of regulatory cytokines that suppress pro-inflammatory pathways in DCs [99].

9.4 Tracking Methods for Dendritic Cells

The promise of an effective DC immunotherapy, especially for the treatment of aggressive cancers, prompts the crucial need for a feasible and clinically relevant method to study the real-time trafficking of these cells in a living organism following administration. The interaction of DCs with T cells can be studied by single-cell imaging tools such as intravital microscopy [100]. Snap-shot ex vivo analyses of lymphatic tissue [101] also provide cross-sectional information regarding DC distribution and immune responses. However, apart from their invasiveness and thus inappropriateness for patient studies, these techniques sample on a limited tissue region and generally cannot be used for longitudinal studies.

Non-invasive technologies such as scintigraphic imaging, computed tomography (CT), positron emission tomography (PET), single photon emission computed tomography (SPECT), magnetic resonance imaging (MRI), ultrasound (US), and optical imaging, e.g., bioluminescence imaging (BLI) have been used for cell tracking in the past decades due to their suitability for longitudinal studies. Non-invasive imaging technologies are applicable to human studies and allow, to varying degrees, long-term investigation. Indeed they provide the possibility to study physiological or pathological processes in intact organisms.

The key factor for successful clinical DC tracking in the future will involve the development of the non-invasive imaging technologies, including hardware, software and probes, to combine these technologies in multimodal approaches [102].

9.4.1 Optical Imaging: Bioluminescence and Fluorescence Tomography

Methods for optical cell tracking include fluorescence reflective imaging, fluorescence molecular tomography and BLI. An advantage of optical imaging over other imaging modalities such as MRI or PET-based methods is that the necessary equipment is small and relatively inexpensive [103]. Optical imaging can also be divided into microscopic and macroscopic methods. Microscopic approaches such as photoacoustic imaging provide high sensitivity and resolution for visualizing lymphatic vessels and superficial lymph nodes [104] but suffer from penetration limitations and restricted areas of visualization making systemic cell migration difficult to assess.

Macroscopic optical imaging techniques such as BLI and fluorescence imaging have been extremely important for semi-quantitative non-invasive imaging of molecular processes in preclinical research but are limited in depth penetration and spatial resolution, two major strengths of other non-invasive imaging modalities, particularly MRI [103, 105]. Nonetheless, these techniques have been valuable to study the long-term homing kinetics of DCs in preclinical models [106]. BLI was also used to identify the best application route of DC vaccines [107]. Non-invasive fluorescence imaging has increasingly been used for DC tracking in the mouse [108–114]. Some of these studies implemented a bimodal approach to study DC migration in the lymph nodes by using both MRI and near-infrared optical imaging [110, 111]. Furthermore, the implementation of multifunctional PLGA (poly(lactic-co-glycolic acid)) nanoparticles [110] or programmed nanoparticles [115] (that contain imaging probes in combination with antigen [110] or immunomodulator [115]) has also been used to simultaneously label and modulate DCs prior to their application in vivo. In several of these experiments, quantum dots were employed as very bright labels with superior physical properties [116] for tracking of DCs [108]. However, safety concerns associated with heavy metals in their composition have hampered their broader use in human studies [117]. Potentially, the development of a new generation of smaller-sized quantum dots may be a future option for visualizing cells in real-time with enhanced sensitivity in vivo [118]. A study employing

BLI and SPECT demonstrated the distribution of injury-responsive DCs, as well as suppression of neuronal death, in an experimental stroke model where DCs were loaded with therapeutic cargo [119].

9.4.2 Nuclear Imaging: Scintigraphy, SPECT, and PET

Two radioisotopes have been used for DC labeling in conjunction with gamma camera scintigraphy, including ^{111}In-oxine and ^{99}mTc-hexamethylpropylene amine oxime (HMPAO). ^{111}In-oxine was introduced in the mid-1970s as a labeling agent for leukocyte scintigraphy [120]. ^{111}In scintigraphy was also one of the first methods used to study the distribution of DCs in small animals; in these studies ex vivo cross-sectional examinations were performed to quantify the number of ^{111}In-labeled DCs in different lymphatic tissue [121]. Several years later planar scintigraphy was employed to image DC migration in living mice [122], and only a few years later the method was employed to image DC migration in melanoma patients [123–126]. The ^{111}In-oxine complex is a non-specific cell labeling agent, and its lipid-solubility enables it to penetrate through the cellular bilayer membrane to attach firmly to cytoplasmic components. Although ^{111}In-oxine-labeling has been successfully used in the field of infection/inflammation scintigraphy, over the years the labeling agent has been replaced by ^{99}mTc-HMPAO, due to favorable physical characteristics, availability, cost, radiation burden and image quality when compared to ^{111}In-oxine [127]. Similarly to ^{111}In-oxine, the ^{99}mTc-HMPAO complex is lipophilic and readily penetrates leukocyte membranes; it enters the cell via passive diffusion and gets converted into a hydrophilic complex intracellularly [128]. In clinical practice ^{99}mTc-HMPAO is also frequently used for brain scintigraphy, since it readily crosses the blood brain barrier and can reflect regional cerebral blood flow [129]. ^{99}mTc-HMPAO may be used with higher activities than ^{111}In-oxine and might have the theoretical advantage of higher sensitivity. However, ^{111}In-oxine has been shown to give better labeling stability in DCs than ^{99}mTc-HMPAO [130]. Both complexes are used for labeling autologous white blood cells in clinical practice, and European guidelines on labeling with ^{99}mTc-HMPAO [131] and ^{111}In-oxine [132] for this purpose have been recently published. 2D scintigraphy is gradually being replaced by 3D SPECT for the visualization of the DC distribution [133].

Even though it can only provide 2D images, planar gamma camera imaging is still in popular use for clinical trials to study the distribution of DC vaccines in patients. Planar scintigraphy is commonly used in combination with anatomical imaging methods, e.g., SPECT and MRI in a bi- or multi-modal approach to add useful information especially with respect to the location of the injected DC [134–137]. These studies have also provided valuable information regarding DC therapy. For example, the first clinical study employing MRI to track DCs reported that ultrasound control of DC delivery may be insufficient to target the lymph nodes accurately [134]. In metastatic cancer patients, DCs delivered intranodally reach deeper lymphatic chains whereas those delivered intradermally reach draining inguinal lymph nodes [135]. Furthermore, a reduction in DC number was shown to improve DC migration [137].

Together with MRI, PET is a dominant imaging modality in the clinic; in combination, these imaging modalities provide anatomic, physiologic, metabolic and functional information [138]. The first PET studies in animals to track DC distribution were performed using planar projection imaging and ^{18}F cell labeling using N-succinimidyl-4-[^{18}F]fluorobenzoate [139]. Although ^{18}F cell labeling for PET studies allows for a quantitative assessment of DC biodistribution in vivo, experiments are restricted to a few hours (4 h) following delivery of ^{18}F-labeled DCs due to the short half-life of ^{18}F ($t_{1/2}$ = 110 min) [139]. Following intravenous (i.v.) injection, ^{18}F-labeled DCs localized in organs with a high capillary network, e.g., lung, spleen, and liver with only a small proportion reaching peripheral lymph nodes. Following s.c. injection, the majority of DCs remained at the injection site, although a significant amount reached the draining lymph node [139]. PET-specific reporter gene methods were also developed for T cell tracking to overcome the weaknesses of short-lived radiotracers such as ^{18}F [140, 141]. More recently, an HSV1 thymidine kinase reporter gene was used to label DCs prior to delivery in an experimental fungal disease model; ^{18}F activity was reported in blood, lung, and thymus even 7 days after intranasal administration of DCs [142]. This study also reported an increased pro-inflammatory effector T cell (TEM) response [142]. More recently, new PET radiotracers with longer half-lives, e.g., ^{64}Cu-PTSM ($t_{1/2}$ = 762 min) were developed for labeling lymphocytes; these cells could be tracked for up to 20 h in mice using micro-PET [143]. Subsequently, ^{64}Cu-PTSM was employed to label DCs prior to delivery in a breast cancer mouse model [144]. In this study, important information was deduced regarding the best delivery routes for DC immunotherapies; intraperitoneal (i.p.) application of DCs was far superior with respect to DC delivery and migration to lymph nodes and spleen than subcutaneous DC application [144]. Similar to studies employing ^{18}F labeling [139], ^{64}Cu-PTSM-labeled DCs homed to organs with a high density capillary network following i.v. delivery, including lungs, then liver, but not to lymph nodes [144]. Interestingly, these data from mice are in line with results obtained from myeloma patients; a study employing three different imaging modalities (^{18}F-FDG PET, ^{111}In-oxine SPECT and ^{64}Cu-PTSM PET) showed accumulation of DCs in the liver, spleen, and lung within 1 h following i.v. delivery [136].

9.4.3 Cell Tracking Using Magnetic Resonance Methods

In contrast to PET, SPECT, and optical imaging modalities, MRI has excellent spatial resolution and compares well with the other non-invasive modalities with respect to sensitivity of cell detection and tissue penetration depth [105]. Similar to PET and SPECT, MRI is an imaging modality where experimental applications can be easily translated to the clinic. It is possible to track cells in a whole organism since it provides simultaneous anatomical and dynamic physiological information. One key advantage of MRI is that it does not necessitate the use of radiotracers to label and detect cells, as is required in PET, SPECT, and scintigraphy imaging. A study in melanoma patients also reported that scintigraphy is not sufficient to follow small numbers of ^{111}In-labeled DCs in vivo [137]. Therefore, further non-invasive methods with better safety profiles need to be developed in order to ensure sufficient sensitivity and unequivocal cellular detection over sufficiently long periods of time.

Several developments have been made within the MRI field in the past decade for cell tracking. Labeling of cells for MRI detection primarily makes use of contrast agents that are incorporated into the cell (via mechanisms of pinocytosis, phagocytosis, or receptor-mediated endocytosis [145]) or by tagging them to cellular surface entities [146, 147]; this renders the cells distinguishable from surrounding cells by MRI.

9.4.3.1 Contrast agents modulating relaxation times

The first cell labeling strategies made use of compounds that typically shorten MRI relaxation times (T_1, T_2, T_2^*). By shortening T_2^*, iron oxide nanoparticles create a strong negative contrast and signal extinction due to localized perturbations in the magnetic field. A variety of iron oxide–based labels are available; these include superparamagnetic iron oxide particles (SPIO), ultrasmall iron oxide agents (USPIO), and micrometer-sized iron oxide particles (MPIO). The latter are particularly applicable for labeling stem cells since they allow long-term developmental, embryonic, and homing studies [148, 149]. SPIO has been used to label DC cancer vaccines in melanoma patients [134]. Similar nanoparticles were used to follow DCs in animal experiments after s.c. and i.v. administration [150, 151]. In analogous experiments using radiotracers and PET or SPECT imaging in mice [144] and patients [136], one study employing SPIO showed that human DCs home to liver and spleen, as well as lymph nodes, in a murine model 24 h after i.v. injection; following s.c. injection, DCs homed to draining lymph nodes but not liver or spleen [151]. However, in this study only long-term cell homing was investigated and shorter time-points were not studied [151], in contrast to previous studies employing ^{64}Cu-PTSM radiotracer [144].

Notwithstanding the high sensitivity of iron oxide particles, one major drawback of these contrast agents is a general difficulty distinguishing negative contrast (voids) they create, due to the presence of labeled cells, from other void artifacts (e.g., blood clots or air bubbles). On the other hand, T_1 shortening agents such as gadolinium (Gd) or manganese (Mn) chelates (e.g., Gd-DTPA or Mn-DPDP) result in an increase in MRI signal (positive contrast). Therefore these contrast agents do not have the signal void problems associated with iron oxide nanoparticles, but are much less sensitive. DCs labeled with a clinical formulation of Gd(III)-HP-DO3A were administered intratumorally and visualized as MRI enhancements in the brain [152]. More recently human DCs labeled with gadolinium and fluorescence-based nanoparticles (Gd-HNP) could be followed and quantified with high sensitivity following i.v. and i.p. delivery in mice [153]. By interfering with intracellular ^{1}H relaxation times, both T_1 and T_2^* contrast agents can only be detected indirectly as signal enhancements or extinctions on ^{1}H MRI scans. Especially for iron oxide nanoparticles there is a general difficulty to distinguish and quantify the labeled cells from surrounding tissue.

9.4.3.2 Fluorine magnetic resonance

In the past decade there has been a growing interest in an alternate MRI nucleus (^{19}F) that could eliminate the above impasse. The potential applications for ^{19}F in MR spectroscopy and imaging were recognized a few years following Lauterbur's original report introducing two-dimensional imaging to NMR and the coining of the term MRI [154]. However, the first pioneering work in which ^{19}F MRI-based methods were used to track cells in vivo was published nearly 30 years later by Ahrens and colleagues [155].

9.4.3.2.1 Basics of ^{19}F MR methods

The elegance of ^{19}F MRI is that ^{19}F in organic bound form is absent in eukaryotic cells, and therefore its introduction into mammals will yield basically background-free images and provides a means for highly selective detection of cells in vivo throughout the organism. Typically nanoparticles rich in ^{19}F are prepared by emulsifying perfluorocarbon compounds (PFCs) and used to label cells including DCs. Another advantage of PFCs is their biological and chemical inertness. ^{19}F MRI methods also provide a means for directly quantifying the number of ^{19}F-labeled cells within specific regions of interest [105]. The similarity between ^{19}F and ^{1}H NMR properties with respect to sensitivity and resonance frequency is also an advantage. The same pulse power settings can be used for both nuclei. After power calibration and adjustments for the ^{1}H channel, the same settings can be applied for the ^{19}F channel. This is particularly helpful when designing new radio frequency coils for dual ^{19}F and ^{1}H MR imaging [156]. In the study introducing ^{19}F MRI for DC labeling, perfluoro-15-crown-5 ether (PFCE) nanoparticles were employed as ^{19}F labels for fetal skin

DCs and bone-marrow-derived DCs [155]. PFCE is biologically stable and provides a high payload of ^{19}F nuclei as it contains 20 equivalent ^{19}F spins. It is nowadays one of the most commonly used PFCs. In their study, Ahrens et al. observed DC migration through the lymphatics 6 h following subcutaneous delivery [155]. Following 3 days after i.v. delivery, a ^{19}F signal could be clearly observed in liver and spleen, and weakly in the lungs [155]. Again, this is in alignment with parallel reports employing radiotracers as DC labels [136, 144] and later reports employing SPIOs [151].

9.4.3.2.2 Sensitivity of ^{19}F MR methods

The main challenge with ^{19}F MRI is sensitivity, and this can be a limitation when low numbers of ^{19}F labeled DCs are to be imaged. In the first ^{19}F MRI study dealing with murine DC labeling, an average of 5.2×10^{12} ^{19}F spins was reported per DC; at 11.7 T a detection of 2×10^5 labeled DCs per voxel was estimated [155]. A more recent publication studying the uptake of clinically applicable ^{19}F label in human DCs reported an increased number of ^{19}F atoms (1.7×10^{13}) per DC [157] and improved detection (2×10^3 labeled DCs per voxel at 7 T), which equates to approximately 1.3×10^3 cells per mm^3 [157].

Table 9.1 Signal sensitivity and cell detection limits

^{19}F (μmol)/10^6 DCs	^{19}F spins/DC	Magnetic Field (T)	Detection (DCs/voxel)	Detection (DCs/mm^3)	Ref.
48	5.2×10^{12}	11.7	2×10^5	—	[155]
150	1×10^{12}	7	—	—	[160]
375	1.7×10^{13}	7	2×10^3	1×10^3	[157]
20	1.6×10^{12}	9.4	1.5×10^5	1.2×10^3	[158]

The number of ^{19}F spins per DC depends on the dose of PFCE employed and on the aminophospholipid composition of the nanoparticle shell [158]. Varying the ^{19}F content in the culture from 4 to 40 μmol per 10^6 cells we observe 0.24×10^{12} to 0.38×10^{13} ^{19}F spins per DC [158]. At 9.4 T, we can detect 1.5×10^5 cells per $5 \times 5 \times 5$ mm^3 voxel; this equates to 1.2×10^3 cells per mm^3 [158]. By comparison, the sensitivity of SPIO-labeling was estimated to be 10^3 DCs per mm^3 at 3 T and 500 DCs per mm^3 at 7 T [159].

Realizing the need to compensate for setbacks in ^{19}F MRI signal sensitivity, we previously aimed to increase the ^{19}F signal per cell by improving cellular uptake; we varied particle size and showed that particles larger than 500 nm are taken up more effectively by DCs [161] thus eliciting an enhanced ^{19}F signal following in vivo delivery (Fig. 9.2). However, larger-sized particles also promote the expression of maturation markers and major histocompatibility complexes on DCs and subsequently increase T cell responses to specific antigen (Fig. 9.2), which is undesirable in situations (e.g., autoimmunity) where the immune system should be kept under control [161].

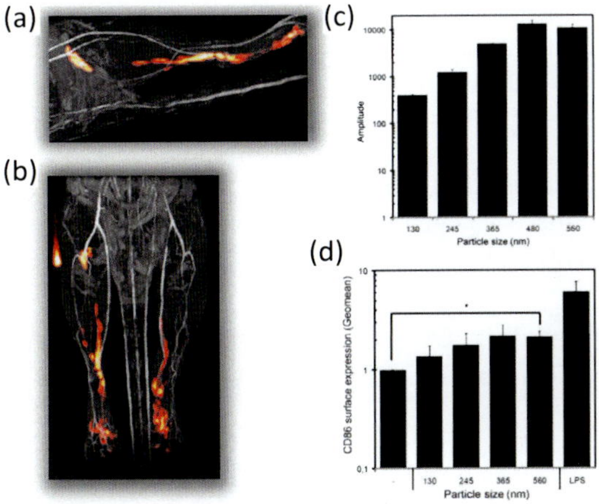

Figure 9.2 Large particle size promotes DC signal in ^{19}F MRI but promotes DC maturation. (a) Sagittal and (b) coronal ^{19}F and ^{1}H MRI overlay showing the migration of DCs through the lymphatic vessel into the popliteal lymph nodes (c). Increased ^{19}F signal of DCs labeled with larger sized PFC particles. (d) Increased expression of the maturation marker CD86 in DCs labeled with larger sized PFC particles. Reprinted with permission from ref. 161.

We observed an increased uptake of nanoparticles in DCs when ^{19}F nanoparticles were enriched with dipalmitoyl phosphatidylethanolamine (DPPE) [158] as shown by both electron microscopy and ^{19}F MRS (Fig. 9.3). Due to increased phase transition temperatures (T_m), saturated phospholipids such as DPPE (T_m = 64°C) promote cellular transfection [162] and stability of nanoparticles [163].

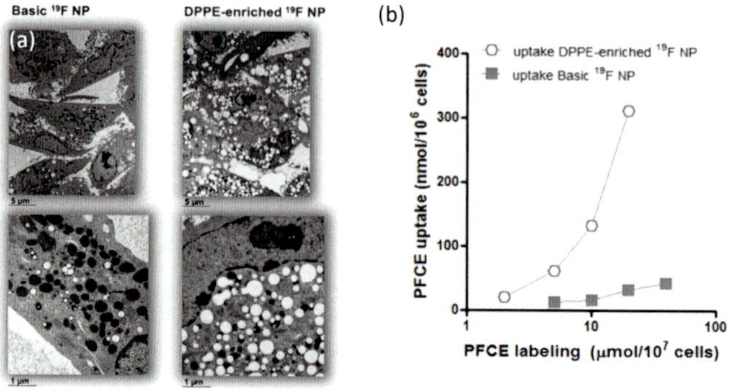

Figure 9.3 Enrichment of nanoparticles with dipalmitoyl phosphoethanolamine (DPPE) promotes their uptake in DCs. (a) Ultrathin sections of DCs labeled with basic and DPPE-enriched PFCE nanoparticles. (b) ^{19}F signal per 10^6 cells following incubation with increasing concentrations of PFCE.

Probes consisting of several symmetrically positioned and equivalent CF_3 groups are desirable. A superfluorinated molecular probe was designed using a simple one-step reaction; the developed probe is ideal as it gives a single intense resonance peak (thus avoiding CS artifacts) while providing a high signal (1.92×10^{13} ^{19}F atoms per DC), as seen in Fig. 9.4 [164].

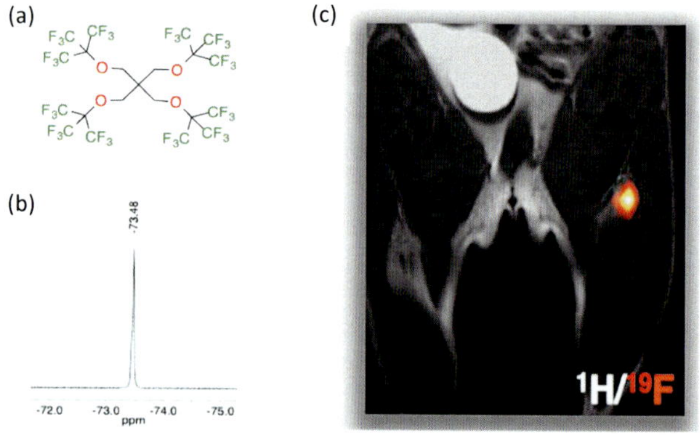

Figure 9.4 Superfluorinated molecular probe. (a) Chemical structure and (b) ^{19}F NMR spectrum of 1,3bis [[1,1,1,3,3,3hexafluoro 2-(trifluoromethyl)propan2yl]oxy] 2,2bis[[1,1,1,3,3,3-hexafluoro 2(trifluoromethyl)propan2yl]oxymethyl]propane. (c) Coronal ^{19}F and ^{1}H MRI overlay showing DCs at the popliteal lymph node following subcutaneous delivery in the vicinity. Reprinted with permission from ref. 164.

9.4.3.2.3 Safety and toxicity

Several in vitro studies utilizing both animal [155, 161] and human [160] DCs have investigated the effects of PFC labeling on the cell viability and function. However, a more detailed assessment is needed in the future to monitor the cellular toxicity in studies investigating different PFC structures and their lipidic carriers as cell labels. PFCs do not appear to be toxic but their persistence in the body remains a major concern. Their lack of toxicity is probably due to their outstanding chemical and biochemical inertness. The C-F bond is the strongest single bond encountered in organic chemistry and gets stronger with increasing number of ^{19}F atoms per carbon [165]. This together with their high gas-dissolving capacities has actually placed PFCs amongst major players as blood substitutes in medicine [166].

There is no apparent metabolic degradation process for PFCs and circulatory half-life is dependent on uptake by the reticuloendothelial system [166]. Initial toxicity studies in animals estimated the median lethal dose (LD_{50}) at 26–35 g of PFC/kg although the volume injected (>100 mL/kg) and the phospholipid composition of the nanoparticle may have a major part in toxicity [166]. The highest dose used for blood reconstitution in human studies was 1.8 g PFC/kg and this was associated with flu-like side effects [167]. Since higher amounts of PFC emulsions have been employed to visualize and monitor inflammatory processes in animal models [156, 168–170], new strategies need to be developed in order to translate the methods for imaging inflammatory disease in humans. A study from the group of Ulrich Flögel reported on the clearance of linear PFCs with predicted short biological half-lives, in order to seek alternative ^{19}F probes for perfluoro-15-crown-5 ether (PFCE) for imaging inflammation [171]. Compared to the estimated biological half-life for PFCE (250 days), perfluorodecalin had the shortest biological half-life (9 days) but did not give rise to sufficient MRI signal. In contrast to the cyclic PFCE, linear PFCs are known to be characterized by complex ^{19}F MR spectra. By employing a reconstruction method that integrates the multiple spectra it was possible to acquire artifact-free ^{19}F MRI of these linear PFCs. Perfluorooctyl bromide (PFOB) and trans-bis-perfluorobutyl ethylene (F-44E) also had significantly shorter clearance times than PFCE, with biological half-lives of 12 days (PFOB) and 28 days (F-44E), respectively, and provided equally good signals in vivo compared to PFCE [171]. We note that the intracellular PFC dose within DCs, labeled ex vivo, lie within the nanogram range and would not be toxic in vivo. Nonetheless, it would be very appealing to study the more complex linear PFCs for DC labeling strategies in conjunction with advanced MRI sequences such as methods using sine-bell acquisition-weighted chemical shift imaging [171].

9.4.3.2.4 Clinical application and future perspectives

In a proof-of-concept study to visualize DCs in cancer patients, Ahrens and colleagues labeled patient DCs with a clinical-grade ^{19}F label (CS-1000, Celsense) [172]. Following intradermal administration in the same patients, ^{19}F-labeled DCs were visualized at their site of delivery using combined ^{19}F and ^{1}H MR imaging at 3.0 T (Fig. 9.5) [172].

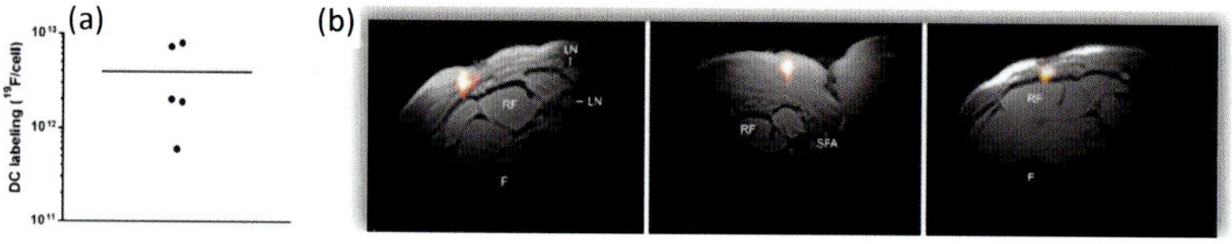

Figure 9.5 ^{19}F MRI of DC therapy in colorectal adenocarcinoma patients. (a) A summary of ^{19}F quantification results in DCs in the patient cohort. The horizontal bar is the average labeling for population (3.9 × 10^{12} ^{19}F/cell). (b) Axial composite ^{19}F/^{1}H images of the right thigh at 4 h post-inoculation in three patients, 53-year-old female (left), 45-year-old female (middle), 61-year-old male (right); ^{19}F-labeled DCs are displayed in "hot-iron" over the ^{1}H anatomy (F = femur, RF = rectus femoris, SFA = superficial femoral artery, LN = inguinal lymph node). Reprinted with permission from ref. 172.

Although DCs could be clearly visualized at their site of delivery, there was no MRI evidence of cell accumulation in the draining lymph nodes (LN) [172], possibly due to cell densities in these regions being on the order of, or below, the cell detection limit threshold for the experimental configuration employed [172].

Albeit considerable advances have been achieved in developing ^{19}F labels and their cellular uptake for in vivo imaging, sensitivity still remains an issue in situations where regions containing low numbers of ^{19}F spins need to be imaged. Although optimization of the ^{19}F labels is one important strategy for boosting ^{19}F signal, averaging is often still required to maximize detection thus lengthening scan time. To optimize SNR and to improve sensitivity in ^{19}F MRI, other alternatives need to be sought, e.g., changes in the pulse sequences. In the past few years, sequences tailored for ^{19}F MRI of complex PFCs have been developed to suppress J modulation effects caused by long-range couplings in order to eliminate chemical shift artifacts; these include multi-spin echo sequences with optimized interpulse delay and chemical shift imaging (CSI) with selective refocusing pulses [171, 173]. Compressed sensing (CS) was adapted for ^{19}F MRI methods to image inflammation. By reconstructing sparse and under-sampled data, this mathematical method significantly reduces scanning time. The first application of CS for ^{19}F MRI was to accurately and reproducibly reconstruct under-sampled ^{19}F spectroscopic datasets from in vivo ^{19}F CSI [174]. Later CS was employed in a RARE (rapid acquisition with relaxation enhancement) image pulse sequence incorporating a predetermined phase encoding gradient table yielding a pseudo-random, non-uniform sampling pattern that emphasized the k-space center; in these studies more than an eightfold reduction in imaging time was achieved without pronounced image degradation and with minimal impact on the image SNR and ^{19}F quantification accuracy [175]. Radio frequency (RF) coils with geometries that lack disruptive lumped elements [176, 177] and provide uniform RF (B1) fields, as well as improved SNR, will be beneficial for tracking DCs in vivo. The development of double-tuned $^{19}F/^{1}H$ RF coils with a solenoid design that make use of only one transmit channel for both nuclei is practical for clinical MR scanners with a limited number of transmit channels [178]. The implementation of cryogenic cooling to RF probes [179] has provided an immense improvement in SNR for proton [180–185] and X-nuclei [186] MRI. The use of very low temperatures (20–30 K), closed cycle cooled RF-coils, and preamplifiers delivers a sensitivity superior to standard room temperature RF-coils with SNR improvements 2–3 times higher than the same coil at RT which translates into an equivalent increase in spatial resolution, equivalent enhancement in cell detection or alternatively a reduction in scanning time by a factor of ~9.

9.5 Conclusion

The key factors that will determine the success of clinical DC tracking in the future will involve a combination of technologies such as cryogenic hardware to boost SNR, pulse sequence developments that reduce scan time, novel labels for selective and long-term cell labeling, as well as a harmonious combination of multi-modal imaging technologies with parallel quantitative approaches such as ^{19}F MR spectroscopy. The goal is to improve signal sensitivity in order to detect even minute amounts of cells, as well as to achieve better spatiotemporal resolution to follow complex lymphatic circuitries and cell kinetics in living organisms.

References

1. Steinman, R. M., and Cohn, Z. A. (1973). Identification of a novel cell type in peripheral lymphoid organs of mice. I. Morphology, quantitation, tissue distribution, *J. Exp. Med.*, **137**, pp. 1142–1162.
2. Geissmann, F., Gordon, S., Hume, D. A., Mowat, A. M., and Randolph, G. J. (2010). Unravelling mononuclear phagocyte heterogeneity, *Nat. Rev. Immunol.*, **10**, pp. 453–460.
3. Bar-On, L., et al. (2010). CX3CR1+ CD8alpha+ dendritic cells are a steady-state population related to plasmacytoid dendritic cells, *Proc. Natl. Acad. Sci. U. S. A.*, **107**, pp. 14745–14750.
4. Germain, R. N., and Jenkins, M. K. (2004). In vivo antigen presentation, *Curr. Opin. Immunol.*, **16**, pp. 120–125.
5. Merad, M., Sathe, P., Helft, J., Miller, J., and Mortha, A. (2013). The dendritic cell lineage: Ontogeny and function of dendritic cells and their subsets in the steady state and the inflamed setting, *Annu. Rev. Immunol.*, **31**, pp. 563–604.
6. Fogg, D. K., et al. (2006). A clonogenic bone marrow progenitor specific for macrophages and dendritic cells, *Science*, **311**, pp. 83–87.
7. Auffray, C., et al. (2009). CX3CR1+ CD115+ CD135+ common macrophage/ DC precursors and the role of CX3CR1 in their response to inflammation, *J. Exp. Med.*, **206**, pp. 595–606.

8. Onai, N., Obata-Onai, A., Schmid, M. A., Ohteki, T., Jarrossay, D., and Manz, M. G. (2007). Identification of clonogenic common Flt3+M-CSFR+ plasmacytoid and conventional dendritic cell progenitors in mouse bone marrow, *Nat. Immunol.*, **8**, pp. 1207–1216.

9. Naik, S. H., et al. (2007). Development of plasmacytoid and conventional dendritic cell subtypes from single precursor cells derived in vitro and in vivo, *Nat. Immunol.*, **8**, pp. 1217–1226.

10. Thomas, R., and Lipsky, P. E. (1996). Presentation of self peptides by dendritic cells. Possible implications for the pathogenesis of rheumatoid arthritis, *Arthritis Rheum.*, **39**, pp. 183–190.

11. Page, G., Lebecque, S., and Miossec, P. (2002). Anatomic localization of immature and mature dendritic cells in an ectopic lymphoid organ: Correlation with selective chemokine expression in rheumatoid synovium, *J. Immunol.*, **168**, pp. 5333–5341.

12. Kingston, D., Schmid, M. A., Onai, N., Obata-Onai, A., Baumjohann, D., and Manz, M. G. (2009). The concerted action of GM-CSF and Flt3-ligand on in vivo dendritic cell homeostasis, *Blood*, **114**, pp. 835–843.

13. Waskow, C., et al. (2008). The receptor tyrosine kinase Flt3 is required for dendritic cell development in peripheral lymphoid tissues, *Nat. Immunol.*, **9**, pp. 676–683.

14. Briseno, C. G., Murphy, T. L., and Murphy, K. M. (2014). Complementary diversification of dendritic cells and innate lymphoid cells, *Curr. Opin. Immunol.*, **29C**, pp. 69–78.

15. Thattai, M., and Van Oudenaarden, A. (2004). Stochastic gene expression in fluctuating environments, *Genetics*, **167**, pp. 523–530.

16. Schaffer, W. M. (1974). Optimal reproductive effort in fluctuating environments, *Am. Nat.*, **108**, pp. 783–790.

17. Germain, R. N. (2012). Maintaining system homeostasis: The third law of Newtonian immunology, *Nat. Immunol.*, **13**, pp. 902–906.

18. Bendall, S. C., et al. (2011). Single-cell mass cytometry of differential immune and drug responses across a human hematopoietic continuum, *Science*, **332**, pp. 687–696.

19. Jaitin, D. A., et al. (2014). Massively parallel single-cell RNA-seq for marker-free decomposition of tissues into cell types, *Science*, **343**, pp. 776–779.

20. Siegal, F. P., et al. (1999). The nature of the principal type 1 interferon-producing cells in human blood, *Science*, **284**, pp. 1835–1837.

21. Cella, M., Jarrossay, D., Facchetti, F., Alebardi, O., Nakajima, H., Lanzavecchia, A., and Colonna, M. (1999). Plasmacytoid monocytes

migrate to inflamed lymph nodes and produce large amounts of type I interferon, *Nat. Med.*, **5**, pp. 919–923.

22. Randolph, G. J., Inaba, K., Robbiani, D. F., Steinman, R. M., and Muller, W. A. (1999). Differentiation of phagocytic monocytes into lymph node dendritic cells in vivo, *Immunity*, **11**, pp. 753–761.
23. Serbina, N. V., Salazar-Mather, T. P., Biron, C. A., Kuziel, W. A., and Pamer, E. G. (2003). TNF/iNOS-producing dendritic cells mediate innate immune defense against bacterial infection, *Immunity*, **19**, pp. 59–70.
24. Geissmann, F., Jung, S., and Littman, D. R. (2003). Blood monocytes consist of two principal subsets with distinct migratory properties, *Immunity*, **19**, pp. 71–82.
25. Schuler, G., and Steinman, R. M. (1985). Murine epidermal Langerhans cells mature into potent immunostimulatory dendritic cells in vitro, *J. Exp. Med.*, **161**, pp. 526–546.
26. Shortman, K., and Naik, S. H. (2007). Steady-state and inflammatory dendritic-cell development, *Nat. Rev. Immunol.*, **7**, pp. 19–30.
27. Steinman, R. M., et al. (2003). Dendritic cell function in vivo during the steady state: A role in peripheral tolerance, *Ann. N. Y. Acad. Sci.*, **987**, pp. 15–25.
28. Anandasabapathy, N., et al. (2011). Flt3L controls the development of radiosensitive dendritic cells in the meninges and choroid plexus of the steady-state mouse brain, *J. Exp. Med.*, **208**, pp. 1695–1705.
29. Steinman, R. M., Turley, S., Mellman, I., and Inaba, K. (2000). The induction of tolerance by dendritic cells that have captured apoptotic cells, *J. Exp. Med.*, **191**, pp. 411–416.
30. Krasteva, M., et al. (1998). Dual role of dendritic cells in the induction and down-regulation of antigen-specific cutaneous inflammation, *J. Immunol.*, **160**, pp. 1181–1190.
31. Hawiger, D., et al. (2001). Dendritic cells induce peripheral T cell unresponsiveness under steady state conditions in vivo, *J. Exp. Med.*, **194**, pp. 769–779.
32. Ludewig, B., Odermatt, B., Landmann, S., Hengartner, H., and Zinkernagel, R. M. (1998). Dendritic cells induce autoimmune diabetes and maintain disease via de novo formation of local lymphoid tissue, *J. Exp. Med.*, **188**, pp. 1493–1501.
33. Gallegos, A. M., and Bevan, M. J. (2004). Central tolerance to tissue-specific antigens mediated by direct and indirect antigen presentation, *J. Exp. Med.*, **200**, pp. 1039–1049.

34. Hadeiba, H., et al. (2012). Plasmacytoid dendritic cells transport peripheral antigens to the thymus to promote central tolerance, *Immunity*, **36**, pp. 438–450.

35. Svensson, M. N., et al. (2013). Fms-like tyrosine kinase 3 ligand controls formation of regulatory T cells in autoimmune arthritis, *PLoS One*, **8**, p. e54884.

36. Kavousanaki, M., Makrigiannakis, A., Boumpas, D., and Verginis, P. (2010). Novel role of plasmacytoid dendritic cells in humans: Induction of interleukin-10–producing treg cells by plasmacytoid dendritic cells in patients with rheumatoid arthritis responding to therapy, *Arthritis Rheum.*, **62**, pp. 53–63.

37. Huang, Y.-M., Xiao, B.-G., Özenci, V., Kouwenhoven, M., Teleshova, N., Fredrikson, S., and Link, H. (1999). Multiple sclerosis is associated with high levels of circulating dendritic cells secreting pro-inflammatory cytokines, *J. Neuroimmunol.*, **99**, pp. 82–90.

38. Serafini, B., Columba-Cabezas, S., Di Rosa, F., and Aloisi, F. (2000). Intracerebral recruitment and maturation of dendritic cells in the onset and progression of experimental autoimmune encephalomyelitis, *Am. J. Pathol.*, **157**, pp. 1991–2002.

39. Kivisakk, P., Imitola, J., Rasmussen, S., Elyaman, W., Zhu, B., Ransohoff, R. M., and Khoury, S. J. (2009). Localizing central nervous system immune surveillance: meningeal antigen-presenting cells activate T cells during experimental autoimmune encephalomyelitis, *Ann. Neurol.*, **65**, pp. 457–469.

40. Greter, M., et al. (2005). Dendritic cells permit immune invasion of the CNS in an animal model of multiple sclerosis, *Nat. Med.*, **11**, pp. 328–334.

41. Whartenby, K. A., et al. (2005). Inhibition of FLT3 signaling targets DCs to ameliorate autoimmune disease, *Proc. Natl. Acad. Sci. U. S. A.*, **102**, pp. 16741–16746.

42. Yogev, N., et al. (2012). Dendritic cells ameliorate autoimmunity in the CNS by controlling the homeostasis of PD-1 receptor+ regulatory T cells, *Immunity*, **37**, pp. 264–275.

43. D'agostino, P. M., Gottfried-Blackmore, A., Anandasabapathy, N., and Bulloch, K. (2012). Brain dendritic cells: biology and pathology, *Acta Neuropathol.*, **124**, pp. 599–614.

44. Idoyaga, J., et al. (2013). Specialized role of migratory dendritic cells in peripheral tolerance induction, *J. Clin. Invest.*, **123**, pp. 844–854.

45. Banchereau, J., and Steinman, R. M. (1998). Dendritic cells and the control of immunity, *Nature*, **392**, pp. 245–252.

46. Economopoulos, V., Noad, J. C., Krishnamoorthy, S., Rutt, B. K., and Foster, P. J. (2011). Comparing the MRI appearance of the lymph nodes and spleen in wild-type and immuno-deficient mouse strains, *PLoS One*, **6**, p. e27508.

47. Gruetter, R. (1993). Automatic, localized in vivo adjustment of all first- and second-order shim coils, *Magn. Reson. Med.*, **29**, pp. 804–811.

48. Caliceti, P., and Veronese, F. M. (2003). Pharmacokinetic and biodistribution properties of poly(ethylene glycol)-protein conjugates, *Adv. Drug Deliv. Rev.*, **55**, pp. 1261–1277.

49. Allen, T. M., Hansen, C., Martin, F., Redemann, C., and Yau-Young, A. (1991). Liposomes containing synthetic lipid derivatives of poly(ethylene glycol) show prolonged circulation half-lives in vivo, *Biochim. Biophys. Acta*, **1066**, pp. 29–36.

50. Smaby, J. M., Brockman, H. L., and Brown, R. E. (1994). Cholesterols interfacial interactions with sphingomyelins and phosphatidylcholines - hydrocarbon chain structure determines the magnitude of condensation, *Biochemistry*, **33**, pp. 9135–9142.

51. Virtanen, J. A., Cheng, K. H., and Somerharju, P. (1998). Phospholipid composition of the mammalian red cell membrane can be rationalized by a superlattice model, *Proc. Natl. Acad. Sci. U. S. A.*, **95**, pp. 4964–4969.

52. Zachowski, A. (1993). Phospholipids in animal eukaryotic membranes: transverse asymmetry and movement, *Biochem. J.*, **294**(Pt 1), pp. 1–14.

53. Bell, D., et al. (1999). In breast carcinoma tissue, immature dendritic cells reside within the tumor, whereas mature dendritic cells are located in peritumoral areas, *J. Exp. Med.*, **190**, pp. 1417–1426.

54. Dunn, G. P., Bruce, A. T., Ikeda, H., Old, L. J., and Schreiber, R. D. (2002). Cancer immunoediting: from immunosurveillance to tumor escape, *Nat. Immunol.*, **3**, pp. 991–998.

55. Wu, G. F., et al. (2011). Limited sufficiency of antigen presentation by dendritic cells in models of central nervous system autoimmunity, *J. Autoimmun.*, **36**, pp. 56–64.

56. Iliff, J. J., et al. (2012). A paravascular pathway facilitates CSF flow through the brain parenchyma and the clearance of interstitial solutes, including amyloid β, *Sci. Translational Med.*, **4**, pp. 147ra111.

57. Hsu, F. J., et al. (1996). Vaccination of patients with B-cell lymphoma using autologous antigen-pulsed dendritic cells, *Nat. Med.*, **2**, pp. 52–58.

58. Banchereau, J., Schuler-Thurner, B., Palucka, A. K., and Schuler, G. (2001). Dendritic cells as vectors for therapy, *Cell*, **106**, pp. 271–274.

59. Palucka, K., and Banchereau, J. (2012). Cancer immunotherapy via dendritic cells, *Nat. Rev. Cancer*, **12**, pp. 265–277.

60. Chapuis, A. G., et al. (2012). Transferred melanoma-specific CD8+ T cells persist, mediate tumor regression, and acquire central memory phenotype, *Proc. Natl. Acad. Sci. U. S. A.*, **109**, pp. 4592–4597.

61. Gattinoni, L., Klebanoff, C. A., and Restifo, N. P. (2012). Paths to stemness: Building the ultimate antitumour T cell, *Nat. Rev. Cancer*, **12**, pp. 671–684.

62. Palucka, A. K., Ueno, H., Fay, J. W., and Banchereau, J. (2007). Taming cancer by inducing immunity via dendritic cells, *Immunol. Rev.*, **220**, pp. 129–150.

63. Palucka, A. K., et al. (2006). Dendritic cells loaded with killed allogeneic melanoma cells can induce objective clinical responses and MART-1 specific CD8+ T-cell immunity, *J. Immunother.*, **29**, pp. 545–557.

64. Kim, J. H., et al. (2007). Phase I/II study of immunotherapy using autologous tumor lysate-pulsed dendritic cells in patients with metastatic renal cell carcinoma, *Clin. Immunol.*, **125**, pp. 257–267.

65. Schadendorf, D., et al. (2006). Dacarbazine (DTIC) versus vaccination with autologous peptide-pulsed dendritic cells (DC) in first-line treatment of patients with metastatic melanoma: a randomized phase III trial of the DC study group of the DeCOG, *Ann. Oncol.*, **17**, pp. 563–570.

66. Small, E. J., et al. (2006). Placebo-controlled phase III trial of immunologic therapy with sipuleucel-T (APC8015) in patients with metastatic, asymptomatic hormone refractory prostate cancer, *J. Clin. Oncol.*, **24**, pp. 3089–3094.

67. Yamanaka, R., and Kajiwara, K., Dendritic cell vaccines, in *Glioma* (Yamanaka, R., ed.), 2012, Springer, New York, pp. 187–200.

68. Dolecek, T. A., Propp, J. M., Stroup, N. E., and Kruchko, C. (2012). CBTRUS statistical report: Primary brain and central nervous system tumors diagnosed in the United States in 2005–2009, *Neuro. Oncol.*, **14** Suppl 5, pp. v1–49.

69. Prins, R. M., et al. (2011). Gene expression profile correlates with T-cell infiltration and relative survival in glioblastoma patients vaccinated with dendritic cell immunotherapy, *Clin. Cancer Res.*, **17**, pp. 1603–1615.

70. Giannoukakis, N., Phillips, B., Finegold, D., Harnaha, J., and Trucco, M. (2011). Phase I (safety) study of autologous tolerogenic dendritic cells in type 1 diabetic patients, *Diab. Care*, **34**, pp. 2026–2032.

71. Mccurry, K. R., Colvin, B. L., Zahorchak, A. F., and Thomson, A. W. (2006). Regulatory dendritic cell therapy in organ transplantation, *Transplant Int.*, **19**, pp. 525–538.

72. Dhodapkar, M. V., Steinman, R. M., Krasovsky, J., Munz, C., and Bhardwaj, N. (2001). Antigen-specific inhibition of effector T cell function in humans after injection of immature dendritic cells, *J. Exp. Med.*, **193**, pp. 233–238.

73. Legge, K. L., Gregg, R. K., Maldonado-Lopez, R., Li, L., Caprio, J. C., Moser, M., and Zaghouani, H. (2002). On the role of dendritic cells in peripheral T cell tolerance and modulation of autoimmunity, *J. Exp. Med.*, **196**, pp. 217–227.

74. Moreau, A., Varey, E., Bouchet-Delbos, L., and Cuturi, M.-C. (2012). Cell therapy using tolerogenic dendritic cells in transplantation, *Transplant. Res.*, **1**, p. 13.

75. Shackleton, M., et al. (2004). The impact of imiquimod, a Toll-like receptor-7 ligand (TLR7L), on the immunogenicity of melanoma peptide vaccination with adjuvant Flt3 ligand, *Cancer Immun.*, **4**, p. 9.

76. Klein, O., Ebert, L.M., Zanker, D., Woods, K., Tan, B.S., Fucikova, J., Behren, A., Davis, I.D., Maraskovsky, E., Chen, W., and Cebon, J. (2013). Flt3 ligand expands CD4+ FoxP3+ regulatory T cells in human subjects, *Eur J Immunol*, **43**, pp. 533-9.

77. Inaba, K., et al. (1992). Generation of large numbers of dendritic cells from mouse bone marrow cultures supplemented with granulocyte/macrophage colony-stimulating factor, *J. Exp. Med.*, **176**, pp. 1693–1702.

78. Xu, Y., Zhan, Y., Lew, A. M., Naik, S. H., and Kershaw, M. H. (2007). Differential development of murine dendritic cells by GM-CSF versus Flt3 ligand has implications for inflammation and trafficking, *J. Immunol.*, **179**, pp. 7577–7584.

79. Sallusto, F., and Lanzavecchia, A. (1994). Efficient presentation of soluble antigen by cultured human dendritic cells is maintained by granulocyte/macrophage colony-stimulating factor plus

interleukin 4 and downregulated by tumor necrosis factor alpha, *J. Exp. Med.*, **179**, pp. 1109–1118.

80. Caux, C., et al. (1996). CD34+ hematopoietic progenitors from human cord blood differentiate along two independent dendritic cell pathways in response to GM-CSF+ TNF alpha, *J. Exp. Med.*, **184**, pp. 695–706.

81. Gilliet, M., et al. (2002). The development of murine plasmacytoid dendritic cell precursors is differentially regulated by FLT3-ligand and granulocyte/macrophage colony-stimulating factor, *J. Exp. Med.*, **195**, pp. 953–958.

82. Greter, M., et al. (2012). GM-CSF controls nonlymphoid tissue dendritic cell homeostasis but is dispensable for the differentiation of inflammatory dendritic cells, *Immunity*, **36**, pp. 1031–1046.

83. Caux, C., Dezutter-Dambuyant, C., Schmitt, D., and Banchereau, J. (1992). GM-CSF and TNF- cooperate in the generation of dendritic Langerhans cells, *Nature*, pp. 258–258.

84. Kalantari, T., et al. (2011). Generation of immunogenic and tolerogenic clinical-grade dendritic cells, *Immunol. Res.*, **51**, pp. 153–160.

85. Banchereau, J., and Palucka, A. K. (2005). Dendritic cells as therapeutic vaccines against cancer, *Nat. Rev. Immunol.*, **5**, pp. 296–306.

86. Whiteside, T. L., et al. (2011). Shipping of therapeutic somatic cell products, *Cytotherapy*, **13**, pp. 201–213.

87. Yilmaz, A., et al. (2006). Differential effects of statins on relevant functions of human monocyte-derived dendritic cells, *J. Leukoc. Biol.*, **79**, pp. 529–538.

88. Yilmaz, A., et al. (2004). HMG-CoA reductase inhibitors suppress maturation of human dendritic cells: New implications for atherosclerosis, *Atherosclerosis*, **172**, pp. 85–93.

89. Waiczies, S., et al. (2007). Geranylgeranylation but not GTP loading determines rho migratory function in T cells, *J. Immunol.*, **179**, pp. 6024–6032.

90. Leuenberger, T., et al. (2014). Modulation of dendritic cell immunobiology via inhibition of 3-hydroxy-3-methylglutaryl-CoA (HMG-CoA) reductase, *PLoS One*, **9**, p. e100871.

91. Li, X.-L., et al. (2013). Atorvastatin-modified dendritic cells in vitro ameliorate experimental autoimmune myasthenia gravis by up-regulated Treg cells and shifted Th1/Th17 to Th2 cytokines, *Mol. Cell. Neurosci.*, **56**, pp. 85–95.

92. Xu, H., et al. (2014). Therapeutic potential of atorvastatin-modified dendritic cells in experimental autoimmune neuritis by decreased

Th1/Th17 cytokines and up-regulated T regulatory cells and NKR-P1+ cells, *J. Neuroimmunol.*, **269**, pp. 28–37.

93. Palucka, K., Ueno, H., Fay, J., and Banchereau, J. (2011). Dendritic cells and immunity against cancer, *J. Intern. Med.*, **269**, pp. 64–73.
94. Aktas, O., et al. (2003). Treatment of relapsing paralysis in experimental encephalomyelitis by targeting Th1 cells through atorvastatin, *J. Exp. Med.*, **197**, pp. 725–733.
95. Paul, F., et al. (2008). Oral high-dose atorvastatin treatment in relapsing-remitting multiple sclerosis, *PLoS One*, **3**, p. e1928–e1928.
96. Waiczies, S., Prozorovski, T., Infante-Duarte, C., Hahner, A., Aktas, O., Ullrich, O., and Zipp, F. (2005). Atorvastatin induces T cell anergy via phosphorylation of ERK1, *J. Immunol.*, **174**, pp. 5630–5635.
97. Bendix, I., et al. (2010). MAPK3 deficiency drives autoimmunity via DC arming, *Eur. J. Immunol.*, **40**, pp. 1486–1495.
98. Saraiva, M., and O'garra, A. (2010). The regulation of IL-10 production by immune cells, *Nat. Rev. Immunol.*, **10**, pp. 170–181.
99. Manicassamy, S., et al. (2009). Toll-like receptor 2-dependent induction of vitamin A-metabolizing enzymes in dendritic cells promotes T regulatory responses and inhibits autoimmunity, *Nat. Med.*, **15**, pp. 401–409.
100. Mempel, T. R., Henrickson, S. E., and Von Andrian, U. H. (2004). T-cell priming by dendriticcells in lymph nodes occurs in three distinct phases, *Nature*, **427**, pp. 154–159.
101. Martin-Fontecha, A., Sebastiani, S., Hopken, U. E., Uguccioni, M., Lipp, M., Lanzavecchia, A., and Sallusto, F. (2003). Regulation of dendritic cell migration to the draining lymph node: Impact on T lymphocyte traffic and priming, *J. Exp. Med.*, **198**, pp. 615–621.
102. Barnett, B. P., et al. (2011). Use of perfluorocarbon nanoparticles for non-invasive multimodal cell tracking of human pancreatic islets, *Contrast Media Mol. Imaging*, **6**, pp. 251–259.
103. Weissleder, R. (2002). Scaling down imaging: Molecular mapping of cancer in mice, *Nat. Rev. Cancer*, **2**, pp. 11–18.
104. Kim, C., Song, K. H., Gao, F., and Wang, L. V. (2010). Sentinel lymph nodes and lymphatic vessels: Noninvasive dual-modality in vivo mapping by using indocyanine green in rats—volumetric spectroscopic photoacoustic imaging and planar fluorescence imaging 1, *Radiology*, **255**, pp. 442–450.
105. Ahrens, E. T., and Bulte, J. W. (2013). Tracking immune cells in vivo using magnetic resonance imaging, *Nat. Rev. Immunol.*, **13**, pp. 755–763.

106. Schimmelpfennig, C. H., et al. (2005). Ex vivo expanded dendritic cells home to T-cell zones of lymphoid organs and survive in vivo after allogeneic bone marrow transplantation, *Am. J. Pathol.*, **167**, pp. 1321–1331.

107. Schweichel, D., et al. (2006). Evaluation of DNA vaccination with recombinant adenoviruses using bioluminescence imaging of antigen expression: Impact of application routes and delivery with dendritic cells, *J. Gene Med.*, **8**, pp. 1243–1250.

108. Noh, Y. W., Lim, Y. T., and Chung, B. H. (2008). Noninvasive imaging of dendritic cell migration into lymph nodes using near-infrared fluorescent semiconductor nanocrystals, *FASEB J.*, **22**, pp. 3908–3918.

109. Christian, N. A., et al. (2009). In vivo dendritic cell tracking using fluorescence lifetime imaging and near-infrared-emissive polymersomes, *Mol. Imaging Biol.*, **11**, pp. 167–177.

110. Noh, Y. W., Jang, Y. S., Ahn, K. J., Lim, Y. T., and Chung, B. H. (2011). Simultaneous in vivo tracking of dendritic cells and priming of an antigen-specific immune response, *Biomaterials*, **32**, pp. 6254–6263.

111. Mackay, P. S., et al. (2011). Multimodal imaging of dendritic cells using a novel hybrid magneto-optical nanoprobe, *Nanomedicine*, **7**, pp. 489–496.

112. Sagar, D., Lamontagne, A., Foss, C. A., Khan, Z. K., Pomper, M. G., and Jain, P. (2012). Dendritic cell CNS recruitment correlates with disease severity in EAE via CCL2 chemotaxis at the blood-brain barrier through paracellular transmigration and ERK activation, *J. Neuroinflammation*, **9**, pp. 245.

113. Biffi, S., et al. (2013). Dendritic polyglycerolsulfate near infrared fluorescent (NIRF) dye conjugate for non-invasively monitoring of inflammation in an allergic asthma mouse model, *PLoS One*, **8**, p. e57150.

114. Chen, Y. C., Wen, S., Shang, S. A., Cui, Y., Luo, B., and Teng, G. J. (2014). Magnetic resonance and near-infrared imaging using a novel dual-modality nano-probe for dendritic cell tracking in vivo, *Cytotherapy*, **16**, pp. 699–710.

115. Heo, M. B., and Lim, Y. T. (2014). Programmed nanoparticles for combined immunomodulation, antigen presentation and tracking of immunotherapeutic cells, *Biomaterials*, **35**, pp. 590–600.

116. Jaiswal, J. K., Mattoussi, H., Mauro, J. M., and Simon, S. M. (2002). Long-term multiple color imaging of live cells using quantum dot bioconjugates, *Nat. Biotechnol.*, **21**, pp. 47–51.

117. Hardman, R. (2006). A toxicologic review of quantum dots: Toxicity depends on physicochemical and environmental factors, *Environ. Health Perspect.*, pp. 165–172.

118. Kosaka, N., Mccann, T. E., Mitsunaga, M., Choyke, P. L., and Kobayashi, H. (2010). Real-time optical imaging using quantum dot and related nanocrystals, *Nanomedicine (Lond)*, **5**, pp. 765–776.

119. Manley, N. C., et al. (2013). Derivation of injury-responsive dendritic cells for acute brain targeting and therapeutic protein delivery in the stroke-injured rat, *PLoS One*, **8**, p. e61789.

120. Thakur, M. L., Lavender, J. P., Arnot, R. N., Silvester, D. J., and Segal, A. W. (1977). Indium-111-labeled autologous leukocytes in man, *J. Nucl. Med.*, **18**, pp. 1014–1021.

121. Kupiec-Weglinski, J. W., Austyn, J. M., and Morris, P. J. (1988). Migration patterns of dendritic cells in the mouse. Traffic from the blood, and T cell-dependent and-independent entry to lymphoid tissues, *J. Exp. Med.*, **167**, pp. 632–645.

122. Suda, T., Callahan, R., Wilkenson, R., Van Rooijen, N., and Schneeberger, E. (1996). Interferon-gamma reduces Ia+ dendritic cell traffic to the lung, *J. Leukoc. Biol.*, **60**, pp. 519–527.

123. De Vries, I. J., et al. (2003). Effective migration of antigen-pulsed dendritic cells to lymph nodes in melanoma patients is determined by their maturation state, *Cancer Res.*, **63**, pp. 12–17.

124. Quillien, V., et al. (2005). Biodistribution of radiolabelled human dendritic cells injected by various routes, *Eur. J. Nucl. Med. Mol. Imaging*, **32**, pp. 731–741.

125. Mackensen, A., Krause, T., Blum, U., Uhrmeister, P., Mertelsmann, R., and Lindemann, A. (1999). Homing of intravenously and intralymphatically injected human dendritic cells generated in vitro from CD34+ hematopoietic progenitor cells, *Cancer Immunol. Immunother.*, **48**, pp. 118–122.

126. Morse, M. A., Coleman, R. E., Akabani, G., Niehaus, N., Coleman, D., and Lyerly, H. K. (1999). Migration of human dendritic cells after injection in patients with metastatic malignancies, *Cancer Res.*, **59**, pp. 56–58.

127. Peters, A. M., et al. (1986). Clinical experience with 99mTc-hexamethylpropylene-amineoxime for labelling leucocytes and imaging inflammation, *Lancet*, **2**, pp. 946–949.

128. Kelbaek, H., and Fogh, J. (1985). Technetium-99m labeling of polymorphonuclear leukocytes: Preparation with two different stannous agents, *J. Nucl. Med.*, **26**, pp. 68–71.

129. Matsuda, H. (2014). Quantification in brain SPECT: Noninvasive cerebral blood flow measurements using 99mTc-labeled tracers, in *PET and SPECT in Neurology* (Dierckx, R. A. J. O., et al., eds.), Springer Berlin Heidelberg, p. 75–91.

130. Blocklet, D., et al. (2003). 111In-oxine and 99mTc-HMPAO labelling of antigen-loaded dendritic cells: In vivo imaging and influence on motility and actin content, *Eur. J. Nucl. Med. Mol. Imaging*, **30**, pp. 440–447.

131. De Vries, E. J., Roca, M., Jamar, F., Israel, O., and Signore, A. (2010). Guidelines for the labelling of leucocytes with 99mTc-HMPAO, *Eur. J. Nucl. Med. Mol. Imaging*, **37**, pp. 842–848.

132. Roca, M., De Vries, E. F., Jamar, F., Israel, O., and Signore, A. (2010). Guidelines for the labelling of leucocytes with 111In-oxine, *Eur. J. Nucl. Med. Mol. Imaging*, **37**, pp. 835–841.

133. Martelli, C., et al. (2012). In vivo imaging of lymph node migration of MNP- and (111)In-labeled dendritic cells in a transgenic mouse model of breast cancer (MMTV-Ras), *Mol. Imaging Biol.*, **14**, pp. 183–196.

134. De Vries, I. J., et al. (2005). Magnetic resonance tracking of dendritic cells in melanoma patients for monitoring of cellular therapy, *Nat. Biotechnol.*, **23**, pp. 1407–1413.

135. Alfaro, C., et al. (2011). Pilot clinical trial of type 1 dendritic cells loaded with autologous tumor lysates combined with GM-CSF, pegylated IFN, and cyclophosphamide for metastatic cancer patients, *J. Immunol.*, **187**, pp. 6130–6142.

136. Prince, H. M., et al. (2008). In vivo tracking of dendritic cells in patients with multiple myeloma, *J. Immunother.*, **31**, pp. 166–179.

137. Aarntzen, E. H., et al. (2013). Targeting of 111In-labeled dendritic cell human vaccines improved by reducing number of cells, *Clin. Cancer Res.*, **19**, pp. 1525–1533.

138. Catana, C., Drzezga, A., Heiss, W.-D., and Rosen, B. R. (2012). PET/MRI for neurologic applications, *J. Nucl. Med.*, **53**, pp. 1916–1925.

139. Olasz, E. B., Lang, L., Seidel, J., Green, M. V., Eckelman, W. C., and Katz, S. I. (2002). Fluorine-18 labeled mouse bone marrow-derived dendritic cells can be detected in vivo by high resolution projection imaging, *J. Immunol. Methods*, **260**, pp. 137–148.

140. Koehne, G., et al. (2003). Serial in vivo imaging of the targeted migration of human HSV-TK-transduced antigen-specific lymphocytes, *Nat. Biotechnol.*, **21**, pp. 405–413.

141. Dubey, P., et al. (2003). Quantitative imaging of the T cell antitumor response by positron-emission tomography, *Proc. Natl. Acad. Sci.*, **100**, pp. 1232–1237.

142. Vilekar, P., Awasthi, V., Lagisetty, P., King, C., Shankar, N., and Awasthi, S. (2010). In vivo trafficking and immunostimulatory potential of an intranasally-administered primary dendritic cell-based vaccine, *BMC Immunol.*, **11**, p. 60.

143. Adonai, N., et al. (2002). Ex vivo cell labeling with 64Cu-pyruvaldehyde-bis(N4-methylthiosemicarbazone) for imaging cell trafficking in mice with positron-emission tomography, *Proc. Natl. Acad. Sci.*, **99**, pp. 3030–3035.

144. Viehl, C. T., et al. (2005). A tat fusion protein-based tumor vaccine for breast cancer, *Ann. Surg. Oncol.*, **12**, pp. 517–525.

145. Ahrens, E., Feili-Hariri, M., Xu, H., Genove, G., and Morel, P. (2003). Receptor-mediated endocytosis of iron-oxide particles provides efficient labeling of dendritic cells for in vivo MR imaging, *Magn. Reson. Med.*, **49**, pp. 1006–1013.

146. Modo, M., Hoehn, M., and Bulte, J. W. (2005). Cellular MR imaging, *Mol. Imaging*, **4**, pp. 143–164.

147. Arbab, A. S., Liu, W., and Frank, J. A. (2006). Cellular magnetic resonance imaging: Current status and future prospects, *Expert. Rev. Med. Devices*, **3**, pp. 427–439.

148. Hinds, K. A., et al. (2003). Highly efficient endosomal labeling of progenitor and stem cells with large magnetic particles allows magnetic resonance imaging of single cells, *Blood*, **102**, pp. 867–872.

149. Shapiro, E. M., Skrtic, S., Sharer, K., Hill, J. M., Dunbar, C. E., and Koretsky, A. P. (2004). MRI detection of single particles for cellular imaging, *Proc. Natl. Acad. Sci. U. S. A.*, **101**, pp. 10901–10906.

150. Baumjohann, D., Hess, A., Budinsky, L., Brune, K., Schuler, G., and Lutz, M. B. (2006). In vivo magnetic resonance imaging of dendritic cell migration into the draining lymph nodes of mice, *Eur. J. Immunol.*, **36**, pp. 2544–2555.

151. Briley-Saebo, K. C., et al. (2010). Longitudinal tracking of human dendritic cells in murine models using magnetic resonance imaging, *Magn. Reson. Med.*, **64**, pp. 1510–1519.

152. Sengar, R. S., Spokauskiene, L., Steed, D. P., Griffin, P., Arbujas, N., Chambers, W. H., and Wiener, E. C. (2009). Magnetic resonance imaging-guided adoptive cellular immunotherapy of central nervous system tumors with a T1 contrast agent, *Magn. Reson. Med.*, **62**, pp. 599–606.

153. Aspord, C., et al. (2013). Paramagnetic nanoparticles to track and quantify in vivo immune human therapeutic cells, *Nanoscale*, **5**, pp. 11409–11415.

154. Holland, G. N., Bottomley, P. A., and Hinshaw, W. S. (1977). F-19 magnetic-resonance imaging, *J. Magn. Reson.*, **28**, pp. 133–136.

155. Ahrens, E. T., Flores, R., Xu, H., and Morel, P. A. (2005). In vivo imaging platform for tracking immunotherapeutic cells, *Nat. Biotechnol.*, **23**, pp. 983–987.

156. Waiczies, H., et al. (2013). Visualizing brain inflammation with a shingled-leg radio-frequency head probe for $^{19}F/^{1}H$ MRI, *Sci. Rep.*, **3**, p. 1280.

157. Bonetto, F., Srinivas, M., Heerschap, A., Mailliard, R., Ahrens, E. T., Figdor, C. G., and De Vries, I. J. (2011). A novel (19)F agent for detection and quantification of human dendritic cells using magnetic resonance imaging, *Int. J. Cancer*, **129**, pp. 365–373.

158. Waiczies, S., Lepore, S., Sydow, K., Drechsler, S., Ku, M.C., Martin, C., Lorenz, D., Schutz, I., Reimann, H.M., Purfurst, B., Dieringer, M.A., Waiczies, H., Dathe, M., Pohlmann, A., and Niendorf, T. (2015). Anchoring dipalmitoyl phosphoethanolamine to nanoparticles boosts cellular uptake and fluorine-19 magnetic resonance signal, *Sci Rep*, **5**, pp. 8427.

159. Verdijk, P., et al. (2007). Sensitivity of magnetic resonance imaging of dendritic cells for in vivo tracking of cellular cancer vaccines, *Int. J. Cancer*, **120**, pp. 978–984.

160. Helfer, B. M., et al. (2010). Functional assessment of human dendritic cells labeled for in vivo (19)F magnetic resonance imaging cell tracking, *Cytotherapy*, **12**, pp. 238–250.

161. Waiczies, H., et al. (2011). Perfluorocarbon particle size influences magnetic resonance signal and immunological properties of dendritic cells, *PLoS One*, **6**, p. e21981.

162. Ramezani, M., Khoshhamdam, M., Dehshahri, A., and Malaekeh-Nikouei, B. (2009). The influence of size, lipid composition and bilayer fluidity of cationic liposomes on the transfection efficiency of nanolipoplexes, *Colloids Surf. B Biointerfaces*, **72**, pp. 1–5.

163. Buchanan, K. D., Huang, S., Kim, H., Macdonald, R. C., and Mcpherson, D. D. (2008). Echogenic liposome compositions for increased retention of ultrasound reflectivity at physiologic temperature, *J. Pharm. Sci.*, **97**, pp. 2242–2249.

164. Tirotta, I., et al. (2014). A superfluorinated molecular probe for highly sensitive in vivo 19F-MRI, *J. Am. Chem. Soc.*, **136**, pp. 8524–8527.

165. Banks, R., and Tatlow, J. (1986). A guide to modern organofluorine chemistry, *J. Fluorine Chem.*, **33**, pp. 227–346.

166. Riess, J. G. (2001). Oxygen carriers ("blood substitutes") raison d'Etre, chemistry, and some physiology blut ist ein ganz besondrer saft, *Chem. Rev.*, **101**, pp. 2797–2920.

167. Keipert, P. E. (1995). Use of oxygent™, a perfluorochemical-based oxygen carrier, as an alternative to intraoperative blood transfusion, *Artif. Cells, Blood Substit. Biotechnol.*, **23**, pp. 381–394.

168. Flögel, U., et al. (2008). In vivo monitoring of inflammation after cardiac and cerebral ischemia by fluorine magnetic resonance imaging, *Circulation*, **118**, pp. 140–148.

169. Ebner, B., Behm, P., Jacoby, C., Burghoff, S., French, B. A., Schrader, J., and Flögel, U. (2010). Early assessment of pulmonary inflammation by 19F MRI in vivo, *Circ. Cardiovasc. Imaging*, **3**, pp. 202–210.

170. Van Heeswijk, R. B., Pilloud, Y., Flögel, U., Schwitter, J., and Stuber, M. (2012). Fluorine-19 magnetic resonance angiography of the mouse, *PLoS One*, **7**, p. e42236.

171. Jacoby, C., et al. (2014). Probing different perfluorocarbons for in vivo inflammation imaging by 19F MRI: Image reconstruction, biological half-lives and sensitivity, *NMR Biomed.*, **27**, pp. 261–271.

172. Ahrens, E. T., Helfer, B. M., O'hanlon, C. F., and Schirda, C. (2014). Clinical cell therapy imaging using a perfluorocarbon tracer and fluorine-19 MRI, *Magn. Reson. Med.*, **72**, pp. 1696–1701.

173. Giraudeau, C., et al. (2012). High sensitivity 19F MRI of a perfluorooctyl bromide emulsion: Application to a dynamic biodistribution study and oxygen tension mapping in the mouse liver and spleen, *NMR Biomed.*, **25**, pp. 654–660.

174. Kampf, T., et al. (2010). Application of compressed sensing to in vivo 3D 19F CSI, *J. Magn. Reson.*, **207**, pp. 262–273.

175. Zhong, J., Mills, P. H., Hitchens, T. K., and Ahrens, E. T. (2013). Accelerated fluorine-19 MRI cell tracking using compressed sensing, *Magn. Reson. Med.*, **69**, pp. 1683–1690.

176. Ku, M. C., et al. (2013). GDNF mediates glioblastoma-induced microglia attraction but not astrogliosis, *Acta Neuropathol.*, **125**, pp. 609–620.

177. Kriegl, R., et al. (2014). Novel inductive decoupling technique for flexible transceiver arrays of monolithic transmission line resonators, *Magn. Reson. Med.*, **73**, pp. 1669–1681.

178. Otake, Y., Soutome, Y., Hirata, K., Ochi, H., and Bito, Y. (2014). Double-tuned radiofrequency coil for F and H imaging, *Magn. Reson. Med. Sci.*, **13**, pp. 199–205.

179. Wright, A. C., Song, H. K., and Wehrli, F. W. (2000). In vivo MR micro imaging with conventional radiofrequency coils cooled to 77 degrees K, *Magn. Reson. Med.*, **43**, pp. 163–169.

180. Nouls, J. C., Izenson, M. G., Greeley, H. P., and Johnson, G. A. (2008). Design of a superconducting volume coil for magnetic resonance microscopy of the mouse brain, *J. Magn. Reson.*, **191**, pp. 231–238.

181. Ratering, D., Baltes, C., Nordmeyer-Massner, J., Marek, D., and Rudin, M. (2008). Performance of a 200-MHz cryogenic RF probe designed for MRI and MRS of the murine brain, *Magn. Reson. Med.*, **59**, pp. 1440–1447.

182. Baltes, C., Radzwill, N., Bosshard, S., Marek, D., and Rudin, M. (2009). Micro MRI of the mouse brain using a novel 400 MHz cryogenic quadrature RF probe, *NMR Biomed.*, **22**, pp. 834–842.

183. Baltes, C., Bosshard, S., Mueggler, T., Ratering, D., and Rudin, M. (2011). Increased blood oxygen level-dependent (BOLD) sensitivity in the mouse somatosensory cortex during electrical forepaw stimulation using a cryogenic radiofrequency probe, *NMR Biomed.*, **24**, pp. 439–446.

184. Wagenhaus, B., et al. (2012). Functional and morphological cardiac magnetic resonance imaging of mice using a cryogenic quadrature radiofrequency coil, *PLoS ONE*, **7**, p. e42383.

185. Waiczies, H., Millward, J. M., Lepore, S., Infante-Duarte, C., Pohlmann, A., Niendorf, T., and Waiczies, S. (2012). Identification of cellular infiltrates during early stages of brain inflammation with magnetic resonance microscopy, *PLoS One*, **7**, p. e32796.

186. Elabyad, I. A., Kalayciyan, R., Shanbhag, N. C., and Schad, L. R. (2014). First in vivo potassium-39 MRI at 9.4 T using conventional copper radio frequency surface coil cooled to 77 K, *Biomed. Eng. IEEE Trans.*, **61**, pp. 334–345.

187. Niendorf, T., Pohlmann, A., Reimann, H.M., Waiczies, H., Peper, E., Huelnhagen, T., Seeliger, E., Schreiber, A., Kettritz, R., Strobel, K., Ku, M.-C., and Waiczies, S. (2015). Advancing Cardiovascular, Neurovascular and Renal Magnetic Resonance Imaging in Small Rodents Using Cryogenic Radiofrequency Coil Technology, *Frontiers in Pharmacology*, **6**.

Chapter 10

Neural Stem Cells

Markus Aswendt, Philipp Boehm-Sturm, and Mathias Hoehn

Max Planck Institute for Metabolism Research,
In-vivo-NMR Laboratory, Gleuelerstrasse 50,
50931 Cologne, Germany

markus.aswendt@sf.mpg.de

10.1 Introduction

In vivo ^{19}F MRI is an advanced MRI technique which allows quantitative and unambiguous tracking of ^{19}F-labeled cells in the living organism without radioactive isotope labeling. Compared to MRI contrast agents, such as iron oxide particles, ^{19}F MRI lacks confounding background, e.g., from blood vessels, as the ^{19}F signal is exclusively generated by the ^{19}F-loaded cells. Non-invasive quantification becomes feasible based on the linear relation between ^{19}F signal measured and ^{19}F nuclei present. Most importantly, preclinical data can be translated into ^{19}F MRI protocols for the human, as some commercial ^{19}F agents are already clinically authorized. The prominent feature of ^{19}F MRI is the non-invasive and quantitative tracking of stem cells in the living organism which contributes to a better characterization and understanding of the role of stem cells during regenerative processes.

Fluorine Magnetic Resonance Imaging
Edited by Ulrich Flögel and Eric Ahrens

ISBN 978-981-4745-31-4 (Hardcover), 978-981-4745-32-1 (eBook)
www.panstanford.com

A key player of brain regeneration are neural stem cells (NSCs), which reside during adulthood in specific brain niches and possess the unique feature among brain cells to retain the capacity to proliferate and give rise to new neurons and glia, for example, in response to injuries.

Regenerative medicine for neurological diseases like Parkinson's disease and stroke, in which certain brain regions are lost, is focusing on the role of endogenous as well as transplanted stem cells as a novel treatment strategy. With ^{19}F MRI, researchers now have a tool in hand to visualize how and when engrafted NSCs migrate and repopulate the injured brain regions in vivo. This represents a clear advantage compared to conventional, invasive studies which depend on subgroups of animals for each individual time point to reveal recovery mechanisms as the benefit of the engrafted stem cells. In contrast, ^{19}F MRI requires a lower number of laboratory animals used for a single experiment while providing the intra-individual time profile.

This book chapter introduces (i) the application of ^{19}F MRI for imaging NSCs in regenerative therapies, (ii) the strategy of efficient labeling NSCs with ^{19}F MRI contrast agents, (iii) the in vitro and in vivo imaging protocols, and, finally, (iv) the ^{19}F MRI-specific post-processing and validation steps for imaging cell grafts in the rodent brain.

10.2 Neural Stem Cells Used for Cell Therapy

10.2.1 Definition

Neural stem cells comprise a specific primordial and the least committed cell type of the central nervous system (CNS). In contrast to the more limited progenitor and precursor cells, a single NSC is defined to be (i) multipotent (differentiation in neurons, astrocytes, and oligodendrocytes is possible), (ii) able to repopulate a developing or degenerated region of the CNS, and (iii) able to self-renew (to produce daughter cells with identical properties) [1]. NSCs reside throughout adulthood in at least two neurogenic niches in mammals (Fig. 10.1), including the lateral ventricle wall (subventricular zone, SVZ) and the dentate gyrus of the hippocampus (subgranular zone, SGZ) [2]. In rodents they possess a very high proliferative rate of 10,000 to 30,000 (SVZ) and 9,000 cells (SGZ) per day [3, 4]. NSCs differentiate into neuronal precursor cells, so called neuroblasts, which migrate from the SVZ and SGZ towards the olfactory bulb (rodents), and striatum (human) and granular zone of the hippocampus (rodents/human), respectively [5–7]. These cells provide a source of new neurons, a continuous process which is called neurogenesis. Neurogenesis is regulated according to external stimuli, e.g., learning, activity, environmental enrichment and olfactory sensors [8–11] as well as pathophysiological triggers. Most importantly, NSCs respond to severe CNS injuries, e.g., stroke, traumatic brain injury and Alzheimer's disease [12–14] as part of an endogenous regeneration cascade.

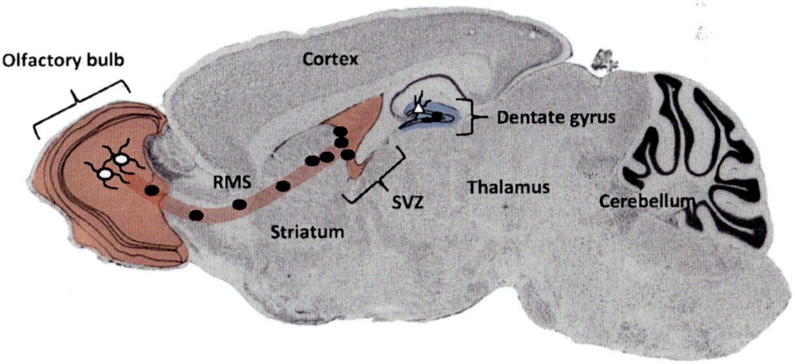

Figure 10.1 Illustration of neurogenesis in the adult mouse brain. NSCs in the subventricular zone (SVZ), adjacent to the lateral ventricle, generate neuroblasts which migrate through the rostral migratory stream (RMS) to the olfactory bulb, where they differentiate into interneurons (pathway highlighted in red). In the dentate gyrus of the hippocampus (blue-colored), NSCs reside in the subgranular zone and produce new neurons for the granule cell layer. Illustration prepared based on a sagittal image of a histological Nissl staining and atlas data provided by ©2014 Allen Institute for Brain Science. Allen Mouse Brain Atlas [Internet]. Available from: http://mouse.brain-map.org/.

Although endogenous neurogenesis is insufficient to replace lost neurons and provide regeneration of neurological disorders, it has stimulated the experimental work on NSC culture and transplantation. The breakthrough of in vitro NSC culture was set by the neurosphere model from primary isolated brain cells by Reynolds and Weiss [15]. Also important was the fundamental work on two essential growth factors, epidermal growth factor (EGF) and fibroblast growth factor (FGF) [16], that enabled enable in vitro NSCs to remain in a proliferative and undifferentiated state. Stable cell lines from these primary cultures were first obtained by oncogene-mediated immortalization [17, 18]. However, were cells differ from naïve NSCs [19]. Stem cell culture condition could be improved to maintain stable cell lines without immortalization for up to 2 years [20], e.g. by differentiation of embryonic stem (ES) cells [21–23]. More recently, cellular reprogramming was introduced which is used to generate patient-specific induced pluripotent (iPS) cells from fibroblasts that can be further differentiated, e.g., into neurons and astrocytes [24]. Furthermore, fibroblasts can be directly converted into induced NSCs (iNSCs) [25]. Because of their unique cellular properties, NSCs are believed to be the ideal source for novel therapy strategies aimed at cell replacement or neuroprotection of brain disorders [26].

10.2.2 Mechanisms of Action in Therapy

The endogenous regeneration potential to reconstitute lost brain tissue and neuronal network connections appears to be highly constricted, as many neurological diseases are irreversible and progressive (e.g., Huntington's disease, multiple sclerosis, Parkinson's, and stroke). Furthermore, the risk of many neurological disorders is age-related, leading to an expected increased occurrence with increasing life expectancy. In the case of stroke, which represents a major cause of morbidity and mortality worldwide, nearly three quarters of all strokes occur in people over the age of 65. The risk of having a stroke more than doubles each decade after the age of 55 (U.S. Centers for Disease Control and Prevention). Stroke is caused by an embolic/thrombotic vessel blockage or a hemorrhagic bleeding which results in oxygen deprivation and subsequent brain cell death. Presently, the clinical therapy is strictly limited to a small subset of patients qualifying for administration of recombinant tissue plasminogen activator (rtPA) for recanalization of the vessels within a strict time limit of up to 4.5 h after stroke onset [27].

Human stroke is experimentally mimicked in rats and mice by occluding the middle cerebral artery (MCAO) [28]. This results in a substantial reduction of blood supply to the cortex and parts of the subcortical regions, leading to a progressive loss of neurons in cortical and subcortical regions. Endogenous NSCs respond to stroke and migrate from the SVZ towards the lesion, differentiate and form new neurons in the area affected by the insult [12, 29–31]. However, NSCs have a finite ability to regenerate and reconstitute a neural lesion [32]. Stem cell therapy aims to support the endogenous repair mechanism by applying exogenously generated, at best patient-specific, stem cells. Preclinical data on human stroke models in rodents indicate that stem cell therapy enhances brain-repair processes and improves functional recovery [33–35]. Furthermore, clinical trials are already in progress (e.g., a phase I study in Guildford, UK, Id-nr. NCT01151124). Besides studying the route, timing, and site of delivery, various types of stem cells are under investigation, including ES cells, NSCs, and stem cells isolated from bone marrow, umbilical cord blood and adipose tissue. Recently, human induced pluripotent stem cells were reported to serve as a patient-specific source for autologous transplantations [36].

Cell replacement therapy initially aimed to replace damaged tissue by spontaneous differentiation into the appropriate cell type. However, this attempt is impeded by the differentiation efficiency, very low neural circuit integration and the general issue of graft survival in vivo. More recently, the observed beneficial effects are explained by a combination of immunosuppressive [37], neuroprotective [38], and neurotrophic effects [39]. Understanding the mechanisms underlying stem cell induced regeneration is in the early stages and heavily dependent on the development of non-invasive imaging techniques with the aim of elucidating these processes under in vivo conditions.

10.2.3 ^{19}F MRI of NSCs

MRI provides the best spatial resolution (down to 100 μm in vivo) in soft tissues compared to nuclear imaging and optical imaging. In addition to anatomical details, specific MRI pulse sequences provide tissue perfusion, structural connectivity (diffusion MRI), brain activity (fMRI) and functional connectivity (resting state fMRI) [40]. Thus, MRI is the method of choice to monitor the complex pathophysiological changes in brain diseases such as stroke [41]. MRI is non-invasive and non-radioactive and thus increasingly becoming the imaging method of choice in routine clinical practice.

However, imaging specific cell types remains a challenge, since MRI signal generation relies on the ubiquitous hydrogen while other nuclei, e.g., ^{23}Na [42] and ^{13}C [43] lack the necessary sensitivity. Thus, for many preclinical models, including NSC therapy of stroke, paramagnetic lanthanide complexes or superparamagnetic iron oxide particles have been used to label cells prior to injection or implantation [44]. Paramagnetic agents can be further functionalized to respond to (patho-) physiological processes in specific neuronal cell types [45, 46]. MRI sensitivity increases with field strength, leading to a reported minimal detection limit of 100 labeled embryonic stem cells in the rat brain at 17.6 T [47]. However, confounding contrast on MR images, e.g., from blood vessels and macrophages, diminishes cell tracking specificity.

An alternative MRI-visible nucleus is the ^{19}F isotope, which possesses the second highest gyromagnetic ratio next to hydrogen and has negligible natural abundance in the body and no MRI background [48]. During the past decade, various ^{19}F cell labels have been optimized to provide efficient ^{19}F MRI-based cell tracking in vivo [49]. Notably, the detection of ^{19}F cell labels is fundamentally different from paramagnetic and superparamagnetic contrast agents. Instead of measuring the effect on T1, T2 and T_2^* relaxation times, the ^{19}F cell label acts like a tracer, and the ^{19}F spin density is directly proportional to the acquired MR signal. Ex vivo labeling of dendritic cells with a ^{19}F contrast agent by Ahrens and colleagues provided the first in vivo demonstration of ^{19}F cell tracking [50]. ^{19}F MRI equipment and processing was adopted to the mouse brain application, e.g., for visualization of amyloid beta plaques and inflammation after stroke [51, 52]. Ruiz-Cabello et al. and Boehm-Sturm et al. were then the first to show the feasibility of ^{19}F labels for in vivo tracking of murine and human NSCs, respectively [53, 54]. In the following section, we introduce ^{19}F cell labels and describe the requirements for efficient and safe labeling of NSCs.

10.3 Labeling NSCs for in vivo Tracking Using ^{19}F MRI

10.3.1 ^{19}F Cell Labels

The prerequisite for ^{19}F MRI-based cell tracking is a sufficient level of intracellular ^{19}F to reach in vivo detection sensitivity. Therefore, ^{19}F cell labels are required to carry a large number of ^{19}F atoms, which ideally respond to a single ^{19}F resonance frequency and are furthermore non-toxic to the cells [48]. The majority of studies use emulsions of perfluorocarbons (PFCs), which are derived from hydrocarbons but have all hydrogen atoms replaced by ^{19}F atoms. These C-F polymers are characterized by very strong intra-molecular bonds and are both lipo- and hydrophilic, which leads to a very high chemical and biological inertness [55]. PFCs are not metabolized by enzymes in vivo [56], are stable at lysosomal pH [56], and are finally cleared via the reticuloendothelial system, and excreted via the lungs and the skin [57, 58]. PFCs can dissolve large amounts of gases, which made them already clinically useful as artificial blood substitutes via partial liquid ventilation [56, 59]. Due to the specific amphiphatic character, PFCs tend to spontaneously form stable and well-organized monolayers [60]. In order to use PFCs for biological applications, an emulsion must be prepared. Emulsions are droplets of immiscible fluorocarbon with a lipid monolayer coating that stabilizes the droplet in the continuous water phase. Additional substances, e.g., fluorescent dyes for fluorescence imaging, can be chemically coupled to the PFC coating. Perfluorocarbons used for labeling thus far include perfluoro-15-crown-5-ether (PFCE), perfluorooctylbromide (PFOB) and linear perfluoropolyethers (PFPEs) (Fig. 10.2a) [49]. In general, the ^{19}F spectra of linear structures are more complex; only macrocyclic structures like PFCE possess a single resonance frequency. Other ^{19}F particles used for cell labeling consist of a polymer in a conformation that only exposes hydrophilic regions ("Micelles") or a perfluorocarbon core with a polymer coat ("Particles") (Fig. 10.2b) [49]. Fluorinated small molecules have been functionalized for targeting enzyme activity [61], gene expression [62] and protein binding [63].

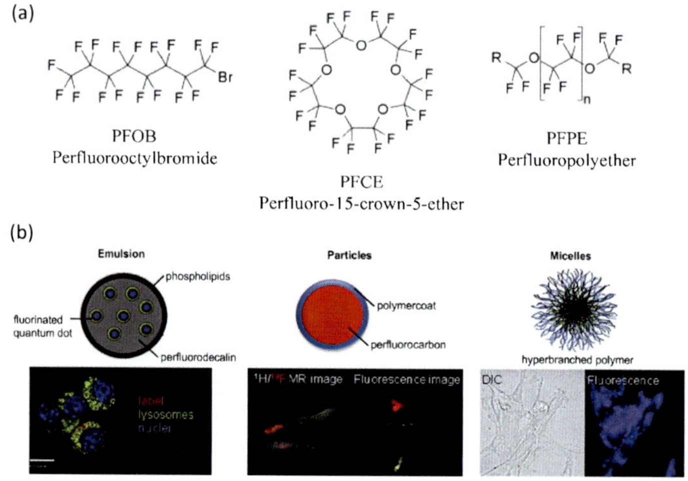

Figure 10.2 Schematic overview of different ^{19}F cell labels. (a) Chemical structures of perfluorocarbons used for ^{19}F MRI cell tracking: perfluoro-15-crown-5 ether (PFCE), perfluorooctyl bromide (PFOB) and linear perfluoropolyethers (PFPE). Chemical structures reprinted from [55]. (b) Emulsions are composed of fluorocarbon droplets ensheathed by a lipid monolayer coating (approximately 200 nm in diameter). The addition of fluorescent dyes or quantum dots provides fluorescence verification of cell labeling as shown in a representative micrograph with quantum dots in red, lysosomes in green, nuclei in blue (10 μm scale bar). Particles can be synthesized in various sizes (200–2,000 nm) and consist of a perfluorocarbon core with a polymer coat, to which additional components, such as fluorescent dyes can be covalently added [48]. Bimodal ^{19}F/fluorescence particles are visualized in the micrograph after injection into the mouse footpad by ^{19}F MRI (left) and fluorescence imaging (right). Micelles are smaller in size (20–30 nm diameter) and consist of a polymer with hydrophobic hyperbranched core and amphiphilic fluoropolymer arms [64]. The inset shows micrographs of labeled cells with a diffusion interference contrast (DIC) image on the left and a fluorescence image on the right, where the micelles are blue. Figure adapted with permission from [49].

A summary of studies using ^{19}F cell labels to track NSCs is reported in Table 10.1. Several key parameters, which have been identified for iron oxide and lanthanide MRI contrast agents, hold true for ^{19}F cell labels as well, including: (i) efficient and intracellular labeling, (ii) potential toxicity and harmful alterations of cellular processes, (iii) potential leakage and transfer to host cells, and (iv) dilution of label with proliferation [65]. The non-invasive tracking of NSCs is strictly sensitivity limited, which is undermined by cell divisions and subsequent label dilution over time.

Table 10.1 ^{19}F MRI studies of murine and human NSC lines and reported labeling protocols, efficacy, and label safety tests

Stem cell type	Immortalized human neural stem cell lines CTX0E03 and STROC05	Non-immortalized human fetal striatal NSCs	Immortalized murine neural stem cells C17.2	Murine NSCs derived by differentiation of D3 ES cells
^{19}F compound	PFPE (CS-1000-DM-Red, Celsense, Inc.)	PFPE (CS-1000, Celsense Inc.)	Custom-made PFCE	PFPE (CS-1000 or CS-1000 green, Celsense Inc.)
Labeling protocol	5 mg/ml (with 10% HSA) for a labeling duration of 24 h was	Poly-HEMA coating, 100 µl/ml PFCE for 36 h in culture medium, 2 × washing with	Incubation in culture medium 2.4–4.8 mM, 3 × washing with PBS	25 µl/ml 42 h incubation in
Labeling efficacy	1.96×10^{12} ^{19}F spins/cell	3.70×10^{12} ^{19}F spins/cell	Not defined	$\sim 10^{12}$ ^{19}F spins/cell
Toxicity test	MTT	Trypan blue	MTS	Trypan blue, proliferation, migration
Stem cell tests	Differentiation (Nestin, GFAP, beta-III tubulin)	Differentiation (GFAP, beta-III tubulin)	—	Differentiation (GFAP, DCX)
Ref.	[66]	[54]	[53]	[67]

10.3.2 Optimization of ^{19}F Cellular Uptake

Notably, labeling protocols are specific to ^{19}F label and stem cell type. Therefore, for each experimental study, the optimal labeling conditions need to be titrated. Reproducible cell uptake of the label is essential for reliable quantification of ^{19}F atoms/cell by NMR spectroscopy. Therefore, a first set of experiments should include labeling protocols in which the ^{19}F label concentration, incubation time, cell culture flask coating and cell density is varied. Finding the optimal protocol can be time-consuming, but is essential to solve labeling challenges. For example, free-floating neurosphere cultures strongly adhere to common flasks after addition of a PFC emulsion which can be solved by addition of serum to the culture medium [66] or by additional coating of the flask surface [54]. Similar to iron oxide particles, cell loading can be improved by addition of transfection agents or attachment of cell-specific targeting peptides which was already reported for non-stem cells [49]. However, certain formulations of PFPEs facilitate uptake without transfection agents in a concentration-dependent manner (Fig. 10.3b). Such a label dilution series is best performed with subsequent measurements of cell viability and ^{19}F uptake. By correlating the number of cells surviving the treatment with the intracellular ^{19}F concentration, one should select the protocol yielding the highest ^{19}F uptake, while maintaining cells viable. In case of NSCs, it is highly recommended to select labeling protocols that have no impact on stem cell properties.

10.3.3 Cell Characterization

Several commercial assays are available to determine the ratio of live versus dead cells, as well as intracellular metabolic activity. The easy-to-use Trypan blue exclusion assay is most often used to verify the viability rate (number of viable cells). By comparing the viability of labeled versus unlabeled cells, one can estimate the survival rate, i.e., the amount of cells surviving contrast agent uptake [45]. Additional measures of cell migration and proliferation complete the characterization of ^{19}F-labeled cells (Fig. 10.3a).

PFCs tend to populate the cell plasma very densely, which could have an effect on intracellular filaments necessary for migratory processes [49]. However, migration of engrafted NSCs from the injection site towards the lesion must be preserved in order to facilitate the desired beneficial therapeutic effect.

Of note, in case of iron oxide particles, Cromer Berman et al. described migratory deficits [68].

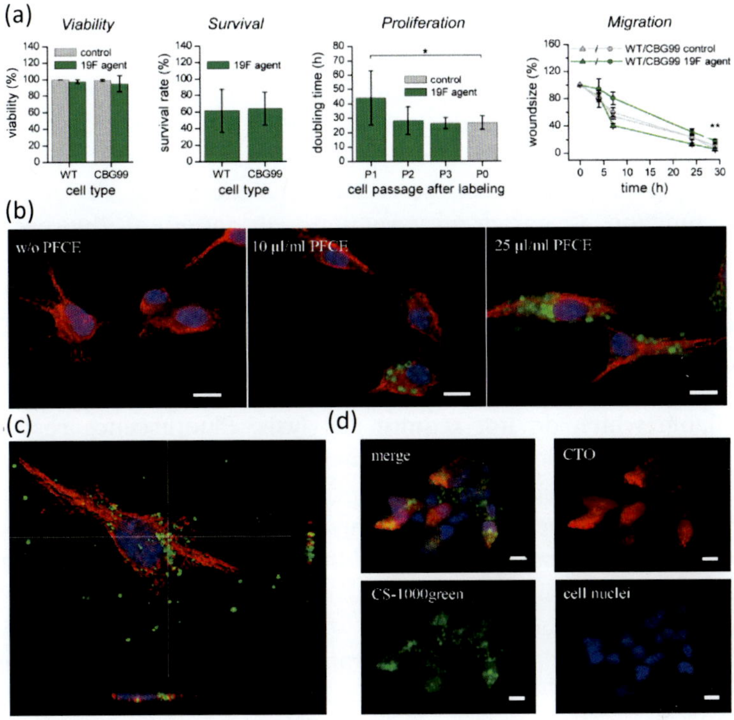

Figure 10.3 Intracellular labeling of murine NSCs with PFPE nanoemulsion. (a) Representative validation data to characterize key cell features after labeling with PFPE, including viability, survival rate, proliferation and migration. (b) Uptake of PFPE is concentration-dependent. Micrographs showing adherent murine NSCs labeled with FITC-coupled PFPE in green, Nestin immunostaining in red and cell nuclei in blue (scale bar 10 µm). (c) Confocal microscopy of the cells labeled with 25 µl/ml PFPE-FITC identifies intracellular location of ^{19}F particles. (d) Fluorescence images of multi-labeled cells counterstained with Hoechst (blue) are shown in the merged image displaying co-localization of CellTracker Orange (CTO, red) and the ^{19}F agent CS-green (green) in the cytoplasm (scale bar 10 µm). Reprinted with permission from [67].

NSC migration can be assayed in vitro by simple microscopy-based scratch (wound) tests [67] or transwell migration assays.

Changes in proliferation can be observed by quantification of Ki67 immunostainings, which mark cells having undergone cell division during the culturing period [66]. Through several passages the number of labeled versus unlabeled viable cells can be used to characterize short- and long-term effects on the proliferation rate/doubling time [67]. In vitro ^{19}F spectroscopy should be included in these proliferation tests to characterize potential leakage and ^{19}F label separation to the daughter cells.

A precise measure of the actual influence of the ^{19}F label on cell function is based on enzymatic formation of fluorescent products. In addition to commonly used MTT (3-(4,5-dimethylthiazol-2-yl)-2,5-diphenyltetra-zolium bromide) and MTS (3-(4,5-dimethylthiazol-2-yl)-5-(3-carboxy-methoxyphenyl)-2-(4-sulfophenyl)-2H-tetrazolium), which are converted into an insoluble purple formazan, novel resazurin-based assays are available which do not require cell lysis. Fluorescence readout of multiple samples can be obtained using a spectrophotometer and easily up-scaled, e.g., to test multiple labeling protocols.

In order to verify intracellular uptake of ^{19}F label, electron microscopy is useful [50], but this is a complex and time consuming technique. More flexibility is achieved with a fluorescent dye-coupled ^{19}F label or simultaneous labeling with a stable intracellular fluorescent tracer for (confocal) fluorescence microscopy (Fig. 10.3c). Long-term labeling of stem cells is achieved with chemical probes like chloromethylfluorescein diacetate (CMFDA or CellTracker) (Fig. 10.3d) and long-chain carbocyanine dyes (DiI or DiO) [69]. However, the stability of covalently bound fluorescent dyes should be carefully controlled [70] to provide a long term label [71]. Admixing fluorescent dyes to the ^{19}F label does not ensure equal distribution of both tracers after cell division or stem cell differentiation.

It should be noted that NSCs in vivo reside only in tightly controlled niches surrounded by supportive cells, and the in vitro culture remains an artificial model to maintain proliferation and undifferentiated state. It is necessary to avoid impacting the specific stem cell phenotype by the ^{19}F cell label in order to ensure the regenerative potential in vivo. Proper verification of the differentiation potential of labeled versus unlabeled cells is necessary. Immunocytochemical staining for markers for NSCs (Nestin, Sox2), neurons (DCX, betaIII-tubulin) and glial cells (GFAP,

S100beta, O4) is used to visualize and quantify the multipotency of NSCs. More precise measures of ^{19}F label effects on the genotype should also be considered using, for example, RT-PCR and microarrays for specific marker sets of the neural lineage including stem cell markers.

10.4 In vitro and in vivo ^{19}F MRI of NSCs

10.4.1 Cell Preparation and Implantation

As a prerequisite for in vivo cell quantification, the average number of ^{19}F spins per cell has to be determined with ^{19}F MR spectroscopy (MRS). For this, the PFC-loaded cells are pipetted several times through a small tip to yield single cell suspensions, counted, and an internal MRS reference standard is added, for example potassium fluoride (KF) or trifluoroacetic acid (TFA). Alternatively, a second tube with an external standard can be placed next to the tube with the cells. The ratio of peaks from the cells and the reference standard is then multiplied with the (known) number of ^{19}F spins in the reference standard and divided by the number of cells.

10.4.2 Imaging Hardware and Pulse Sequences

The cell numbers used in NSC implantation studies are generally lower than in other models of cell therapy. For example the number of dendritic cells implanted for tumor vaccination is about one order of magnitude higher ($\sim10^5$ versus $\sim10^6$ cells per injection). This leads to a need for a hardware setup and pulse sequences that maximize sensitivity. Therefore, the use of fast pulse sequences, such as fast gradient echo (FLASH), turbo spin echo (TSE or RARE), or balanced steady state pulse sequences (SSFP) is recommended. The choice mainly depends on the relaxation times of the ^{19}F label. The disadvantage of using fast pulse sequences is their vulnerability to image artifacts, e.g., blurring.

^{19}F MRI profits from the new generation of very high field MRI systems, and the highest field system available should be used. The only exception may be with hyperpolarized ^{19}F labels for which a long T1 relaxation time is important for long lifetimes of the polarization state [72], but this is still a very small field.

Generally, small surface coils or arrays of these are more sensitive than larger volume resonator coils. However, surface coils have the major disadvantage that the image intensity generated by cells close to the surface coil is higher than from the same number of cells far away from the surface coil. This is due to the non-homogenous magnetic field of these coils, the B1 field. The inhomogeneity makes quantification of cell numbers from different regions or between animals quite difficult. If the B1 field of the transmitting and receiving coil is known in each image voxel, the ^{19}F images can be corrected using the Bloch equations. However, in a cell tracking experiment, an in vivo acquisition of a ^{19}F B1 map is usually too time-consuming since the ^{19}F concentration is too low. If the coil is tunable to both the ^{19}F and the ^{1}H resonance frequency it can be assumed that the B1 fields on both resonance frequencies are in a fixed relation. This can be verified by acquiring B1 field maps for each resonance frequency on a phantom containing both ^{1}H and ^{19}F, e.g., a mixture of TFA and water. In this case, a ^{1}H B1 map can be acquired, converted to a ^{19}F B1 map using the relation found in the phantom, and the ^{19}F image can be corrected for inhomogeneity using the attenuation found when modifying the B1 field in the Bloch equations. Studies have shown the feasibility of such a strategy when using a ^{1}H/^{19}F transmit-receive surface coil [73, 74].

Apart from the MRI system, a specific advantage of ^{19}F MRI hardware in the brain over whole-body measurements is the possibility to stereotactically constrain the head, e.g., using ear- and tooth-bars. This reduces motion artifacts and lengthy MRI scans due to triggering and facilitates precise co-registration of imaging time points for longitudinal analysis.

10.4.3 Estimating Cell Detection Limit

Experimentally, sensitivity of a given hardware setup and pulse sequence is reflected in the minimum ^{19}F concentration (^{19}F spins per voxel) that generates a significant signal-to-noise ratio (SNR) in the MR image. Generally, an SNR of ~3 can visually be detected from the background noise. When the cells are labeled homogenously, the ^{19}F concentration scales linearly with the concentration of cells. It is thus possible with a set of simple in vitro experiments to estimate a cell detection limit of a given hardware setup and type of pulse sequence by imaging a dilution series of either the ^{19}F label or labeled cells. A linear fit of SNR over concentration crosses the minimum SNR (e.g. SNR = 3) at the minimum detectable concentration. For conventional pulse sequences, the image resolution and imaging time can then be adjusted before the in vivo experiment, using the well-known relation between SNR, voxel volume (V), bandwidth per voxel (BW), number of averages (NA), and ^{19}F concentration (C):

$$\text{SNR} = \alpha^* C^* V^* \sqrt{\text{NA/BW}},$$

Here, α is the slope found in fitting the signal from the dilution series and encapsulates all hardware and pulse sequence-specific parameters.

10.5 Validation

10.5.1 Determining the Location of Transplanted Cells and the ^{19}F Cell Label by Histology

Histological validation of the fate of NSCs in regenerative therapies is essential to understand the mode of action and potential benefits for the outcome of neurological diseases such as stroke. However, no native histological staining for ^{19}F is available, as for example the Prussian blue staining to detect iron oxide particles. Therefore, fluorescent dyes or quantum dots coupled to the ^{19}F moieties are essential for microscopic tracing of the ^{19}F cell label on fixed brain tissue sections. We have used a fluorescent reporter expressed by the NSCs, the intracellular dye Celltracker Orange and an immunocytochemical staining specific for engrafted NSCs in order to verify whether cell label still co-locates with the NSCs after a prolonged period (Fig. 10.4a). A precise histological validation is necessary for the exact interpretation of the ^{19}F MRI data and can be used for example to identify chemical shift-induced position changes of the ^{19}F signal. As reported by Bible and colleagues [66], the location of ^{19}F MRI signal corresponded well to human NSCs transplanted into the lesion cavity of stroke mice. However, histology revealed that approximately 20% of ^{19}F-labeled cells were host cells [66]. In line with this observation, we could identify unspecific uptake of

CellTracker Orange in a similar experimental setting. The double-labeled murine NSCs were transplanted next to the lesion cavity, which was determined with anatomical MRI. Assuming a similar dilution of ^{19}F label and CellTracker Orange we can conclude from the histological analysis 14 days post transplantation that the intracellular dye of the engrafted murine NSCs is located in phagocytotic macrophages surrounding the graft site (Fig. 10.4b). This accumulation of intracellular label outside the graft could lead to an overestimation of the number of labeled NSCs (Fig. 10.4c) [67].

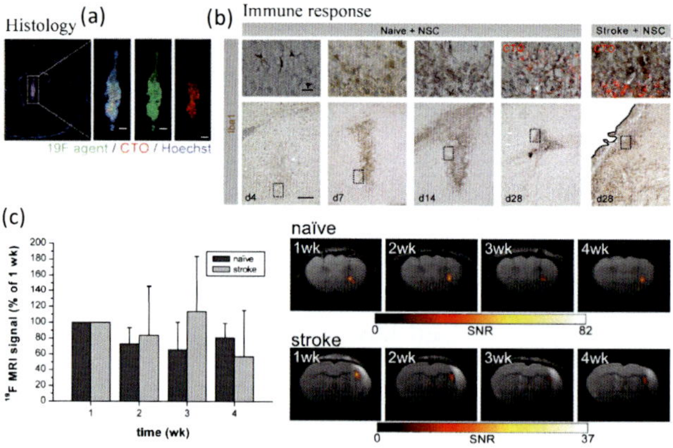

Figure 10.4 Histological validation of in vivo ^{19}F MRI data. (a) Microscopic images of the NSC graft located in the striatum. The zoom-in (right) highlights the intracellular fluorescent dye CellTracker Orange (CTO) in red, CS-1000 green in green and the overlay with cell nuclei in blue (scale bar 200 μm). (b) Innate immune reaction was visualized by staining for Iba1, a marker of microglia and macrophages. Iba1-positive cell accumulation was detected from day 7, with increasing numbers in the stroke-affected hemisphere compared to naïve animals. At day 28, CTO was located in immune cells (white arrows), thus indicating the phagocytosis of grafted cells or released labeling agents. Sections shown in the upper row (60 × magnification, scale bar 20 μm) correspond to boxes in the graft border zone in overview images displayed below (10 × magnification, scale bar 100 mm. (c) Quantitative analysis of ^{19}F SNR in naïve (no stroke) and stroke animals and representative SNR maps highlighting the stroke (white rim) and location of the ^{19}F label (red). We found the ^{19}F signal to be persistent with more scatter in the stroke group but no difference was noted in naïve controls. Reprinted with permission from [67].

10.5.2 Multimodal Approaches: The Better Imaging?

The key advantage of ^{19}F MRI over other methods is the possibility to unambiguously and quantitatively visualize the graft in 3D in deep tissues. The combination with other MRI sequences provides detailed anatomical assignment of the detected ^{19}F cell clusters.

In order to obtain structural and functional cell graft information in vivo, ^{19}F MRI can be combined with optical imaging. In particular, luciferase reporter genes for non-invasive bioluminescence imaging are highly interesting due to the ATP-dependent enzyme reaction, which results in a non-invasive cell viability readout. The reporter gene is stably inserted into the NSC genome, e.g., by lentiviral vector-mediated transduction. Both imaging modalities, ^{19}F MRI and BLI, can be obtained on the same animal, repetitively. Successful application of such a protocol of complementary ^{19}F/^{1}H MRI and BLI has already been demonstrated by Boehm-Sturm and colleagues, who applied both techniques for longitudinal tracking of NSCs in stroke mice over 4 weeks [66]. We found it extremely useful to correlate the BLI data with ^{19}F MRI. Interestingly, the bioluminescence signal of murine NSCs engrafted next to the stroke lesion or a corresponding area in the healthy mouse, decreased rapidly during the first two weeks. However, the ^{19}F MRI signal from the intracellular PFPE label in the grafted cells did not change significantly over 4 weeks. The benefit of the multimodal approach lies in the correlation of both results, which in this case, provide (1) biological information about the cell number and viability (BLI) and (2) the amount of ^{19}F label at the implantation site (^{19}F MRI). Histology confirmed persistent ^{19}F label surrounding the initial graft location, partly taken up by phagocytic immune cells. Multimodal imaging as described here will help to better define the fate of engrafted stem cells in the context of cell therapy.

10.6 Summary

^{19}F MRI is a very appealing approach for cell tracking and investigations of longitudinal cell fate. Apart from a somewhat reduced sensitivity when compared with iron oxide labeled cells, the ^{19}F approach has several advantages. In particular, the safe exclusion of confounding factors for image contrast interpretation, and the ability to quantify cell numbers, will lead to a more sophisticated investigative strategy. This generation of a detailed in vivo cell-biological interpretation of the graft can be further improved when ^{19}F MRI is combined with cell vitality imaging as provided by in vivo bioluminescence imaging.

Acknowledgments

This work was financially supported by grants from the EU-FP7 programs TargetBraln (HEALTH-F2-2012-279017) and BrainPath (PIAPP-GA-2013-612360).

References

1. M. A. Parker, J. K. Anderson, D. A. Corliss, V. E. Abraria, R. L. Sidman, K. I. Park, Y. D. Teng, D. A. Cotanche, and E. Y. Snyder, Expression profile of an operationally-defined neural stem cell clone, *Exp. Neurol.*, vol. 194, no. 2, pp. 320–332, Aug. 2005.
2. F. H. Gage, Neurogenesis in the adult brain, *J. Neurosci.*, vol. 22, no. 3. pp. 612–613, 2002.
3. C. Lois and A. Alvarez-Buylla, Long-distance neuronal migration in the adult mammalian brain, *Science*, vol. 264, no. 5162, pp. 1145–1148, May 1994.
4. H. A. Cameron and R. D. McKay, Adult neurogenesis produces a large pool of new granule cells in the dentate gyrus, *J. Comp. Neurol.*, vol. 435, no. 4, pp. 406–17, Jul. 2001.
5. A. Alvarez-Buylla, Mechanism of migration of olfactory bulb interneurons, *Semin. Cell Dev. Biol.*, vol. 8, no. 2, pp. 207–213, Apr. 1997.
6. A. Ernst, K. Alkass, S. Bernard, M. Salehpour, S. Perl, J. Tisdale, G. Possnert, H. Druid, and J. Frisén, Neurogenesis in the striatum of the adult human brain, *Cell*, vol. 156, no. 5, pp. 1072–1083, Feb. 2014.
7. F. H. Gage, Mammalian neural stem cells, *Science*, vol. 287, no. 5457, pp. 1433–1438, Feb. 2000.
8. T. J. Shors, G. Miesegaes, A. Beylin, M. Zhao, T. Rydel, and E. Gould, Neurogenesis in the adult is involved in the formation of trace memories, *Nature*, vol. 410, no. 6826, pp. 372–376, Mar. 2001.

9. A. G. Dayer, A. A. Ford, K. M. Cleaver, M. Yassaee, and H. A. Cameron, Short-term and long-term survival of new neurons in the rat dentate gyrus, *J. Comput. Neurol.*, vol. 460, no. 4, pp. 563–572, Jun. 2003.

10. G. Kempermann, H. G. Kuhn, and F. H. Gage, More hippocampal neurons in adult mice living in an enriched environment, *Nature*, vol. 386, no. 6624, pp. 493–495, Apr. 1997.

11. H. van Praag, B. R. Christie, T. J. Sejnowski, and F. H. Gage, Running enhances neurogenesis, learning, and long-term potentiation in mice, *Proc. Natl. Acad. Sci. U. S. A.*, vol. 96, no. 23, pp. 13427–13431, Nov. 1999.

12. A. Arvidsson, T. Collin, D. Kirik, Z. Kokaia, and O. Lindvall, Neuronal replacement from endogenous precursors in the adult brain after stroke, *Nat. Med.*, vol. 8, no. 9, pp. 963–970, Sep. 2002.

13. X.-H. Chen, A. Iwata, M. Nonaka, K. D. Browne, and D. H. Smith, Neurogenesis and glial proliferation persist for at least one year in the subventricular zone following brain trauma in rats, *J. Neurotrauma*, vol. 20, no. 7, pp. 623–631, Jul. 2003.

14. K. Jin, V. Galvan, L. Xie, X. O. Mao, O. F. Gorostiza, D. E. Bredesen, and D. A. Greenberg, Enhanced neurogenesis in Alzheimer's disease transgenic (PDGF-APPSw, Ind) mice, *Proc. Natl. Acad. Sci. U. S. A.*, vol. 101, no. 36, pp. 13363–13367, Sep. 2004.

15. B. A. Reynolds and S. Weiss, Generation of neurons and astrocytes from isolated cells of the adult mammalian central nervous system, *Science*, vol. 255, no. 5052, pp. 1707–1710, 1992.

16. A. L. Vescovi, E. A. Parati, A. Gritti, P. Poulin, M. Ferrario, E. Wanke, P. Frölichsthal-Schoeller, L. Cova, M. Arcellana-Panlilio, A. Colombo, and R. Galli, Isolation and cloning of multipotential stem cells from the embryonic human CNS and establishment of transplantable human neural stem cell lines by epigenetic stimulation, *Exp. Neurol.*, vol. 156, no. 1, pp. 71–83, Mar. 1999.

17. E. F. Ryder, E. Y. Snyder, and C. L. Cepko, Establishment and characterization of multipotent neural cell lines using retrovirus vector-mediated oncogene transfer, *J. Neurobiol.*, vol. 21, no. 2, pp. 356–375, Mar. 1990.

18. E. Y. Snyder, D. L. Deitcher, C. Walsh, S. Arnold-Aldea, E. A. Hartwieg, and C. L. Cepko, Multipotent neural cell lines can engraft and participate in development of mouse cerebellum, *Cell*, vol. 68, no. 1, pp. 33–51, Jan. 1992.

19. R. Mi, Y. Luo, J. Cai, T. L. Limke, M. S. Rao, and A. Höke, Immortalized neural stem cells differ from nonimmortalized cortical neurospheres and cerebellar granule cell progenitors, *Exp. Neurol.*, vol. 194, no. 2, pp. 301–319, Aug. 2005.

20. T. Kallur, Human Neural Stem Cells: Region-specific properties and prospects for cell therapy, Lund University, 2008. https://lup.lub.lu.se/search/publication/1150540.

21. P. Koch, T. Opitz, J. A. Steinbeck, J. Ladewig, and O. Brüstle, A rosette-type, self-renewing human ES cell-derived neural stem cell with potential for in vitro instruction and synaptic integration, *Proc. Natl. Acad. Sci. U. S. A.*, vol. 106, no. 9, pp. 3225–3230, Mar. 2009.

22. V. Tropepe, S. Hitoshi, C. Sirard, T. W. Mak, J. Rossant, and D. Van Der Kooy, Direct neural fate specification from embryonic stem cells: A primitive mammalian neural stem cell stage acquired through a default mechanism, *Neuron*, vol. 30, no. 1, pp. 65–78, 2001.

23. L. Conti, S. M. Pollard, T. Gorba, E. Reitano, M. Toselli, G. Biella, Y. Sun, S. Sanzone, Q.-L. Ying, E. Cattaneo, and A. Smith, Niche-independent symmetrical self-renewal of a mammalian tissue stem cell, *PLoS Biol.*, vol. 3, no. 9, p. e283, Sep. 2005.

24. K. Takahashi, K. Tanabe, M. Ohnuki, M. Narita, T. Ichisaka, K. Tomoda, and S. Yamanaka, Induction of pluripotent stem cells from adult human fibroblasts by defined factors, *Cell*, vol. 131, no. 5, pp. 861–872, Nov. 2007.

25. M. Thier, P. Wörsdörfer, Y. B. Lakes, R. Gorris, S. Herms, T. Opitz, D. Seiferling, T. Quandel, P. Hoffmann, M. M. Nöthen, O. Brüstle, and F. Edenhofer, Direct conversion of fibroblasts into stably expandable neural stem cells, *Cell Stem Cell*, vol. 10, no. 4, pp. 473–479, Apr. 2012.

26. O. Lindvall and Z. Kokaia, Stem cells for the treatment of neurological disorders, *Nature*, vol. 441, no. 7097, pp. 1094–1096, Jun. 2006.

27. S. M. Davis and G. A. Donnan, 4.5 hours: The new time window for tissue plasminogen activator in stroke, *Stroke.*, vol. 40, no. 6, pp. 2266–2267, Jun. 2009.

28. R. J. Traystman, Animal models of focal and global cerebral ischemia, *ILAR J.*, vol. 44, no. 2, pp. 85–95, Jan. 2003.

29. G. E. Goings, V. Sahni, and F. G. Szele, Migration patterns of subventricular zone cells in adult mice change after cerebral cortex injury, *Brain Res.*, vol. 996, no. 2, pp. 213–226, Jan. 2004.

30. R. Zhang, Z. Zhang, C. Zhang, L. Zhang, A. Robin, Y. Wang, M. Lu, and M. Chopp, Stroke transiently increases subventricular zone cell

division from asymmetric to symmetric and increases neuronal differentiation in the adult rat, *J. Neurosci.*, vol. 24, no. 25, pp. 5810–5815, Jun. 2004.

31. R. L. Zhang, Z. G. Zhang, L. Zhang, and M. Chopp, Proliferation and differentiation of progenitor cells in the cortex and the subventricular zone in the adult rat after focal cerebral ischemia, *Neuroscience*, vol. 105, no. 1, pp. 33–41, Jan. 2001.
32. K. U. Hunter, C. Yarbrough, J. Ciacci, and K. Urbaniak Hunter, Stem cells in the treatment of stroke, in *Frontiers in Brain Repair*, vol. 671, no. 6 (R. Jandial, ed.), New York, NY: Springer New York, 2010, pp. 105–116.
33. Z. G. Zhang and M. Chopp, Neurorestorative therapies for stroke: Underlying mechanisms and translation to the clinic, *Lancet Neurol.*, vol. 8, no. 5, pp. 491–500, May 2009.
34. F. Locatelli, A. Bersano, E. Ballabio, S. Lanfranconi, D. Papadimitriou, S. Strazzer, N. Bresolin, G. P. Comi, and S. Corti, Stem cell therapy in stroke, *Cell. Mol. life Sci.*, vol. 66, no. 5, pp. 757–772, Mar. 2009.
35. T. Bliss, R. Guzman, M. Daadi, and G. K. Steinberg, Cell transplantation therapy for stroke, *Stroke*, vol. 38, no. 2 Suppl, pp. 817–826, Feb. 2007.
36. K. Oki, J. Tatarishvili, J. Wood, P. Koch, S. Wattananit, Y. Mine, E. Monni, D. Tornero, H. Ahlenius, J. Ladewig, O. Brüstle, O. Lindvall, and Z. Kokaia, Human-induced pluripotent stem cells form functional neurons and improve recovery after grafting in stroke-damaged brain, *Stem Cells*, vol. 30, no. 6, pp. 1120–1133, Jul. 2012.
37. O. Einstein and T. Ben-Hur, The changing face of neural stem cell therapy in neurologic diseases, *Arch. Neurol.*, vol. 65, no. 4, pp. 452–456, Apr. 2008.
38. J. Li, J. Imitola, E. Y. Snyder, and R. L. Sidman, Neural stem cells rescue nervous purkinje neurons by restoring molecular homeostasis of tissue plasminogen activator and downstream targets, *J. Neurosci.*, vol. 26, no. 30, pp. 7839–7848, 2006.
39. N. Kamei, N. Tanaka, Y. Oishi, T. Hamasaki, K. Nakanishi, N. Sakai, and M. Ochi, BDNF, NT-3, and NGF released from transplanted neural progenitor cells promote corticospinal axon growth in organotypic cocultures, *Spine (Phila. Pa. 1976).*, vol. 32, no. 12, pp. 1272–1278, 2007.
40. M. Hoehn and M. Aswendt, Structure-function relationship of cerebral networks in experimental neuroscience: Contribution of magnetic

resonance imaging, *Exp. Neurol.*, vol. 242, no. April, pp. 65–73, Apr. 2013.

41. T. Q. Duong, Multimodal MRI of experimental stroke, *Transl. Stroke Res.*, vol. 3, no. 1, pp. 8–15, Dec. 2011.

42. R. Ouwerkerk, Sodium magnetic resonance imaging: From research to clinical use, *J. Am. Coll. Radiol.*, vol. 4, no. 10, pp. 739–741, Oct. 2007.

43. T. Schaeffter and H. Dahnke, Magnetic resonance imaging and spectroscopy, *Handb. Exp. Pharmacol.*, no. 185 Pt 1, pp. 75–90, Jan. 2008.

44. M. Hoehn, D. Wiedermann, C. Justicia, P. Ramos-Cabrer, K. Kruttwig, T. Farr, and U. Himmelreich, Cell tracking using magnetic resonance imaging, *J. Physiol.*, vol. 584, no. Pt 1, pp. 25–30, Oct. 2007.

45. M. Aswendt, E. Gianolio, G. Pariani, R. Napolitano, F. Fedeli, U. Himmelreich, S. Aime, and M. Hoehn, In vivo imaging of inhibitory, GABAergic neurons by MRI, *Neuroimage*, vol. 62, no. 3, pp. 1685–1693, Sep. 2012.

46. U. Himmelreich, S. Aime, T. Hieronymus, C. Justicia, F. Uggeri, M. Zenke, and M. Hoehn, A responsive MRI contrast agent to monitor functional cell status, *Neuroimage*, vol. 32, no. 3, pp. 1142–1149, Sep. 2006.

47. A. Stroh, C. Faber, T. Neuberger, P. Lorenz, K. Sieland, P. M. Jakob, A. Webb, H. Pilgrimm, R. Schober, E. E. Pohl, and C. Zimmer, In vivo detection limits of magnetically labeled embryonic stem cells in the rat brain using high-field (17.6 T) magnetic resonance imaging, *Neuroimage*, vol. 24, no. 3, pp. 635–645, Feb. 2005.

48. M. Srinivas, A. Heerschap, E. T. Ahrens, C. G. Figdor, and I. J. M. de Vries, (19)F MRI for quantitative in vivo cell tracking, *Trends Biotechnol.*, vol. 28, no. 7, pp. 363–370, Jul. 2010.

49. M. Srinivas, P. Boehm-Sturm, C. G. Figdor, I. J. de Vries, and M. Hoehn, Labeling cells for in vivo tracking using (19)F MRI, *Biomaterials*, vol. 33, no. 34, pp. 8830–8840, Dec. 2012.

50. E. T. Ahrens, R. Flores, H. Xu, and P. A. Morel, In vivo imaging platform for tracking immunotherapeutic cells, *Nat. Biotechnol.*, vol. 23, no. 8, pp. 983–987, Aug. 2005.

51. M. Higuchi, N. Iwata, Y. Matsuba, K. Sato, K. Sasamoto, and T. C. Saido, ^{19}F and 1H MRI detection of amyloid beta plaques in vivo, *Nat. Neurosci.*, vol. 8, no. 4, pp. 527–533, Apr. 2005.

52. U. Flögel, Z. Ding, H. Hardung, S. Jander, G. Reichmann, C. Jacoby, R. Schubert, and J. Schrader, In vivo monitoring of inflammation

after cardiac and cerebral ischemia by fluorine magnetic resonance imaging, *Circulation*, vol. 118, no. 2, pp. 140–148, Jul. 2008.

53. J. Ruiz-Cabello, P. Walczak, D. A. Kedziorek, V. P. Chacko, A. H. Schmieder, S. A. Wickline, G. M. Lanza, and J. W. M. Bulte, In vivo "hot spot" MR imaging of neural stem cells using fluorinated nanoparticles, *Magn. Reson. Med.*, vol. 60, no. 6, pp. 1506–1511, Dec. 2008.
54. P. Boehm-Sturm, L. Mengler, S. Wecker, M. Hoehn, and T. Kallur, In vivo tracking of human neural stem cells with ^{19}F magnetic resonance imaging, *PLoS One*, vol. 6, no. 12, p. e29040, Jan. 2011.
55. J. M. Janjic and E. T. Ahrens, Fluorine-containing nanoemulsions for MRI cell tracking, *Wiley Interdiscip. Rev. Nanomed. Nanobiotechnol.*, vol. 1, no. 5, pp. 492–501, 2009.
56. M. P. Krafft and J. G. Riess, Perfluorocarbons: Life sciences and biomedical uses—Dedicated to the memory of Professor Guy Ourisson, a true RENAISSANCE man, *J. Polym. Sci. Part A Polym. Chem.*, vol. 45, no. 7, pp. 1185–1198, Apr. 2007.
57. O. Castro, A. E. Nesbitt, and D. Lyles, Effect of a perfluorocarbon emulsion (Fluosol-DA) on reticuloendothelial system clearance function, *Am. J. Hematol.*, vol. 16, no. 1, pp. 15–21, Jan. 1984.
58. D. R. Spahn, Blood substitutes. Artificial oxygen carriers: Perfluorocarbon emulsions, *Crit. Care*, vol. 3, no. 5, pp. R93–R97, Jan. 1999.
59. K. C. Lowe, Engineering blood: Synthetic substitutes from fluorinated compounds, *Tissue Eng.*, vol. 9, no. 3, pp. 389–399, Jun. 2003.
60. M. P. Krafft and J. G. Riess, Highly fluorinated amphiphiles and colloidal systems, and their applications in the biomedical field. A contribution, *Biochimie*, vol. 80, no. 5–6, pp. 489–514.
61. S. Mizukami, R. Takikawa, F. Sugihara, Y. Hori, H. Tochio, M. Wälchli, M. Shirakawa, and K. Kikuchi, Paramagnetic relaxation-based ^{19}F MRI probe to detect protease activity, *J. Am. Chem. Soc.*, vol. 130, no. 3, pp. 794–795, Jan. 2008.
62. J. Yu, Z. Ma, Y. Li, K. S. Koeneman, L. Liu, and R. P. Mason, Synthesis and evaluation of a novel gene reporter molecule: Detection of beta-galactosidase activity using ^{19}F NMR of a fluorinated vitamin B6 conjugate+, *Med. Chem.*, vol. 1, no. 3, pp. 255–262, May 2005.
63. Y. Takaoka, T. Sakamoto, S. Tsukiji, M. Narazaki, T. Matsuda, H. Tochio, M. Shirakawa, and I. Hamachi, Self-assembling nanoprobes that display

off/on ^{19}F nuclear magnetic resonance signals for protein detection and imaging, *Nat. Chem.*, vol. 1, no. 7, pp. 557–561, Oct. 2009.

64. W. Du, Z. Xu, A. M. Nyström, K. Zhang, J. R. Leonard, and K. L. Wooley, ^{19}F- and fluorescently labeled micelles as nanoscopic assemblies for chemotherapeutic delivery, *Bioconjug. Chem.*, vol. 19, no. 12, pp. 2492–2498, Dec. 2008.

65. E. Küstermann, U. Himmelreich, K. Kandal, T. Geelen, A. Ketkar, D. Wiedermann, C. Strecker, J. Esser, S. Arnhold, and M. Hoehn, Efficient stem cell labeling for MRI studies, *Contrast Media Mol. Imaging*, vol. 3, no. 1, pp. 27–37, 2008.

66. E. Bible, F. Dell'acqua, B. Solanky, A. Balducci, P. M. Crapo, S. F. Badylak, E. T. Ahrens, and M. Modo, Non-invasive imaging of transplanted human neural stem cells and ECM scaffold remodeling in the stroke-damaged rat brain by ^{19}F-and diffusion-MRI, *Biomaterials*, vol. 33, no. 10, pp. 2858–2871, May 2012.

67. P. Boehm-Sturm, M. Aswendt, A. Minassian, S. Michalk, L. Mengler, J. Adamczak, L. Mezzanotte, C. Löwik, and M. Hoehn, A multi-modality platform to image stem cell graft survival in the naïve and stroke-damaged mouse brain, *Biomaterials*, vol. 35, no. 7, pp. 2218–2226, Feb. 2014.

68. S. M. Cromer Berman, Kshitiz, C. J. Wang, I. Orukari, A. Levchenko, J. W. M. Bulte, and P. Walczak, Cell motility of neural stem cells is reduced after SPIO-labeling, which is mitigated after exocytosis, *Magn. Reson. Med.*, no. 1, pp. 1–8, Feb. 2012.

69. M. Aswendt, J. Adamczak, and A. Tennstaedt, A review of novel optical imaging strategies of the stroke pathology and stem cell therapy in stroke, *Front. Cell. Neurosci.*, vol. 8, no. August, pp. 1–10, Aug. 2014.

70. J. M. Janjic, M. Srinivas, D. K. K. Kadayakkara, and E. T. Ahrens, Self-delivering nanoemulsions for dual fluorine-19 MRI and fluorescence detection, *J. Am. Chem. Soc.*, vol. 130, no. 9, pp. 2832–2841, Mar. 2008.

71. M. Srinivas, M. S. Turner, J. M. Janjic, P. A. Morel, D. H. Laidlaw, and E. T. Ahrens, In vivo cytometry of antigen-specific t cells using ^{19}F MRI, *Magn. Reson. Med.*, vol. 62, no. 3, pp. 747–753, Oct. 2009.

72. U. Bommerich, T. Trantzschel, S. Mulla-Osman, G. Buntkowsky, J. Bargon, and J. Bernarding, Hyperpolarized ^{19}F MRI: Parahydrogen-induced polarization and field variation enable ^{19}F MRI at low spin density, *Phys. Chem. Chem. Phys.*, vol. 12, no. 35, pp. 10309–10312, 2010.

73. M. Srinivas, P. Boehm-Sturm, M. Aswendt, E. D. Pracht, C. G. Figdor, I. J. de Vries, and M. Hoehn, In vivo ^{19}F MRI for cell tracking, *J. Vis. Exp.*, no. 81, p. e50802, Jan. 2013.

74. P. Boehm-Sturm, E. D. Pracht, M. Aswendt, N. Henn, and M. Hoehn, B1 correction for quantitative in vivo ^{19}F magnetic resonance imaging with surface coils, *Proc. Int. Soc. Magn. Reson. Med.*, 2012. Melbourne, Australia.

Part 5
Pharmacology

Chapter 11

Fluorinated Natural Compounds and Synthetic Drugs

Thoralf Niendorf, Yiyi Ji, and Sonia Waiczies

Berlin Ultrahigh Field Facility, Max Delbruck Centre for Molecular Medicine, Robert Rössle Str. 10, Berlin 13353, Germany

thoralf.niendorf@mdc-berlin.de

11.1 Introduction

The driving force behind organofluorine chemistry is that carbon–fluorine bonds impart unique characteristics to the molecules they form. Although not common in nature, organofluorines are encountered on a daily basis in items spanning from refrigerants and household cooking pans, through weatherproof clothing and film-forming foams, to aeronautic-compatible lubricants and medicinal products. Organofluorine chemistry has helped shape some of the remarkable achievements of the 20th century. Amidst the developments in the medical field are applications of organofluorine compounds as oxygen–carrying blood reconstitution products, blockbuster drugs, revolutionary anesthetics, as well as imaging agents. In this chapter, we describe the advantages of fluorination for the improving pharmacodynamics and pharmacokinetics of medicinal products. Furthermore, the

Fluorine Magnetic Resonance Imaging
Edited by Ulrich Flögel and Eric Ahrens

ISBN 978-981-4745-31-4 (Hardcover), 978-981-4745-32-1 (eBook)
www.panstanford.com

opportunities arising when introducing fluorine (^{19}F) atoms to lead medicinal compounds with regard to their quantification in vivo (bioavailability) will be discussed and non-invasive $^{19}F/^{1}H$ magnetic resonance spectroscopy (MRS) and imaging (MRI) techniques for monitoring the fate of ^{19}F-containing drugs will be described.

11.2 Organofluorine Compounds

The origins of organic ^{19}F chemistry date back to the 1890s with the pioneering work by Frédéric Swarts, who discovered chlorine-fluorine halogen-exchange (HALEX) upon mixing organochlorine compounds with inorganic fluorides; this substitution is known as the Swarts reaction [1]. The first 40 years of discoveries by Swarts paved the way for technological breakthroughs in the early 20th century in the field of aliphatic ^{19}F chemistry, including, for example, the synthesis of chlorofluorocarbon refrigerants (CFCs, Freons®) [2] as well as multi-purpose polytetrafluoroethylene (Teflon®, Goretex®) and perfluoro-octanesulfonamide (Scotchgard®) compounds [3]. Some of the organohalogens, especially chlorofluorocarbons (CFCs), DDT (dichloro-diphenyltrichloroethane), and dioxins, were found to be persistent pollutants in our environment, and due to their impact on nature (including ozone depletion) were listed as organic pollutants by the Stockholm Convention [4]. In some cases they were replaced by alternative organofluorines with shorter lifetimes in vivo and within the atmosphere [5].

11.2.1 Naturally Occurring Organofluorine Compounds

Halogen-containing organic compounds such as organofluorines are often believed to be solely the products of an industrial society. However, large proportions of organohalogens also occur in nature [6]. Although most of the naturally occurring organohalogens were unknown upon their discovery, a significant number were found to be identical to synthetic organohalogens, e.g., chloroform, chlorophenols, dioxins, CFCs. By using halogen salts and organic matter, nature appears to produce halogenated compounds similarly to those synthesized in the lab; some of these compounds actually predate the beginning of life on Earth [6].

Natural organohalogens may be formed abiotically as a result of geothermal reactions in forest fires and volcanoes. Mount Etna, an active volcano on the east coast of Sicily, is one of the largest sources of hydrogen fluoride (HF), an important source for organofluorine synthesis. Mount Etna releases 2.2 kg of HF per second [7]. Organohalogens may also be produced by living organisms (microorganisms, plants, marine organisms, higher animals) usually as part of a survival mechanism for the species; for example, the poison-arrow frog *Epipedobates tricolor* secretes the organochlorine epibatidine, a potent poisonous alkaloid [8].

Although fluorine is the most abundant halogen in the earth's crust, organofluorines are the least abundant organohalogens found in nature [9]. Nonetheless, a number of microorganisms, plants and marine invertebrates have been identified as being natural sources of organofluorines [10]. Fluoroacetate (Fig. 1.1) was the first organofluorine compound identified; it was isolated from the shrub *Dichapetalum cymosum* indigenous to South Africa [11] but it is ubiquitous to several other plants (>40 species) all over the world, particularly in Africa, Australia and South America [12]. Fluoroacetate is highly toxic to animals since it converts to fluorocitrate [13] which is a potent inhibitor of citrate synthesis within the energy-producing Krebs cycle [14]. The biosynthetic fluorination pathway leading to fluoroacetate involves a very exceptional enzymatic process; the enzyme fluorinase (S-adenosyl-L-methionine-fluoride adenosyltransferase) catalyses the conversion of fluoride and S-adenosylmethionine to 5′-fluoro-5′-deoxyfluoroadenosine [15, 16]. Other toxic organofluorines include the ω-fluorinated fatty acids that were isolated from the African shrub, *Dichapetalum toxicarium* [17]; originally ω-fluorooleic acid ($C_{18:1F}$) and ω-fluoropalmitic acid ($C_{16:0F}$) were identified [17], although eventually other ω-substituted fluoro acids [18] as well as threo-18-fluoro-9,10-dihydroxystearic acid [19] were isolated. The toxicity of even numbered ω-fluorofatty acids has been attributed to their conversion into fluoroacetate via β-oxidation [20].

Nucleocidin (Fig. 11.1), an adenosine-like nucleoside, was the first naturally occurring organofluorine compound identified that displayed medicinal properties; it exhibited a broad spectrum of antibiotic activity but was too toxic for clinical applications.

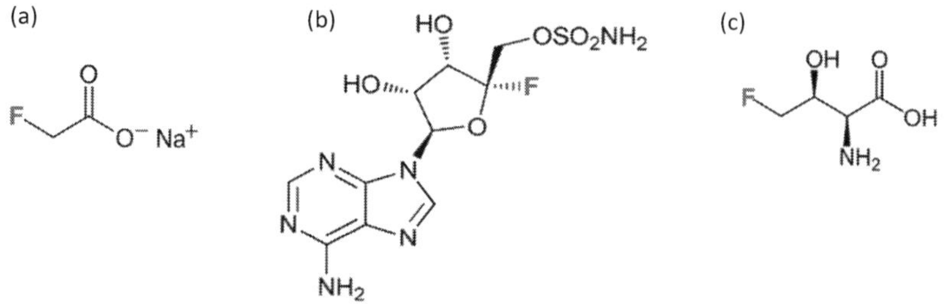

Figure 11.1 Natural organofluorine products. (a) fluoroacetate (b) nucleocidin (c) 4-fluorothreonine.

Nucleocidin was isolated from *Streptomyces calvus*, a microorganism derived from an Indian soil sample [21]. The ^{19}F atom was identified within its empirical formula only several years after its isolation [22]. Plants, such as *Acacia georginae*, also take up fluorides to synthesize volatile fluoroorganics [23], including fluoroacetone [24]. Apart from fluoroacetate, other fluorometabolites appear to be synthesized in nature in the presence of fluorides. The bacterial strain *Streptomyces cattleya*, which is important for the synthesis of the β-lactam antibiotic thienamycin, synthesizes 4-fluorothreonine [25], as confirmed by asymmetric stereoselective synthesis [26] and ^{19}F MRS [27]. Fluorinated natural products have also been isolated from marine organisms. Initial studies with the sponge *Phakellia fusca*, collected from the Xisha Islands in the South China Sea, identified 5-fluorouracil alkaloids, some of which have been synthesized as antitumor compounds [28].

Since halogenation in the lab is hazardous because it requires elemental halogens, other alternatives have been sought. The idea of enzymatic halogenation as a source of future biocatalysis was eventually developed [29]. The concept evolved following reports of an enzymatic reaction in the marine microorganism *Streptomyces cattleya* catalyzing the biosynthesis of fluoroacetate in fermentation broths [25] via the action of fluorinase on fluoride and S-adenosylmethionine [15, 16].

11.2.2 Redesign and Scale-Up of Natural Synthesis

At the turn of this century, genetic engineering of natural biocatalysts has emerged as a method to access libraries of structurally related natural products. This biotechnology development offered a feasible alternative to complex and costly schemes for the synthesis of natural organic product analogs [30]. Application of biosynthetic mutants has been a powerful strategy for scaling up production of organofluorine compounds, including alkaloid derivatives and complex aminocyclitol natural products [31]. Using a mutasynthetic approach, several fluorinated tropane alkaloids could be produced in the lab by feeding fluorinated phenyllactic acid analogs to transformed *Datura stramonium* root cultures [32]. There is a growing interest in studying nature's large and yet unexplored diversity in order to identify unique enzymes that can catalyze organofluorine reactions for biotechnology products [29]. Marine sources of organofluorine enzymes will be likely candidates because of the extreme conditions (pressure, temperature, salinity) that are not encountered in terrestrial habitats [29]. In particular, coral reef ecosystems are likely targets for enzyme discovery due to the enormous biodiversity that includes corals, sponges, algae, and bacteria. In the future it is also highly likely that more biogenic and abiogenic organofluorines will be identified on this planet and outer space. Since hydrogen fluoride has also been detected in interstellar space [33], there is also an increased possibility that organofluorine compounds will be amidst the organic chemicals found in deep space.

11.2.3 Organofluorine Synthesis

Considering the low availability of naturally occurring organofluorine compounds, development of the field of organofluorine synthesis has been essential. Organofluorine compounds can be synthesized either by (i) nucleophilic fluorination using negatively charged fluoride ions or by (ii) electrophilic fluorination using elemental fluorine or reagents containing nitrogen–fluorine (N–F) or oxygen–fluorine (O–F) bonds. As mentioned above, fluorination chemistry effectively had its beginning near the end of the 19th century with the crucial discoveries of Frédéric Swarts concerning indirect HALEX fluorination to convert chlorocarbons to chlorofluorocarbons and fluorocarbons using SbF3 [1]. Nucleophilic fluorination has been commonly used to synthesize organofluorine compounds and typically involves application of fluoride ions. Towards the middle of the 20th century, an electrochemical fluorination method was developed (Simons process), resulting in replacement of hydrogen by fluorine atoms (perfluorination) [34–36]. Incidentally these methods were carried out only a few years following Plunkett's discovery of tetrafluoroethylene polymers (Teflon®) [3].

Another well-established fluorination method involves direct fluorination using elemental fluorine (F_2). However, due to the massive reactivity of F_2, direct fluorination of organic compounds is uncommon [37]. On the other hand, direct fluorination methods were used to synthesize perfluorocarbons (PFCs) in the vapor phase [38], and these PFCs were eventually used in the Manhattan Project due to their compatibility with uranium hexafluoride. More than a decade later, highly turbulent jets of diluted organic reactant were mixed with F_2 in the presence of circulating reactor gases [39]. Half a century later a method of pumping F_2 through mini-reactors was introduced to safely fluorinate organic compounds [40]. Methods of electrophilic fluorination other than those employing F_2 gas are also commonly employed to form C–F bonds. Sources of electrophilic fluorine can be obtained from compounds such as hypofluorites [41, 42], fluoroxy compounds [43] or N-fluoropyridinium salts [44]. The latter as well as N-fluorobenzenesulfonimides are largely used in the preparation of fluorinated glucocorticosteroids [45, 46].

11.2.4 Advantages of Incorporating Fluorine to Bioactive Molecules

Introduction of ^{19}F atoms to lead organocompounds affects their physicochemical characteristics and as a result their pharmacological properties [47]. Fluorine is the most electronegative atom (Pauling electronegativity = 3.98 [48]) and the smallest substituent that can replace hydrogen in a molecule (van der Waals radius of 1.47 Å [49]). Although the size of the fluorine atom is closer to oxygen (1.52 Å) than hydrogen (1.20 Å), the conformational effects (steric bulkiness) of fluorinated substituents closely mimic those of hydrogen atoms; despite slight differences in size, they add only a limited extra steric demand at receptor sites [50]. Many fluorinated molecules are in fact recognized by macromolecular receptor sites as the natural non-fluorinated analogue [51–53]. Steric size is determined from the A value, which is defined as the free energy difference between axial and equatorial substitution [54]. The conformational equilibria of the different fluorinated substituents can be studied using dynamic ^{19}F MR spectroscopy [50]. Since the ^{19}F atom has the highest electronegativity within the periodic table, fluorinated substituents have strong electron-withdrawing properties. Of note, fluorine and fluorinated substituents, e.g., $-CF_3$ groups, have similarly high relative electron-withdrawing power [55]. The specific properties of ^{19}F, i.e., electronegativity and size, alter the physicochemical properties of fluorinated drug molecules, ultimately influencing their pharmacokinetics and pharmacodynamics.

11.2.4.1 Changes in polarity

Fluorination increases the polarity of organic compounds, and this increases their interaction and binding affinity to their substrates. Single C–F substituents, particularly on an aromatic ring, can exhibit polarity influences that promote binding of the modified drug to its substrate [56]. For instance, a single fluorine-substitution in thrombin inhibitors resulted in fivefold increase in binding to thrombin [57]. Historically, introduction of a fluorine atom to promote therapeutic efficacy was first reported with glucocorticosteroids in 1954 [58]. Introduction of fluorine at the 9α-position of the natural hydrocortisone increased the binding affinity to glucocorticoid receptors and retarded the oxidation of the proximal 11-OH group, thereby improving therapeutic efficacy (anti-inflammatory activity) by an order of magnitude [58].

11.2.4.2 Influence on lipophilicity

The introduction of one or more fluorine atoms into a compound also alters its lipophilicity [52, 59], which plays a key role for absorption, transport and receptor binding of bioactive compounds. Lipophilicity is increased following aromatic fluorination, poly/perfluorination and fluorination in the vicinity of basic nitrogen atoms, since fluorine reduces the basicity of the nitrogen functionality [52, 59], which facilitates the passage of the drugs through biological membranes. As a result of increased lipophilicity and blood–brain barrier penetration, the fluorinated forms of pyridinium oximes were more therapeutic at reactivating acetylcholinesterase (AChE) levels that were depleted following exposure to life-threatening paraoxon organophosphorus pesticides [60]. However, excessive lipophilicity also results in reduced water solubility leading to incomplete absorption following oral administration [56, 59, 61]. In some cases, fluorination decreases lipophilicity, e.g., following monofluorination or trifluoromethylation of saturated alkyl groups or following single hydrogen/fluorine substitution in close proximity to oxygen [52, 59].

11.2.4.3 Changes in the acid dissociation constant (pK_a)

Due to fluorine's strong electronegativity, fluorinated substituents also give rise to strong influences on acids and bases. Changes in pK_a depend on the position of the fluorine substituent relative to the acidic or basic group [62]. A change in pK_a often has a strong effect on both the pharmacokinetics and binding affinity of a molecule [63]. This can be challenging when the increase in basicity introduced by a substituent to enhance binding affinity results in low bioavailability due to the compound's limited ability to pass through membranes. For each lead compound, the level and extent of fluorine substitution needs to be determined in order to achieve the ideal pK_a value. For example, monofluorination of 3-(3-(piperidin-1-yl)propyl)indoles results in a sufficient reduction in pK_a to increase the bioavailability of potent 5-HT(hydroxytryptamine)1D receptor ligands without significantly reducing the binding affinity and potency of the compound [64]. Difluorination, however, reduces the basicity of the piperidines beyond a critical threshold, at which point binding to the receptors is compromised [64].

11.2.4.4 Influence on metabolic stability

Fluorination also improves the metabolic stability of pharmacological compounds. Follow drug administration, several mechanisms contribute to metabolism, degradation, and/or elimination of these compounds, which include rapid oxidative processes by liver enzymes, e.g., cytochrome P450 enzymes. Low metabolic stability limits the bioavailability of the compound. One strategy of promoting stability is to block the metabolically labile site with a fluorine substituent without impairing binding to the target protein [65]. By introducing fluorine in the para position of the lead structure of potent cholesterol absorption inhibitors, metabolic attack via benzylic hydroxylation was prevented [66]. The higher dissociation energy of a C–F bond (105.4 kcal/mol) compared to the C–H bond (98.8 kcal/mol) gives an enhanced oxidative and thermal stability to fluorinated analogues [51, 56, 59]. Fluorination can also protect a molecule from hydrolytic metabolism, by effecting reaction rates and stability of the intermediates. This is particularly important for oral administration of drugs that are sensitive to acidic media on the stomach [67, 68].

Taken together, fluorination alters the physicochemical behavior of organic compounds and is in fact a commonly employed process in medicinal chemistry and drug design to improve absorption, metabolic stability, bioavailability, and binding affinity [61, 69, 70].

11.2.5 Organofluorine Compounds in Medicinal Chemistry

Medicinal chemistry developed during a similar epoch as organofluorine chemistry but it was not until the late 1940s that both fields crossed paths. It was indeed after half a century of Swarts' first works [71] that ^{19}F chemistry started being implemented for pharmaceutical purposes. The notion of medicinal chemistry evolved from the time of Ehrlich, when as a medical student he first postulated that molecules will specifically bind to cell receptors via a lock-key mechanism to selectively target microorganisms or transformed cancerous cell-types and not host cells [72]. Half a century later the first fluorinated drugs were synthesized, including the glucocorticosteroidal anti-inflammatory drug fluorocortisone [58] and the antineoplastic agent 5-fluorouracil (5-FU) [73]. Both of these drugs are still listed as essential medicines by the World Health Organization [74]. 5-FU alkaloids were incidentally isolated from marine organisms nearly 50 years later after their first synthesis [28]. The first studies by Heidelberger and colleagues on 5-FU [73] were, however, motivated by previous observations showing the profound biological effects obtained when fluorine was substituted for hydrogen in several classes of compounds (e.g., fluoroacetate [13], carcinogenic fluoroamines [75], fluorinated glucocorticosteroids [58, 76]) and to some extent anti-cancer nucleic acid analogues [77].

11.2.5.1 Fluorine in the pharmaceutical industry

The isolation of naturally occurring organofluorine compounds with medicinal and toxic properties [21, 22], as well as the first syntheses of pharmacological fluorine compounds [73, 78], instigated further efforts in incorporating ^{19}F into medicinal chemistry. Within the decades following the first synthesis of fluorinated glucocorticosteroids [78] and anti-cancer 5-FU [73], several ^{19}F-containing pharmaceutical compounds have been synthesized for medicinal purposes. Major developments include incorporation of ^{19}F into inhalation anesthesia, anti-inflammatory compounds (steroidal and non-steroidal), CNS drugs, as well as anticancer medications. Synthesis of organofluorine compounds has been of particular importance in the development of enzymatic inhibitors ultimately leading to protease inhibitors for the treatment of human immunodeficiency virus (HIV). The first use of fluorinated aldehydes and ketones as inhibitors of hydrolases demonstrated inhibition of acetylcholinesterase [79]. Subsequently, fluoroketones were shown to inhibit proteases operating by widely varying mechanisms [80, 81]. These earlier achievements led to the development of further protease inhibitors [82–84] and subsequently to a new class of structurally complex fluorinated inhibitors of HIV proteases [85].

Today, ^{19}F continues to be an important element during lead optimization in drug discovery. It is estimated that up to 20% of pharmaceuticals prescribed contain a fluorine atom. In 2010 it was reported that 30% of the foremost 30 blockbuster drugs by sales (according to U.S. Pharmaceutical Sales) contained a fluorine atom [86]. From statistics accumulated from the beginning of 2014 this trend decreased to 23% of the leading 30 blockbuster drugs (Table 11.1), probably due to the steady increase in sales for biologicals, such as fusion proteins and monoclonal antibodies, between 2010 and 2014 [87].

Table 11.1 List of fluorinated drugs within the top 100 list of prescribed drugs in US (update 2014) [87]

Drug group	Active ingredient	Activity	Structure	^{19}F atoms
Lipid-lowering drugs	Rosuvastatin	HMG-CoA reductase inhibitor	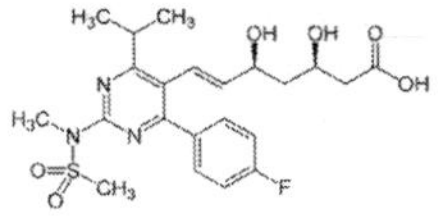	1
	Ezetimibe	Niemann-Pick C1-Like 1 protein inhibitor	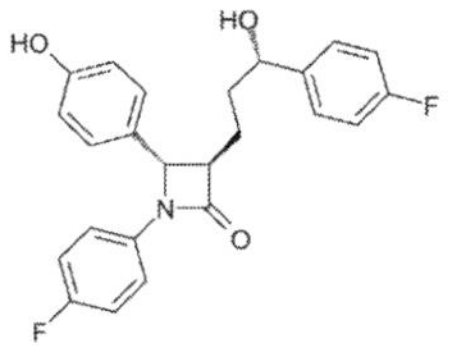	2
Anti-hyperglycemic drugs	Sitagliptin	Dipeptidyl peptidase-4 (DPP-4) inhibitor	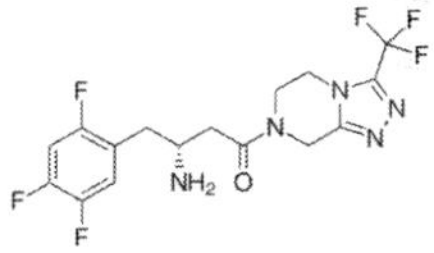	6
Calcimimetic drugs	Cinacalcet	Allosteric activator of calcium-sensing receptor	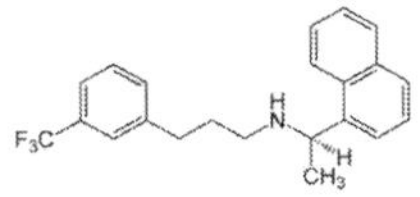	3
Anti-psychotic drugs	Paliperidone	Dopamine antagonist 5-HT2A antagonist	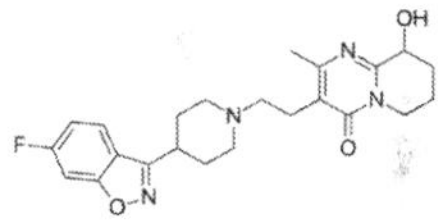	1
Anti-acid drugs	Dexlanso-prazole	(H^+,K^+)-ATPase proton pump inhibitor	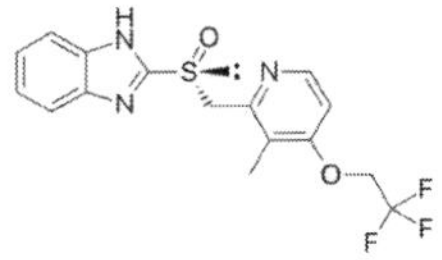	3
Anti-inflammatory drugs	Fluticasone	Glucocortico-steroid	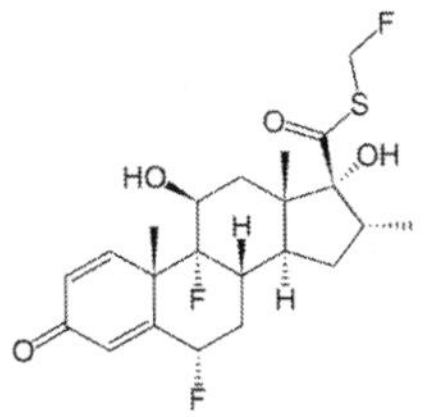	3

(*Continued*)

Table 11.1 (*Continued*)

Drug group	Active ingredient	Activity	Structure	^{19}F atoms
	Celecoxib	Cyclooxygenase-2 selective non-steroidal anti-inflammatory drug	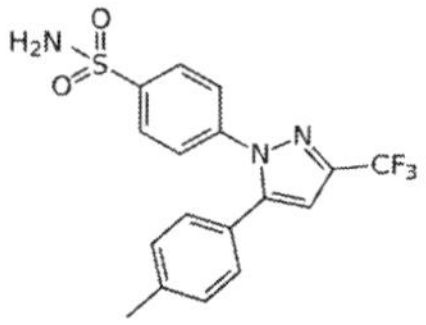	3
Anti-retroviral drugs	Emtricitabine	Nucleoside analogue reverse transcriptase inhibitor	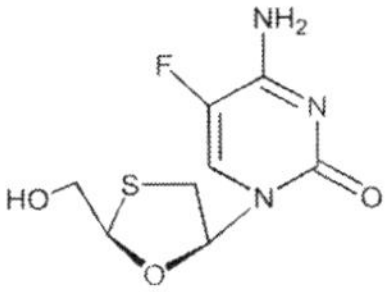	1
	Efavirenz	Non-nucleoside reverse transcriptase inhibitor	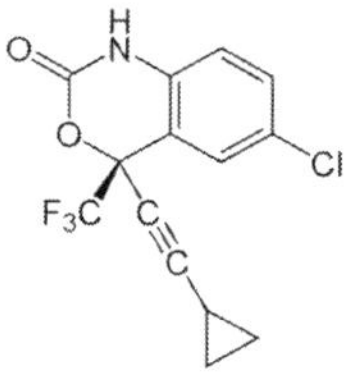	3
	Raltegravir	Integrase inhibitor	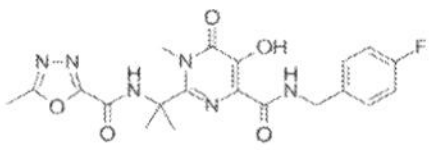	3
	Elvitegravir	Integrase inhibitor	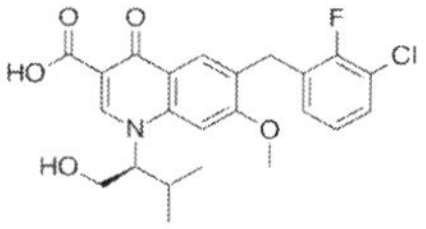	1
Chemo-therapeutic agents	Capecitabine	5-fluorouracil prodrug	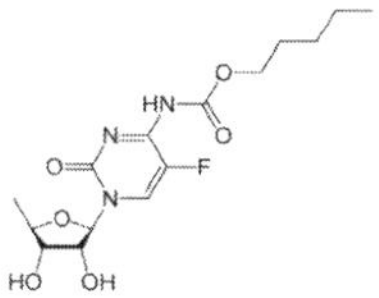	1

11.3 Fluorine MR-Based Spectroscopy

High-resolution MRS has been invaluable for delivering structural information regarding synthetic and natural compounds in medicinal chemistry. ^{19}F is among the most studied nuclei in MRS because of its unique properties and the variety of molecules containing fluorine. The lack of background signal as a result of the scarcity of fluorine in the body as well as the excellent sensitivity to changes in the microenvironment have been exploited to design and apply diverse reporter molecules and pharmacological designs. ^{19}F has the advantage of 100 percent abundance and a gyromagnetic ratio and sensitivity similar to ^{1}H. The chemical shift range for ^{19}F is very large compared to that of hydrogen; organofluorine compounds span a range of over 350 ppm [88]. Thus resonances of multiple ^{19}F nuclei within organofluorine compounds can be easily separated even under rather inhomogenous magnetic field conditions, and the spectra are usually first order [89].

At the turn of the century, ^{19}F MRS emerged as a screening tool to facilitate structure-activity relationships during drug discovery [90]. ^{19}F MRS has evolved into a versatile, reliable, and efficient tool for performing biochemical binding assays. The lone pair electrons on ^{19}F outermost shell can participate in non-bonded interactions with the local environment, rendering ^{19}F chemical shifts extremely sensitive to even very small changes in attractive intermolecular forces such as van der Waals contacts in proteins [91]. The geometrical complexity of molecular structures gives rise to changes in local magnetic fields. These alterations in-turn result in a chemical shift (δ) of the frequency of NMR transitions between molecules [92]. Chemical shift is typically expressed in parts-per-million (ppm) and is defined as the difference between the resonance frequencies of the nuclei of interest and that of a reference compound divided by the operating frequency of the NMR spectrometer. Fluorotrichloromethane ($CFCl_3$) is typically employed as internal reference, where its resonance frequency set to zero ppm, to measure the chemical shift of ^{19}F-containing active compounds [89]. ^{19}F MRS has been used to monitor duplex and triplex formation during DNA conformational changes [93]. For this, the alkaloid 5-fluorodeoxyuridine (5-FU) was used as signal transmitter (^{19}F NMR reporter) via its incorporation into oligodeoxynucleotides (ODN) (Fig. 11.2a). Single stranded ODNs containing the ^{19}F reporter display several resonances at −165.7 ppm due to the formation of higher order structures (Fig. 11.2b). Addition of a non-fluorinated ODN duplex results in the appearance of a new signal at −166.9 ppm attributed to the triplex formation and its disappearance at 50°C (Fig. 11.2b). ^{19}F NMR studies have also been performed to study protein structure and function.

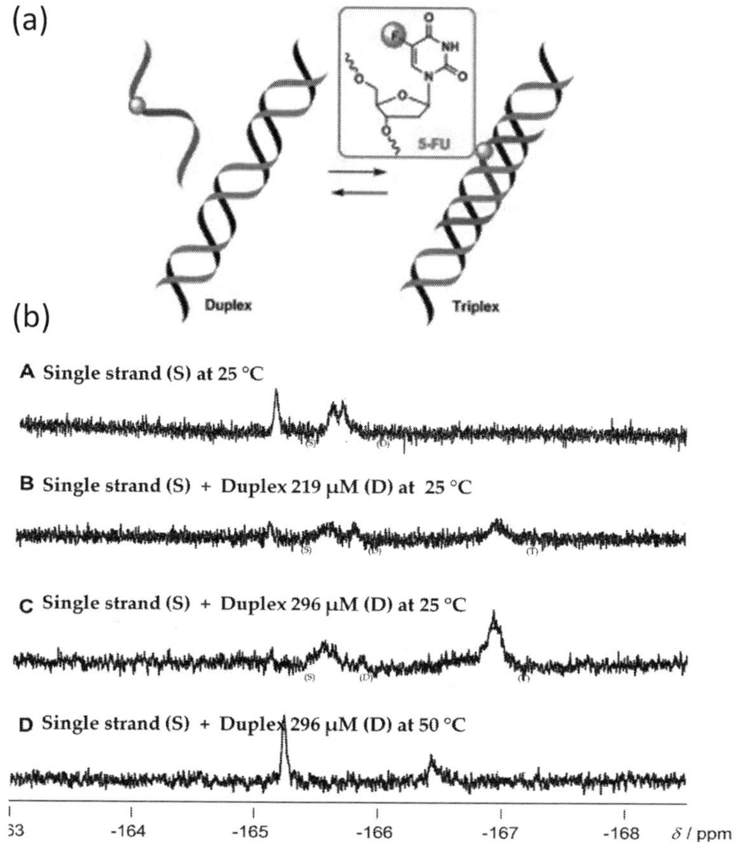

Figure 11.2 Incorporation of the alkaloid 5-FU as ^{19}F reporter to study DNA conformational changes. (a) Sketch showing incorporation of 5-FU labeled single strand (blue) with DNA duplex strand to form triplex. (b) ^{19}F NMR spectra of single stranded (S) oligodeoxynucleotide (upper); single strand (S) in the presence of lower concentration DNA duplex at 25°C (middle) and single strand (S) in the presence of higher concentration DNA duplex (lower) all at 25°C. Reprinted with permission from Ref. 93.

In these applications, the protein needs to be ^{19}F-labeled; one method is to incorporate ^{19}F-labeled amino acids to microbial cultures using orthogonal tRNA/aminoacyl-tRNA synthetase to achieve site-specific incorporation in the target protein [94]. In a later study, a conformationally restricted amino acid was designed as a ^{19}F label to overcome electronic and steric perturbations commonly observed in CF_3 analogies of serine and threonine [95, 96]. Using site-specific ^{19}F NMR, protein labeled with ^{19}F can be studied in complexes with various pharmacological ligands; chemical shifts of strategically located ^{19}F labels in the cytoplasmic region of the β2-adrenergic receptor (b2AR) studied in association with β2 agonists and antagonists revealed that agonist binding primarily shifts the equilibrium toward the G protein–specific active state of the transmembrane helix VI of the receptor [97]. ^{19}F compounds have also been developed to investigate physiological changes (gene activity, pH, ion concentrations, oxygen tension, hypoxia, vascular flow and volume) in vivo [98].

11.3.1 Pharmacokinetic Studies Employing ^{19}F MR Spectroscopy

MRS is also an important tool to investigate the pharmacological profile of drugs and their byproducts in excised tissue samples. Liquid chromatography/mass spectrometry (LC/MS) is another method of choice to obtain quantitative and qualitative information for new drug compounds in fecal and urine samples. The main disadvantage of MS for studying drug pharmacokinetics, however, is the interference from endogenous components that leads to ion suppression as well as the requirement of metabolite standards that are not necessarily available during the early stages of drug discovery. Indeed, studies showed that ^{19}F MRS could provide more information regarding identification and quantification of fluorine-containing compounds and metabolites following administration (Fig. 11.3) [99].

^{19}F MRS of urine samples from rats, mice and humans treated with 5-FU prodrug capecitabine (CAP, N4-pentyloxycarbonyl-5′-deoxy-5-fluorocytidine) identified new metabolites (e.g., 5-fluorocytosine, 5-fluoro-6-hydroxycytosine, fluoride ion, 2-fluoro-3-hydroxypropionic acid and fluoroacetate) that were previously not identified by liquid chromatography [99]. Shown in Fig. 11.3 are the metabolites detected in the urine of patients orally administered with a daily dose of 1.9 g of CAP [99].

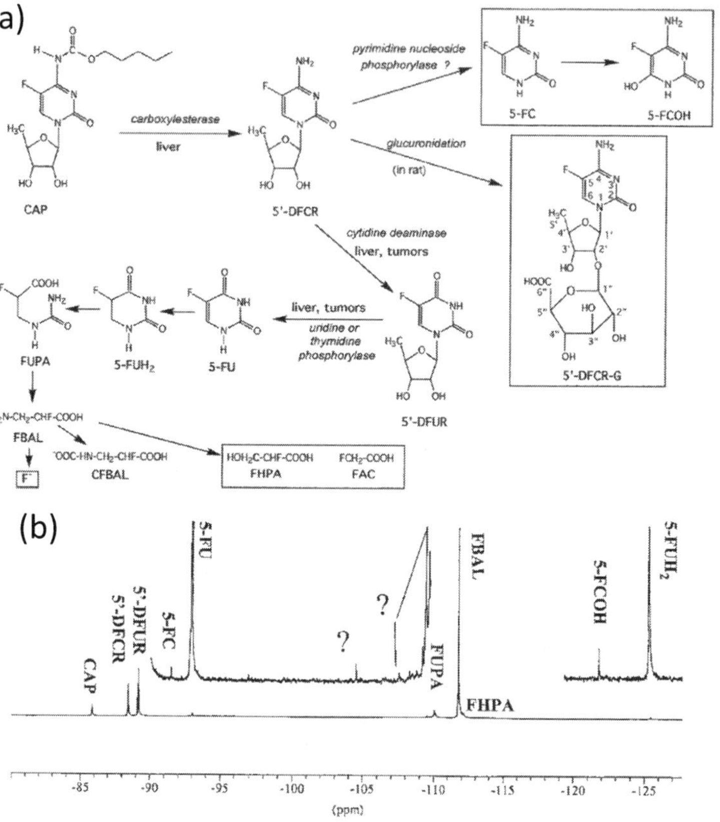

Figure 11.3 Metabolic fate of the oral 5-FU pro-drug capecitabine (CAP). (a) Metabolic pathway of capecitabine. Metabolites of capecitabine that had not been previously identified by liquid chromatography are shown in boxes (5-FC: 5-fluorocytosine, 5-FCOH: 5-fluoro-6-hydroxycytosine, FHPA: 2-fluoro-3-hydroxypropionic acid, FAC: fluoroacetate). (b) Typical ^{1}H-decoupled ^{19}F NMR spectra of urine samples collected from patients 12 h after the first daily dose of CAP. Reprinted with permission from Ref. 99.

11.3.2 Methods of Studying ^{19}F Drugs in vivo

In vivo drug pharmacokinetics have been typically investigated using radiolabeling (e.g., with ^{14}C or ^{18}F labels) in combination with positron emission tomography (PET). Already from first PET studies for ^{18}F-labeled 5-FU it was possible to determine concentrations of drug and metabolites from the ^{18}F tracer uptake in different regions, e.g., vasculature, tumor tissue and liver [100–102]. The activity of the radiotracer is measured as standardized uptake values (SUV) (Eq. 11.1).

$$SUV = \frac{[\text{radiotracer}]_{\text{tissue}}\ (\text{kBq/g})}{\text{injected radiotracer dose (kBq)}} \times \text{body weight (g)} \quad (11.1)$$

Apart from a restricted observation window in PET studies due to the short half-life of radiotracers (e.g., ^{18}F $t_{1/2}$ = 110 min), one drawback of radiolabeling pharmaceuticals is that no distinction is made between the drug applied and metabolites generated from it, since the total ^{18}F pool is measured. Furthermore, radiolabeled compounds are often not immediately available in early-stage drug discovery due to the high cost and lengthy synthesis. These limitations can be overcome by employing ^{19}F MRS to study the fate of ^{19}F-containing drugs in vivo.

11.3.2.1 From in vitro to animal and human ^{19}F MRS studies

The MR properties of fluorine are extremely favorable for the purpose of studying the fate of fluorinated drugs in vivo. Through physicochemical parameters such as chemical shift, scalar coupling, chemical exchange and relaxation processes, the fate and function of ^{19}F-containing drugs (e.g., 5-fluorouracil, fluoxetine, gemcitabine) and anesthetics (e.g., isoflurane, methoxyflurane) can be studied by ^{19}F MRS. ^{19}F MRS studies were initially performed in tumor cell cultures [103, 104], in perfused liver [105] and in animal models [106–108] to resolve 5-FU and its fluoronucleotide metabolites. At around the same time, in vivo ^{19}F MRS studies were also carried out with fluorinated ether anesthetics (methoxyflurane, isoflurane, enflurane, fluroxene, and hexafluoroethane) in the rat brain; the anesthetic potency was shown to correlate to ^{19}F spin-spin relaxation times probably due to a relationship between drug immobilization and drug affinity within a specific chemical environment [109]. More recently ^{19}F MRS was performed in the eye of New Zealand white rabbits to study the elimination half-lives and clearance of the corticosteroid triamcinolone acetonide phosphate (TAP) following intravitreal and subconjunctival injections [110]. This study demonstrates a much higher rate of drug signal decay after subconjunctival compared to intravitreal injection in vivo (1.4 vs. 0.09 h^{-1}) and underscores the utility of ^{19}F MRS in studying fluid dynamics and blood circulation of the eye in association with pharmacokinetic studies [110].

Following the first studies in animals [107, 108], in vivo ^{19}F MRS studies were also extended to patients receiving fluorinated drugs. In patients receiving 5-FU as chemotherapy, signals corresponding to 5-FU and the metabolite fluoro-β-alanine (FBal) were detected in the liver [111]. A number of fluorinated psychoactive drugs have also been investigated in vivo by ^{19}F MRS; these compounds are rich in fluorine atoms and are expected to provide a high signal-to-noise ratio (SNR) (Fig. 11.4). Psychotropic compounds such as fluphenazine, fluoxetine and fluvoxamine (Fig. 11.4) are characterized by a trifluoromethyl group that is an important structural component of many bioactive compounds due to its polar influence and effect on lipophilicity [89]. Previously ^{19}F spectra of the neuroleptic fluphenazine could be acquired in the frontal lobe, basal ganglia and occipital lobe of a chronic schizophrenic patient using a surface coil (10 cm in diameter) [112]. Brain pharmacokinetics and tissue distribution of fluvoxamine and fluoxetine were more recently carried out using ^{19}F chemical shift imaging (CSI) at a magnetic field strength of B_0 = 3 T [113]. The metabolism of the 5-FU pro-drug capecitabine [114] and the non-steroidal anti-inflammatory drug (NSAID) niflumic acid [115] was also investigated using in vivo ^{19}F MRS of the liver at 3 T.

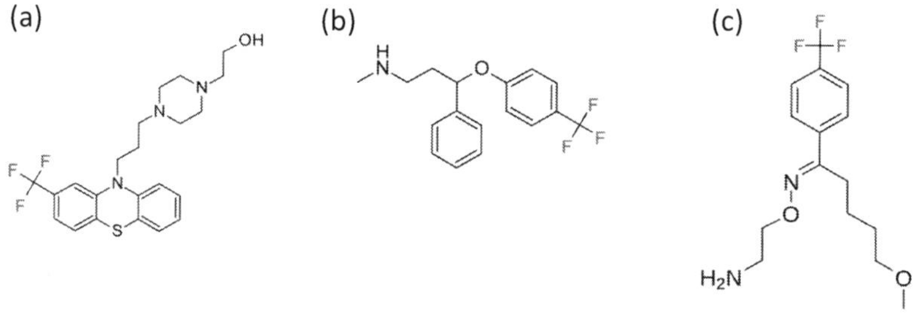

Figure 11.4 Chemical structure of fluorinated psychotropic medications: (a) Fluphenazine, (b) fluoxetine, and (c) fluvoxamine.

11.3.2.2 ^{19}F MR imaging studies of fluorinated drugs

^{19}F MR represents a very useful tool not only for spectroscopic applications, but also for imaging purposes, in order to determine the metabolic fate of ^{19}F drugs in vivo. For studying the levels of fluvoxamine and fluoxetine after drug withdrawal, ^{19}F MRS and CSI at 3 T [116] was used and resulted in the unexpected finding that drugs or possibly closely related metabolites remain trapped in some tissue several days (up to 100 days) after stopping treatment [116]. However, a limitation of ^{19}F CSI at lower field strengths ($B_0 \leq 7$ T) is that sufficient SNR can only be achieved at low spatial resolution within a reasonable acquisition time. The partial volume effect occurs in imaging when the region of interest is smaller than three times the full-width at half maximum (FWHM) of the reconstructed image resolution [117]. Furthermore CSI requires advanced MR sequences such as the sine-squared acquisition-weighted technique and a reconstruction method that integrates multiple spectra [118].

11.3.2.3 ^{19}F MRI of fluorinated drugs at ultrahigh magnetic field strength

To address the sensitivity and spatial resolution constraints of ^{19}F MR it is conceptually appealing to pursue in vivo ^{19}F MR at ultrahigh magnetic field strengths ($B_0 \geq 7.0$ T). To approach this goal, we recently demonstrated the feasibility of hand and wrist ^{19}F in vivo MRI at 7.0 T following topical application of the NSAID 2-{[3-(Trifluoromethyl) phenyl]amino}benzoic acid (flufenamic acid). For this purpose, a wrist and hand volume radiofrequency (RF) coil was constructed to monitor the fate of this compound (Fig. 11.5). Prior to the in vivo application, careful specific absorption rate (SAR) assessment was performed using electromagnetic field simulations which showed that the maximum 10 g local SAR of 16.2 W/kg (@4 W stimulated power) was well within limits governed by the IEC 60601-2-33 guidelines [117]. In vivo ^{19}F MRI was conducted using a 3D gradient-echo imaging technique modified for ^{19}F application using the following parameters: repetition time (TR) = 15 ms, echo time (TE) = 1 ms, matrix size = 48 × 48, FOV = 100 × 100 mm^2, slab thickness = 80 mm, number of slices = 16, slice thickness = 5 mm, 64 averages, total acquisition time (TA) = 12:20 min. Proton images were acquired using a T_1-weighted fast spin-echo based technique (TR = 400 ms, TE = 9 ms, matrix size = 384 × 384, FOV = 100 × 100 mm^2, number of slices = 16, slices thickness = 5 mm, TA = 1 min). Twenty minutes prior to imaging, the wrist of the volunteer received a 10 g topical application of a cream containing flufenamic acid at a concentration of 1 mmol/l. Figure 11.5 shows in vivo images of the wrist acquired with the dual-tunable ^{19}F/^{1}H-birdcage setup at 7.0 T. Following epicutaneous administration of flufenamic acid to the human wrist [119] and knee [120], we could localize this compound with mm accuracy. Although NSAIDs are known to penetrate human skin, no human in vivo studies have been carried out so far that show the penetration efficiency and route of the drug to the site of inflammation. With this in mind, the first ^{19}F images of fluorine-containing NSAIDs are encouraging and point towards the prospect of applying ^{19}F MRI and NSAID therapy to the field of theranostics for visualizing and measuring the pharmocokinetics and concentration of the therapeutically active compound reaching the inflammatory site in rheumatoid arthritis patients.

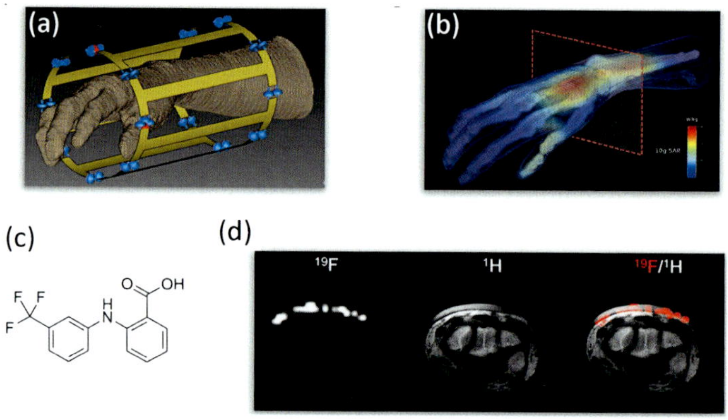

Figure 11.5 Imaging of flufenamic acid in hand and wrist at 7.0 T. (a) Model of the ^{19}F/^{1}H birdcage RF coil with the right arm of a voxel-based human model. (b) 3D maximum intensity projection of the 10 g specific absorption rate (SAR) on the voxel model arm. The transversal slice depicts the position of axial view used for ^{19}F/^{1}H MRI. (c) Chemical structure of flufenamic acid. (d) Left: masked and threshold filtered axial ^{19}F gradient echo (GRE) image; middle: anatomical axial ^{1}H GRE image; right: overlay of the ^{19}F GRE image (red) with the corresponding anatomical ^{1}H GRE image.

11.3.2.4 The future

As 7.0 T MR increasingly becomes used for research, it will advance the capabilities of in vivo ^{19}F MR for the assessment of the metabolic fate of ^{19}F drugs in living organisms. Taking the speed of progress into account, an optimistic practitioner might envision a clinical role for tailored 7.0 T ^{19}F MR applications in the future. It is a vision that continues to inspire basic research and clinical science into ^{19}F MR at 7.0 T. Demonstrable progress in 7.0 T ^{19}F MR technologies and methodology is providing encouragement for the imaging community to tackle solutions of the many outstanding issues. One important development on the hardware horizon is the advent of actively shielded 7.0 T MR systems, which are far more compatible with installations in clinical imaging suites than current models used in basic research requiring hundreds of tons of iron shielding. Magnet length is another important consideration. The recently introduced generation of actively shielded 7.0 T magnets has a length of approximately 255 cm while the passively shielded version has a length of approximately 336 cm. The shorter length is beneficial to relax economic and ergonomic constraints and to improve patient comfort. The requirements of ^{19}F MR at 7.0 T are also likely to pave the way for further advances in RF coil technology, including a broad move to multi-transmit MR systems equipped with eight or more transmission channels [121]. Although the broad spectrum of ^{19}F MR applications makes it challenging to identify a single optimal RF coil configuration, the selected design should meet the following minimum requirements: (a) light weight, (b) capability with multiple patient positions, body habitus, and anatomical variants, (c) ease of clinical use, (d) receptive field large enough to cover the anatomy of choice, (e) sufficient sensitivity to acquire images in clinically acceptable scan times, (f) low amplifier noise profile for parallel and accelerated imaging and (g) uncompromised patient comfort. While much of the pioneering work on novel RF coil array configurations customized for ^{19}F MR is currently occurring at 7.0 T, recognition of the benefits of these designs may result in an eventual migration to lower field strengths common in today's clinical practice.

In addition to RF coil developments, further technical developments are required to image the biodistribution and bioavailability of fluorinated drugs in vivo. Spectroscopy imaging such as CSI or single-voxel based imaging will be necessary for a quantitative analysis of the drug availability in specific regions in vivo. Reconstruction methods that integrate the multiple spectra of complex drug structures will also be necessary. As discussed in other chapters, the goal is to increase signal strength and sensitivity, which is the "currency" for image resolution and scanning time.

11.4 Conclusion

Fluorine MRS has proven to be an essential tool for studying pharmacodynamics, via molecular structure and function, as well as pharmacokinetic (absorption, distribution, metabolism) properties of new compounds during drug discovery. Future in vivo methods using ^{19}F MRS will involve a combination of emerging technologies such as cryogenic hardware [122] and ultrahigh magnetic field strengths [123] to boost sensitivity along with spatio-temporal resolution, in addition to MR pulse sequence developments that reduce scan time and reconstruction methods that will create images out of complex spectral data. The goal is to improve signal sensitivity in order to detect minute traces of drug. In short, while today's ^{19}F in vivo MR techniques remain in a state of creative flux, productive engagement in this area continues to drive further developments.

References

1. Swarts, F. (1892). Étude sur le fluochloroforme, *Bulletins de l'Académie royale des sciences, des lettres et des beauxarts de Belgique*, **24**, pp. 474–484.
2. Midgley Jr, T., and Henne, A. L. (1930). Organic fluorides as refrigerants, *Ind. Eng. Chem.*, **22**, pp. 542–545.
3. Plunkett, R. J. (1941). *Tetrafluoroethylene polymers*, Google Patents.
4. Annex, I. (2002). Stockholm convention on persistent organic pollutants, *America*, **8**.
5. Blowers, P., and Lownsbury, J. M. (2010). Carbon dioxide emission implications if hydrofluorocarbons are regulated: A refrigeration case study, *Environ. Sci. Technol.*, **44**, pp. 1526–1529.

6. Gribble, G. W. (2004). Amazing organohalogens, *Am. Sci.*, **92**, pp. 342–349.

7. Francis, P., Burton, M. R., and Oppenheimer, C. (1998). Remote measurements of volcanic gas compositions by solar occultation spectroscopy, *Nature*, **396**, pp. 567–570.

8. Angerer, K. (2011). Frog tales–on poison dart frogs, epibatidine, and the sharing of biodiversity, *Innovation: Eur. J. Soc. Sci. Res.*, **24**, pp. 353–369.

9. O'hagan, D., and Harper, D. B. (1999). Fluorine-containing natural products, *J. Fluorine Chem.*, **100**, pp. 127–133.

10. Gribble, G. W. (1998). Naturally occurring organohalogen compounds, *Acc. Chem. Res.*, **31**, pp. 141–152.

11. Marais, J. (1944). Dicrotaline: The toxic alkaloid from *Crotalaria dura* (Wood & Evans) and *Crotalaria globifera* (E. Mey), *Onderstepoort J. Vet. Res.*, **20**, pp. 61–65.

12. Hall, R. (1972). The distribution of organic fluorine in some toxic tropical plants, *New Phytol.*, **71**, pp. 855–871.

13. Buffa, P., and Peters, R. (1949). The in vivo formation of citrate induced by fluoroacetate and its significance, *J. Physiol.*, **110**, pp. 488–500.

14. Peters, R., Wakelin, R., and Buffa, P. (1953). Biochemistry of fluoroacetate poisoning the isolation and some properties of the fluorotricarboxylic acid inhibitor of citrate metabolism, *Proc. R. Soc. London. Ser. B Biol. Sci.*, **140**, pp. 497–506.

15. Murphy, C. D., O'hagan, D., and Schaffrath, C. (2001). Identification of a PLP-dependent threonine transaldolase: A novel enzyme involved in 4-fluorothreonine biosynthesis in streptomyces cattleya, *Angew. Chem. Int. Ed.*, **40**, pp. 4479–4481.

16. O'hagan, D., Schaffrath, C., Cobb, S. L., Hamilton, J. T., and Murphy, C. D. (2002). Biochemistry: Biosynthesis of an organofluorine molecule, *Nature*, **416**, pp. 279–279.

17. Ward, P., and Hall, R. (1964). Fluoro-fatty acids in the seeds of Dichapetalum toxicarium, *Nature*, **201**, pp. 611–612.

18. Hamilton, J. T., and Harper, D. B. (1997). Fluoro fatty acids in seed oil of *Dichapetalum toxicarium*, *Phytochemistry*, **44**, pp. 1129–1132.

19. Harper, D. B., Hamilton, J. T. G., and Ohagan, D. (1990). Identification of threo-18-fluoro-9,10-dihydroxystearic acid-a novel omega-fluorinated fatty-acid from dichapetalum-toxicarium seeds, *Tetrahedron Letters*, **31**, pp. 7661–7662.

20. Pattison, F. L. M. (1959). Toxic aliphatic fluorine compounds, *Toxic aliphatic fluorine compounds.*

21. Thomas, S., et al. (1957). Nucleocidin, a new antibiotic with activity against Trypanosomes, *Antibiot. Ann. 1956–1957*, pp. 716–721.

22. Morton, G. O., Lancaster, J. E., Van Lear, G. E., Fulmor, W., and Meyer, W. E. (1969). Structure of nucleocidin. III. revised structure, *J. Am. Chem. Soc.*, **91**, pp. 1535–1537.

23. Peters, R. A., and Shorthouse, M. (1967). Observations on the metabolism of fluoride in Acacia georginae and some other plants, *Nature*, **216**, pp. 80–81.

24. Peters, R. A., and Shorthouse, M. (1971). Identification of a volatile constituent formed by homogenates of acacia georginae exposed to fluoride, *Nature*, **231**, pp. 123–124.

25. Sanada, M., et al. (1986). Biosynthesis of fluorothreonine and fluoroacetic acid by the thienamycin producer, Streptomyces cattleya, *J. Antibiot.*, **39**, pp. 259–265.

26. Amin, M., Harper, D., Moloney, J., Murphy, C., and Howard, J. K. (1997). A short highly stereoselective synthesis of the fluorinated natural product (2S, 3S)-4-fluorothreonine, *Chem. Commun.*, **15**, pp. 1471–1472.

27. Reid, K. A., Bowden, R. D., Dasaradhi, L., Amin, M. R., and Harper, D. B. (1995). Biosynthesis of fluorinated secondary metabolites by Streptomyces cattleya, *Microbiology*, **141**, pp. 1385–1393.

28. Xu, X.-H., Yao, G.-M., Li, Y.-M., Lu, J.-H., Lin, C.-J., Wang, X., and Kong, C.-H. (2003). 5-Fluorouracil derivatives from the sponge Phakellia fusca, *J. Nat. Prod.*, **66**, pp. 285–288.

29. Herrera-Rodriguez, L. N., Meyer, H.-P., Robins, K. T., and Khan, F. (2011). Perspectives on biotechnological halogenation, *Chim. Oggi/ Chem. Today*, **29**, p. 47.

30. Lane, A. L., and Moore, B. S. (2011). A sea of biosynthesis: Marine natural products meet the molecular age, *Nat. Prod. Rep.*, **28**, pp. 411–428.

31. Almabruk, K. H., Lu, W., Li, Y., Abugreen, M., Kelly, J. X., and Mahmud, T. (2013). Mutasynthesis of fluorinated pactamycin analogues and their antimalarial activity, *Org. Lett.*, **15**, pp. 1678–1681.

32. O'hagan, D., J. Robins, R., Wilson, M., W. Wong, C., Berry, M., and Zabetakis, I. (1999). Fluorinated tropane alkaloids generated by directed biosynthesis in transformed root cultures of Datura stramonium, *J. Chem. Soc. Perkin Trans.*, **1**, pp. 2117–2120.

33. Neufeld, D. A., Zmuidzinas, J., Schilke, P., and Phillips, T. G. (1997). Discovery of interstellar hydrogen fluoride, *Astrophys. J. Lett.*, **488**, p. L141.

34. Simons, J. H., and Harland, W. J. (1949). The electrochemical process for the production of fluorocarbons 3. from hydrogen fluoride-soluble organic substances, *J. Electrochem. Soc.*, **95**, pp. 55–59.

35. Simons, J. H., Francis, H. T., and Hogg, J. A. (1949). The electrochemical process for the production of fluorocarbons 2. the electrolysis of solutions of organic substances in liquid hydrogen fluoride, *J. Electrochem. Soc.*, **95**, pp. 53–55.

36. Simons, J. H. (1949). The electrochemical process for the production of fluorocarbons. 1. The generalized procedure and its use with nitrogen compounds, *J. Electrochem. Soc.*, **95**, pp. 47–47.

37. Toon, R. (2011). Fluorine, an obsession with a tragic past, *Educ. Chem.*, **48**, p. 148.

38. Fukuhara, N., and Bigelow, L. A. (1941). The action of elementary fluorine upon organic compounds. XI. The vapor phase fluorination of benzene, *J. Am. Chem. Soc.*, **63**, pp. 2792–2795.

39. Tyczkowski, E. A., and Bigelow, L. A. (1955). The action of elementary fluorine upon organic compounds. XIX. A new jet fluorination reactor, *J. Am. Chem. Soc.*, **77**, pp. 3007–3008.

40. Chambers, R. D., Fox, M. A., and Sandford, G. (2005). Elemental fluorine Part 18. Selective direct fluorination of 1,3-ketoesters and 1,3-diketones using gas/liquid microreactor technology, *Lab Chip*, **5**, pp. 1132–1139.

41. Lerman, O., Tor, Y., and Rozen, S. (1981). Acetyl hypofluorite as a taming carrier of elemental fluorine for novel electrophilic fluorination of activated aromatic rings, *J. Org. Chem.*, **46**, pp. 4629–4631.

42. Rozen, S., Lerman, O., Kol, M., and Hebel, D. (1985). Electrophilic fluorination of unsaturated systems with the recently developed acetyl hypofluorite, *J. Org. Chem.*, **50**, pp. 4753–4758.

43. Barton, D. H. R., Ganguly, A. K., Hesse, R. H., Loo, S. N., and Pechet, M. M. (1968). Organic reactions of fluoroxy-compounds: Electrophilic fluorination of aromatic rings, *Chem. Commun. (London)*, pp. 806–808.

44. Umemoto, T., Fukami, S., Tomizawa, G., Harasawa, K., Kawada, K., and Tomita, K. (1990). Power- and structure-variable fluorinating agents. The N-fluoropyridinium salt system, *J. Am. Chem. Soc.*, **112**, pp. 8563–8575.

45. Poss, A. J., and Shia, G. A. (1995). γ-Fluorination of unsaturated ketones with N-fluorobenzenesulfonimide, *Tetrahedron Lett.*, **36**, pp. 4721–4724.

46. Poss, A. J., Van Der Puy, M., Nalewajek, D., Shia, G. A., Wagner, W. J., and Frenette, R. L. (1991). *N*-fluoropyridinium pyridine heptafluorodiborate: A useful fluorinating agent, *J. Org. Chem.*, **56**, pp. 5962–5964.

47. Muller, K., Faeh, C., and Diederich, F. (2007). Fluorine in pharmaceuticals: Looking beyond intuition, *Science*, **317**, pp. 1881–1886.

48. Pauling, L. (1960). *The Nature of the Chemical Bond and the Structure of Molecules and Crystals: An Introduction to Modern Structural Chemistry*: Cornell University Press.

49. Bondi, A. (1964). van der Waals volumes and radii, *J. Phys. Chem.*, **68**, pp. 441–451.

50. Carcenac, Y., Tordeux, M., Wakselman, C., and Diter, P. (2006). Experimental determination of the conformational free energies (A values) of fluorinated substituents in cyclohexane by dynamic ^{19}F NMR spectroscopy. Part 2. Extension to fluoromethyl, difluoromethyl, pentafluoroethyl, trifluoromethylthio and trifluoromethoxy groups, *New J. Chem.*, **30**, pp. 447–457.

51. Filler, R., and Saha, R. (2009). Fluorine in medicinal chemistry: A century of progress and a 60-year retrospective of selected highlights, *Future Med. Chem.*, **1**, pp. 777–791.

52. Smart, B. E. (2001). Fluorine substituent effects (on bioactivity), *J. Fluorine Chem.*, **109**, pp. 3–11.

53. Park, B. K., Kitteringham, N. R., and O'neill, P. M. (2001). Metabolism of fluorine-containing drugs, *Ann. Rev. Pharmacol. Toxicol.*, **41**, pp. 443–470.

54. Winstein, S., and Holness, N. J. (1955). Neighboring carbon and hydrogen. XIX. t-butylcyclohexyl derivatives. Quantitative conformational analysis, *J. Am. Chem. Soc.*, **77**, pp. 5562–5578.

55. Hansch, C., Leo, A., and Taft, R. W. (1991). A survey of Hammett substituent constants and resonance and field parameters, *Chem. Rev.*, **91**, pp. 165–195.

56. Purser, S., Moore, P. R., Swallow, S., and Gouverneur, V. (2008). Fluorine in medicinal chemistry, *Chem. Soc. Rev.*, **37**, pp. 320–330.

57. Olsen, J. A., et al. (2003). A fluorine scan of thrombin inhibitors to map the fluorophilicity/fluorophobicity of an enzyme active site: Evidence for C-F...C=O interactions, *Angew. Chem.*, **115**, pp. 2611–2615.

58. Fried, J., and Sabo, E. F. (1954). 9α-fluoro derivatives of cortisone and hydrocortisone, *J. Am. Chem. Soc.*, **76**, pp. 1455–1456.

59. Böhm, H. J., et al. (2004). Fluorine in medicinal chemistry, *Chembiochem*, **5**, pp. 637–643.

60. Jeong, H. C., Kang, N. S., Park, N. J., Yum, E. K., and Jung, Y. S. (2009). Reactivation potency of fluorinated pyridinium oximes for acetylcholinesterases inhibited by paraoxon organophosphorus agent, *Bioorg. Med. Chem. Lett.*, **19**, pp. 1214–1217.

61. Bartusik, D., and Aebisher, D. (2014). ^{19}F applications in drug development and imaging—a review, *Biomed. Pharmacother.*

62. Baillon, J. G., Mamont, P. S., Wagner, J., Gerhart, F., and Lux, P. (1988). Fluorinated analogues of spermidine as substrates of spermine synthase, *Eur. J. Biochem.*, **176**, pp. 237–242.

63. Rodgers, T., and Rowland, M. (2006). Physiologically based pharmacokinetic modelling 2: Predicting the tissue distribution of acids, very weak bases, neutrals and zwitterions, *J. Pharm. Sci.*, **95**, pp. 1238–1257.

64. Van Niel, M. B., et al. (1999). Fluorination of 3-(3-(piperidin-1-yl) propyl)indoles and 3-(3-(piperazin-1-yl)propyl)indoles gives selective human 5-HT1D receptor ligands with improved pharmacokinetic profiles, *J. Med. Chem.*, **42**, pp. 2087–2104.

65. Van De Waterbeemd, H., Smith, D. A., Beaumont, K., and Walker, D. K. (2001). Property-based design: Optimization of drug absorption and pharmacokinetics, *J. Med. Chem.*, **44**, pp. 1313–1333.

66. Dugar, S., et al. (1996). Metabolism and structure activity data based drug design: Discovery of (−) SCH 53079 an analog of the potent cholesterol absorption inhibitor (−) SCH 48461, *Bioorg. Med. Chem. Lett.*, **6**, pp. 1271–1274.

67. Wang, J., et al. (2013). Fluorine in pharmaceutical industry: Fluorine-containing drugs introduced to the market in the last decade (2001–2011), *Chem. Rev.*, **114**, pp. 2432–2506.

68. Bégué, J.-P., and Bonnet-Delpon, D. (2008). *Bioorganic and Medicinal Chemistry of Fluorine*: John Wiley & Sons.

69. Kirk, K. L. (2006). Fluorine in medicinal chemistry: Recent therapeutic applications of fluorinated small molecules, *J. Fluorine Chem.*, **127**, pp. 1013–1029.

70. Hagmann, W. K. (2008). The many roles for fluorine in medicinal chemistry, *J. Med. Chem.*, **51**, pp. 4359–4369.

71. Baker, W., Hey, D., Smith, F., Campbell, N., Harper, S., and Shoppee, C. (1947). Organic chemistry, *Annu. Rep. Prog. Chem.*, **44**, pp. 82–216.

72. Ehrlich, P. (1899). Croonian lecture: On immunity with special reference to cell life, *Proc. R. Soc. London*, **66**, pp. 424–448.

73. Heidelberger, C., et al. (1957). Fluorinated pyrimidines, a new class of tumour-inhibitory compounds, *Nature*, **179**, 663–666.

74. Organization, W. H. (2014). *The selection and use of essential medicines: Report of the WHO Expert Committee, 2013 (including the 18th WHO model list of essential medicines and the 4th WHO model list of essential medicines for children)*: World Health Organization.

75. Miller, J., Miller, E., and Finger, G. (1953). On the enhancement of the carcinogenicity of 4-dimethylaminoazobenzene by fluoro-substitution, *Cancer Res.*, **13**, pp. 93–97.

76. Tannhauser, P., Pratt, R. J., and Jensen, E. V. (1956). The preparation of 21-fluorosteroids, *J. Am. Chem. Soc.*, **78**, pp. 2658–2659.

77. Stock, C. C. (1954). Experimental cancer chemotherapy, *Adv. Cancer Res.*, **2**, pp. 425–492.

78. Fried, J., and Sabo, E. F. (1953). Synthesis of 17α-hydroxycorticosterone and its 9α-halo derivatives from 11-EPI-17α-hydroxycorticosterone, *J. Am. Chem. Soc.*, **75**, pp. 2273–2274.

79. Brodbeck, U., Schweikert, K., Gentinetta, R., and Rottenberg, M. (1979). Fluorinated aldehydes and ketones acting as quasi-substrate inhibitors of acetylcholinesterase, *Biochim. Biophys. Acta (BBA)—Enzymol.*, **567**, pp. 357–369.

80. Imperiali, B., and Abeles, R. H. (1986). Inhibition of serine proteases by peptidyl fluoromethyl ketones, *Biochemistry*, **25**, pp. 3760–3767.

81. Gelb, M. H. (1986). Fluoro ketone phospholipid analogs: New inhibitors of phospholipase A2, *J. Am. Chem. Soc.*, **108**, pp. 3146–3147.

82. Abouabdellah, A., and Welch, J. T. (1994). Optically active fluorinated β-lactam building blocks: A novel fluorinated retroamide isostere, *Tetrahedron: Asymmetry*, **5**, pp. 1005–1013.

83. Hoffman, R. V., and Tao, J. (1998). A stereocontrolled synthesis of monofluoro ketomethylene dipeptide isosteres, *J. Org. Chem.*, **64**, pp. 126–132.

84. Garrett, G. S., Emge, T. J., Lee, S. C., Fischer, E. M., Dyehouse, K., and Mciver, J. M. (1991). Synthesis of the monofluoro ketone peptide isostere, *J. Org. Chem.*, **56**, pp. 4823–4826.

85. Myers, A. G., Barbay, J. K., and Zhong, B. (2001). Asymmetric synthesis of chiral organofluorine compounds: Use of nonracemic fluoroiodoacetic acid as a practical electrophile and its application

to the synthesis of monofluoro hydroxyethylene dipeptide isosteres within a novel series of HIV protease inhibitors, *J. Am. Chem. Soc.*, **123**, pp. 7207–7219.

86. O'hagan, D. (2010). Fluorine in health care: Organofluorine containing blockbuster drugs, *J. Fluorine Chem.*, **131**, pp. 1071–1081.
87. Sales, U. S. P. (2013). *http://www. drugs.com/stats/top100/2013/q4/sales*; 4th Quaterly Update].
88. Günther, H. (2013). *NMR Spectroscopy: Basic Principles, Concepts and Applications in Chemistry*: John Wiley & Sons.
89. Dolbier, W. R. (2009). *Guide to Fluorine NMR for Organic Chemists*: John Wiley & Sons.
90. Shuker, S. B., Hajduk, P. J., Meadows, R. P., and Fesik, S. W. (1996). Discovering high-affinity ligands for proteins: SAR by NMR, *Science*, **274**, pp. 1531–1534.
91. Campos-Olivas, R., Aziz, R., Helms, G. L., Evans, J. N., and Gronenborn, A. M. (2002). Placement of 19F into the center of GB1: Effects on structure and stability, *FEBS Lett.*, **517**, pp. 55–60.
92. Proctor, W. G., and Yu, F. C. (1950). The dependence of a nuclear magnetic resonance frequency upon chemical compound, *Phys. Rev.*, **77**, pp. 717–717.
93. Tanabe, K., Sugiura, M., and Nishimoto, S.-I. (2010). Monitoring of duplex and triplex formation by ^{19}F NMR using oligodeoxynucleotides possessing 5-fluorodeoxyuridine unit as ^{19}F signal transmitter, *Bioorg. Med. Chem.*, **18**, pp. 6690–6694.
94. Wang, L., Brock, A., Herberich, B., and Schultz, P. G. (2001). Expanding the genetic code of escherichia coli, *Science*, **292**, pp. 498–500.
95. Tkachenko, A. N., Mykhailiuk, P. K., Afonin, S., Radchenko, D. S., Kubyshkin, V. S., Ulrich, A. S., and Komarov, I. V. (2013). A ^{19}F NMR label to substitute polar amino acids in peptides: A CF3-substituted analogue of serine and threonine, *Angew. Chem.*, **125**, pp. 1526–1529.
96. Tkachenko, A. N., Radchenko, D. S., Mykhailiuk, P. K., Afonin, S., Ulrich, A. S., and Komarov, I. V. (2013). Design, synthesis, and application of a trifluoromethylated phenylalanine analogue as a label to study peptides by solid-state ^{19}F NMR spectroscopy, *Angew. Chem. Int. Ed.*, **52**, pp. 6504–6507.
97. Liu, J. J., Horst, R., Katritch, V., Stevens, R. C., and Wüthrich, K. (2012). Biased signaling pathways in β2-adrenergic receptor characterized by ^{19}F-NMR, *Science*, **335**, pp. 1106–1110.

98. Yu, J.-X., Kodibagkar, V. D., Cui, W., and Mason, R. P. (2005). ^{19}F: A versatile reporter for non-invasive physiology and pharmacology using magnetic resonance, *Curr. Med. Chem.*, **12**, pp. 819–848.

99. Desmoulin, F., Gilard, V., Malet-Martino, M., and Martino, R. (2002). Metabolism of capecitabine, an oral fluorouracil prodrug: ^{19}F NMR studies in animal models and human urine, *Drug Metab. Dispos.*, **30**, pp. 1221–1229.

100. Strauss, L. G., and Conti, P. S. (1991). The applications of PET in clinical oncology, *J. Nuclear Med.*, **32**, pp. 623–648.

101. Hohenberger, P., Strauß, L. G., Lehner, B., Frohmüller, S., Dimitrakopoulou, A., and Schlag, P. (1993). Perfusion of colorectal liver metastases and uptake of fluorouracil assessed by H215O and [18F]uracil positron emission tomography (PET), *Eur. J. Cancer*, **29**, pp. 1682–1686.

102. Dimitrakopoulou, A., et al. (1993). Studies with positron emission tomography after systemic administration of fluorine-18-uracil in patients with liver metastases from colorectal carcinoma, *J. Nuclear Med.*, **34**, pp. 1075–1081.

103. Malet-Martino, M.-C., Faure, F., Vialaneix, J.-P., Palevody, C., Hollande, E., and Martino, R. (1986). Noninvasive fluorine-19 NMR study of fluoropyrimidine metabolism in cell cultures of human pancreatic and colon adenocarcinoma, *Cancer Chemother. Pharmacol.*, **18**, pp. 5–10.

104. Keniry, M., Benz, C., Shafer, R. H., and James, T. L. (1986). Noninvasive spectroscopic analysis of fluoropyrimidine metabolism in cultured tumor cells, *Cancer Res.*, **46**, pp. 1754–1758.

105. Cabanac, S., Malet-Martino, M. C., Bon, M., Martino, R., Nedelec, J. F., and Dimicoli, J. L. (1988). Direct ^{19}F NMR spectroscopic observation of 5-fluorouracil metabolism in the isolated perfused mouse liver model, *NMR Biomed.*, **1**, pp. 113–120.

106. Prior, M. J. W., Maxwell, R. J., and Griffiths, J. R. (1990). In vivo ^{19}F NMR spectroscopy of the antimetabolite 5-fluorouracil and its analogues: An assessment of drug metabolism, *Biochem. Pharmacol.*, **39**, pp. 857–863.

107. Griffiths, J. R., et al. (1987). Monitoring cancer therapy by NMR spectroscopya, *Ann. N. Y. Acad. Sci.*, **508**, pp. 183–199.

108. Stevens, A., Morris, P., Iles, R., Sheldon, P., and Griffiths, J. (1984). 5-fluorouracil metabolism monitored in vivo by ^{19}F NMR, *Br. J. Cancer*, **50**, pp. 113.

109. Evers, A. S., Haycock, J. C., and Andre D'avignon, D. (1988). The potency of fluorinated ether anesthetics correlates with their ^{19}F spin-spin

relaxation times in brain tissue, *Biochem. Biophys. Res. Commun.*, **151**, pp. 1039–1045.

110. Liu, X., Kevin Li, S., and Jeong, E.-K. (2010). Ocular pharmacokinetic study of a corticosteroid by ¹⁹ F MR, *Exp. Eye Res.*, **91**, pp. 347–352.

111. Wolf, W., Silver, M. S., Albright, M. J., Weber, H., Reichardt, U., and Sauer, R. (1987). A non-invasive study of drug metabolism in patients as studied by ^{19}F NMR spectroscopy of 5-fluorouracil, *Ann. N. Y. Acad. Sci.*, **508**, pp. 491–493.

112. Durst, P., Schuff, N., Crocq, M.-A., Mokrani, M.-C., and Macher, J.-P. (1990). Noninvasive in vivo detection of a fluorinated neuroleptic in the human brain by ^{19}F nuclear magnetic resonance spectroscopy, *Psychiatry Res.*, **35**, pp. 107–114.

113. Bolo, N. R., Hode, Y., Nedelec, J.-F., Laine, E., Wagner, G., and Macher, J.-P. (2000). Brain pharmacokinetics and tissue distribution in vivo of fluvoxamine and fluoxetine by fluorine magnetic resonance spectroscopy, *Neuropsychopharmacology*, **23**, pp. 428–438.

114. Klomp, D. W. J., Van Laarhoven, H. W. M., Kentgens, A. P. M., and Heerschap, A. (2003). Optimization of localized ^{19}F magnetic resonance spectroscopy for the detection of fluorinated drugs in the human liver, *Magn. Reson. Med.*, **50**, pp. 303–308.

115. Bilecen, D., Schulte, A.-C., Kaspar, A., Küstermann, E., Seelig, J., Elverfeldt, D. V., and Scheffler, K. (2003). Detection of the non-steroidal anti-inflammatory drug niflumic acid in humans: A combined ^{19}F-MRS in vivo and in vitro study, *NMR Biomed.*, **16**, pp. 144–151.

116. Bolo, N. R., Hodé, Y., and Macher, J. P. (2004). Long-term sequestration of fluorinated compounds in tissues after fluvoxamine or fluoxetine treatment: A fluorine magnetic resonance spectroscopy study in vivo, *Magn. Reson. Mater. Phys. Biol. Med.*, **16**, pp. 268–276.

117. International-Electrotechnical-Commission, *Medical electrical equipment.-Part 2-33: Particular requirements for the basic safety and essential performance of magnetic resonance equipment for medical diagnosis* 2010: IEC.

118. Jacoby, C., et al. (2014). Probing different perfluorocarbons for in vivo inflammation imaging by ^{19}F MRI: Image reconstruction, biological half-lives and sensitivity, *NMR Biomed*, **27**, pp. 261–271.

119. Waiczies, H., et al. (2012). Towards theranostics of rheumatoid arthritis: ^{1}H/^{19}F imaging of non-steroidal anti-inflammatory drugs in hand and wrist at 7 Tesla. in *Proc. Intl. Soc. Mag. Reson. Med. (20)*.

120. Ji, Y., Waiczies, H., Winter, L., Neumanova, P., Hofmann, D., Rieger, J., Mekle, R., Waiczies, S., and Niendorf, T. (2015). Eight-channel transceiver RF coil array tailored for $^{1}H/^{19}F$ MR of the human knee and fluorinated drugs at 7.0 T, *NMR in Biomedicine.*

121. Oezerdem, C., Winter, L., Graessl, A., Paul, K., Els, A., Weinberger, O., Rieger, J., Kuehne, A., Dieringer, M., Hezel, F., Voit, D., Frahm, J., and Niendorf, T. (2016). 16-channel bow tie antenna transceiver array for cardiac MR at 7.0 tesla, *Magnetic Resonance in Medicine*, **75**, pp. 2553–2565.

122. Niendorf, T., Pohlmann, A., Reimann, H.M., Waiczies, H., Peper, E., Huelnhagen, T., Seeliger, E., Schreiber, A., Kettritz, R., Strobel, K., Ku, M.-C., and Waiczies, S. (2015). Advancing Cardiovascular, Neurovascular and Renal Magnetic Resonance Imaging in Small Rodents Using Cryogenic Radiofrequency Coil Technology, *Frontiers in Pharmacology*, **6**.

123. Niendorf, T., Paul, K., Oezerdem, C., Graessl, A., Klix, S., Huelnhagen, T., Hezel, F., Rieger, J., Waiczies, H., Frahm, J., Nagel, A.M., Oberacker, E., and Winter, L. (2015). W(h)ither human cardiac and body magnetic resonance at ultrahigh fields? technical advances, practical considerations, applications, and clinical opportunities, *NMR in Biomedicine*, pp. n/a-n/a.

Part 6
Other Biomedical Applications

Chapter 12

Imaging of the Respiratory System

Marcus J. Couch,[a,b] **Alexei V. Ouriadov,**[b] **and Mitchell S. Albert**[a,b]

[a]*Lakehead University, Thunder Bay, Ontario, Canada*
[b]*Thunder Bay Regional Research Institute, Thunder Bay, Ontario, Canada*

albertmi@tbh.net

Pulmonary fluorine-19 (^{19}F) magnetic resonance imaging (MRI) using inhaled inert fluorinated gases can potentially provide high-quality structural and functional images of the lungs. Inert fluorinated gases have the advantages of being nontoxic, abundant, relatively inexpensive, and the technique can be performed on any MRI scanner with broadband imaging capabilities. In this chapter, the development of inert fluorinated gas MRI, current progress in animal and human imaging, and the future prospects of this technique are discussed. The current state of other lung MR imaging techniques, such as hyperpolarized (HP) noble gas MRI and ^{1}H-based MRI techniques, is also briefly discussed in order to provide some context. Overall, inert fluorinated gas MRI has the potential to be clinically adopted as an inexpensive pulmonary imaging modality that can aid in disease diagnosis.

Fluorine Magnetic Resonance Imaging
Edited by Ulrich Flögel and Eric Ahrens

ISBN 978-981-4745-31-4 (Hardcover), 978-981-4745-32-1 (eBook)
www.panstanford.com

12.1 Introduction

Due to the increasing prevalence of chronic respiratory diseases, such as asthma, chronic obstructive pulmonary disease (COPD) and cystic fibrosis, there is a need in the medical community for new technologies that can aid in the diagnosis and management of chronic respiratory diseases [1–3]. Pulmonary function is typically assessed in the clinic using spirometry and plethysmography, where airflow is measured at the mouth during various breathing maneuvers. Spirometry is able to measure flow-related parameters such as FEV_1, which refers to the volume of air exhaled in the first second of a forced expiration [4], while plethysmography is able to measure volume-related parameters such as the functional residual capacity (FRC) [5]. Pulmonary function tests are standard clinical diagnostic tools that have a low cost and are widely available; however, they can only provide a global measurement of lung function that is well known to be fundamentally heterogeneous [6].

A number of imaging techniques have been developed that can potentially provide additional regional information regarding pulmonary structure and function, such as chest X-rays, computed tomography (CT), single photon emission computed tomography (SPECT), positron emission tomography (PET), and MRI [7]. Since there are growing concerns regarding cumulative radiation doses due to imaging procedures that use ionizing radiation [8], MRI is potentially an ideal imaging modality as it is both non-ionizing and non-invasive. The image quality from conventional proton (^1H) MR images of the lung has historically been very poor, due to a low tissue density and other factors. Some recent progress has been made in developing optimized image acquisition strategies [9] and functional lung measurements [10]; however, ^1H MRI is only indirectly sensitive to pulmonary ventilation.

Hyperpolarized (HP) noble gas MRI, using helium-3 (^3He) or xenon-129 (^{129}Xe), is a pulmonary imaging technique that is able to directly visualize the distribution of ventilation in the lungs, as well as provide a measurement of regional ventilation, gas exchange, and lung microstructure [11–14]. Unfortunately, HP gas MRI has a number of special requirements that make it an unrealistic technique for wide-scale clinical implementation, such as the need for scarce and expensive noble gas isotopes, an expensive polarizer, and specially trained personnel. Fluorine-19 MRI of the lungs using inert fluorinated gases may be an inexpensive alternative to HP noble gas MRI that can offer similar structural information and functional biomarkers. One notable advantage of inert fluorinated gases is that they can be mixed with O_2, which allows for safe dynamic imaging over multiple breaths. HP gases, on the other hand, are typically inhaled in anoxic gas mixtures, thereby limiting the scan duration. Inert fluorinated gas MRI has been performed extensively in the lungs of animals since 1982. High resolution 3D images [15], as well as measurements of functional biomarkers, such as the apparent diffusion coefficient (ADC) [16], wash-in/wash-out kinetics [17], and the ventilation/perfusion ratio [18] have been demonstrated in animals. Preliminary work in the lungs of healthy volunteers and patients with lung diseases has been reported recently, and improvements in image quality are ongoing [19, 20]. Overall, inert fluorinated gas MRI may become a viable pulmonary imaging modality that can provide clinically useful information for the diagnosis and management of chronic respiratory diseases.

12.2 ^{1}H MRI of the Lung

Since its discovery in the early 1970s [21], MRI has become an important diagnostic tool in both clinical and research settings. MRI is able to provide images with an excellent soft tissue contrast using a non-invasive and non-ionizing method. Conventional ^{1}H MRI of the lungs has historically been challenging due to a number of well-known factors: a low tissue density in the lungs, magnetic susceptibility differences at air/tissue interfaces leading to a short T_2^*, and respiratory/cardiac motion. Some progress has been made in alleviating these issues, particularly with the use of efficient image acquisition techniques, such as ultra-short echo time (UTE) imaging [9]. UTE can help to overcome the short T_2^* in the lungs by capturing the signal almost immediately after the RF pulse and while sampling *k*-space in a radial fashion. It has recently been demonstrated that pulmonary ^{1}H UTE imaging is reproducible, and that the UTE MR signal intensity correlates with pulmonary function test parameters in patients with COPD [22]. Zero echo time (ZTE) imaging is a variant of UTE, where even shorter echo times are achieved by applying non-selective hard RF pulses while the spatial encoding gradients are already turned on. Preliminary ZTE imaging has shown promise in the lungs of animals and humans [23, 24]. In addition to short TE approaches, techniques that use a short TR may also be able to help address the challenges of conventional ^{1}H MRI of the lung, such as balanced steady-state free precession (SSFP) [25]. Balanced SSFP uses a TR on the order of ~1 ms and recycles magnetization into subsequent pulse sequence repetitions, rather than spoiling the remaining transverse magnetization after the acquisition window. This technique makes efficient use of the pulmonary ^{1}H signal and therefore helps to overcome the low tissue density in the lungs.

In addition to optimized ^{1}H-based MRI techniques that provide structural and anatomical information, it is also possible to obtain some functional lung information with techniques such as O_2-enhanced MRI [26, 27]. Since O_2 is paramagnetic, it tends to shorten longitudinal relaxation times in lung tissue, leading to a signal enhancement in T_1-weighted images. A "ventilation" image can be obtained by subtracting an un-enhanced image, and bright areas correspond to well-ventilated regions of the lung that had a signal enhancement due to the presence of O_2. This technique has the ability to detect ventilation defects due to airway obstructions [28]. O_2-enhanced MRI has also been combined with ^{1}H UTE imaging to provide additional signal enhancements [29, 30]. Other recent efforts have focused on using O_2-enhanced MRI to obtain quantitative information, such as specific ventilation (defined as the ratio of fresh gas entering a lung region divided by the end-expiratory volume of that region). By using alternating breathing cycles of air (21% O_2) and 100% O_2, the signal changes over time can be fitted to a model describing the O_2 kinetics to compute a map of specific ventilation [31].

An alternative to O_2-enhanced MRI for ^{1}H-based functional lung imaging is Fourier-decomposition (FD) MRI, which does not require an exogenous contrast agent [32]. In essence, FD MRI rapidly acquires ^{1}H images in free breathing subjects, and then a pixel-wise Fourier decomposition of the signal intensity yields spectral peaks corresponding to the cardiac and respiratory frequencies. The amplitudes of the spectral peaks are then used to generate ventilation- and perfusion-weighted images. This technique has been shown to be reproducible [33], and it has been validated with comparisons to SPECT/CT [34] and HP ^{3}He MRI [35]. Although ^{1}H-based functional lung imaging techniques are inexpensive and they can be performed on virtually any MRI scanner, they are potentially disadvantaged since they have a poor signal-to-noise ratio (SNR) and they are ultimately only indirect measurements of pulmonary ventilation. For a discussion of other recent advances in lung ^{1}H MRI, the interested reader is referred to a recent review by Miller et al. [36].

12.3 Hyperpolarized Noble Gas MRI

Although some progress has been made in the development of functional ^{1}H-based lung imaging techniques, many MRI researchers have focused on lung imaging using HP ^{3}He and ^{129}Xe, a technique that can offer high resolution and high-quality images that are directly sensitive to ventilation and lung function [11–14]. Through optical pumping [37], the MR signal from ^{3}He and ^{129}Xe can be increased by up to 100,000 times above thermal equilibrium levels, thereby allowing for direct imaging of the distribution of the HP noble gas in the lung airspaces following inhalation. HP noble gas lung MRI was first demonstrated in 1994 by Albert et al., and ^{129}Xe MR images of excised mouse lungs were reported (Fig. 12.1) [38].

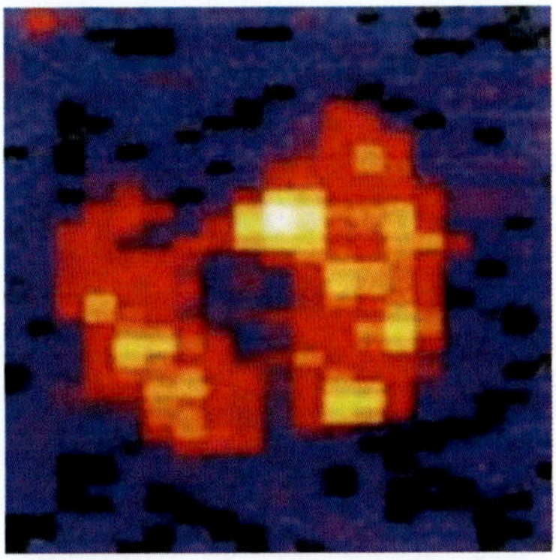

Figure 12.1 First biological hyperpolarized ^{129}Xe MR image acquired in excised mouse lungs. Image reproduced with permission from Albert et al. (1994). *Nature,* 370, 199–201.

Born out of the desire to understand the mechanism of general anesthesia in the brain, ^{129}Xe was chosen due to its anesthetic ability, NMR sensitivity, and its ability to dissolve into blood and be carried to tissues and organs of interest [39]. Naturally, imaging the lungs seemed like a practical starting point, since inhalation is the simplest route of administration and the signal in the gas phase would be much higher than in the dissolved phase. Although some preliminary work in animals and humans has continued to explore functional MRI of ^{129}Xe in the brain [40–42], the majority of the HP noble gas literature published in the past 20 years has focused on imaging the lungs [12, 13]. Figure 12.2 shows current examples of HP ^{3}He MR images acquired in a healthy individual and three patients with pulmonary diseases: asthma, moderate COPD, and severe COPD [43]. Although each image in itself bears a striking structural distinctiveness, many functional imaging techniques have been developed to provide additional information, such as ventilation defect volume (VDV) measurements [44], airway tree measurements [45], ADC measurements [46], and mapping the alveolar partial pressure of O_2 (P_AO_2) [47].

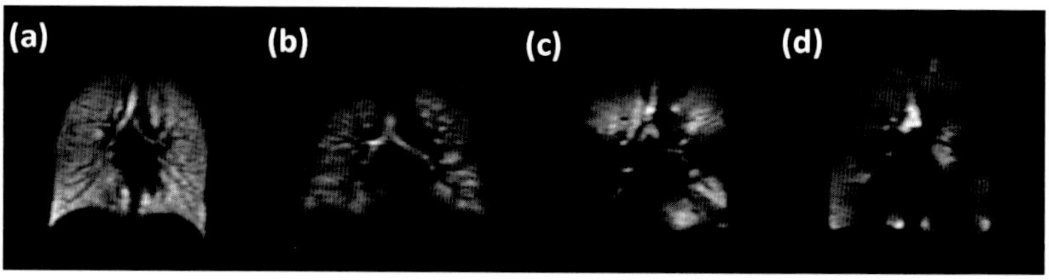

Figure 12.2 HP ^{3}He MR images acquired from four individuals: (a) healthy, (b) asthma, (c) moderate COPD, and (d) severe COPD. Images were reproduced with permission from Couch et al. (2014). *NMR Biomed.*, 27, 1525–1534.

Unlike conventional MRI, the T_1 relaxation time of HP gases characterizes the decay of the HP magnetization, rather than a time constant for regrowth to thermal equilibrium. Since the HP magnetization is non-recoverable, image acquisitions must carefully preserve the reservoir of HP magnetization so that a near-constant signal level is maintained throughout the acquisition, such as in a variable flip angle (VFA) approach [48]. Historically, the use of ^{3}He has been more common in the literature, due to the available polarizer technology and its gyromagnetic ratio being 2.7 times greater than that of ^{129}Xe, leading to a greater signal strength (see Table 12.1 for a comparison of the physical properties of ^{3}He and ^{129}Xe). Unfortunately, ^{3}He is well known to be an extremely rare and expensive isotope, and its only means of production is through the radioactive decay of tritium; the cost of ^{3}He has further increased due to government sequestering of ^{3}He for use in neutron detectors and national security [49]. For this reason, ^{129}Xe MRI has been receiving increasing attention recently as it is cheaper and more widely available, and improvements in polarizer technology are ongoing [50, 51]. Recent work has focused on validating ^{129}Xe against ^{3}He for HP MRI [52], measuring ^{129}Xe ADC [53], diffusion anisotropy [54], and probing gas exchange [55]. It should also be noted that the anesthetic properties of xenon has necessitated some careful studies with respect to safety and tolerability [56]. Regardless of whether ^{3}He or ^{129}Xe is used for imaging, both gases require a polarizer to prepare and process the gases, and only a handful of research sites in the world currently have access to this expensive polarizer technology. HP gas imaging also requires an MRI scanner with broadband multinuclear capabilities and dedicated RF coils. Other current perspectives on HP noble gas MRI have been reviewed recently by Lilburn et al. [12].

12.4 Properties of Inert Fluorinated Gas MRI

Due to the high cost of noble gas isotopes and the limited availability of polarizer technology, ^{19}F MRI of the lungs using inhaled inert fluorinated gases is emerging as a potential alternative to HP ^{3}He and ^{129}Xe MRI of the lungs [19, 20]. Preliminary work has shown that inert fluorinated gas MRI can provide high-quality images of the distribution of lung ventilation and can also yield quantitative functional biomarkers that are similar to HP noble gas MRI [43]. A discussion of the physical properties of inert fluorinated gases will be helpful in understanding how pulmonary ^{19}F MRI is possible. Inert gases such as tetrafluoromethane (CF_4), sulfur hexafluoride (SF_6), hexafluoroethane (C_2F_6), and perfluoropropane (C_3F_8 or PFP) can be used in ^{19}F MRI. Fluorinated gases are non-toxic and safe for human inhalation, as they are commonly used in pulmonary function tests, such as SF_6 (in small concentrations) in the multiple inert gas elimination technique (MIGET) [57]. This technique has the advantage of using gases that are more abundant and less expensive compared to HP gases. Like HP gas MRI, inert fluorinated gas imaging also requires a broadband MRI scanner and dedicated RF coils; however, inert fluorinated gases do not need to be hyperpolarized prior to their use in MRI, which eliminates the need for polarizer technology that is currently very expensive. Therefore, inert fluorinated gas MRI is more likely to be clinically adopted as a routine pulmonary imaging modality.

A comparison of the physical properties of ^{129}Xe, ^{3}He, PFP, SF_6, and C_2F_6 is shown in Table 12.1 [15, 18, 19, 58–64]. It can immediately be seen that ^{19}F has a very high gyromagnetic ratio and a high natural abundance; in addition, inert fluorinated gases have a negligible solubility [63], and the T_1 relaxation times of fluorinated gases in the lungs are substantially shorter than HP gases [18, 19, 59, 64]. In terms of HP gas MRI, a long T_1 can be beneficial for imaging within a breath-hold, and changes in T_1 allow for P_AO_2 mapping [47]. In the context of inert fluorinated gas MRI, a short T_1 is advantageous, since averaging is necessary to overcome a thermal polarization that has much less available magnetization than HP gases. Since the ^{19}F longitudinal magnetization recovers quickly, short TRs and high excitation flip angles can be used. For example, assuming that TR = T_1, the Ernst angle of 68° can be used for imaging. HP gases, on the other hand, have a pool of non-recoverable longitudinal magnetization that is typically probed with low flip angles.

Table 12.1 Comparison of ^{129}Xe, ^{3}He and ^{19}F properties

Parameter	^{129}Xe	^{3}He	^{19}F (PFP)	^{19}F (SF_6)	^{19}F (C_2F_6)
Nuclear Spin, I	1/2	1/2	1/2		
Gyromagnetic ratio, γ, [MHz rad T^{-1}]	74.521	203.782	251.662		
Natural Abundance [%]	26.4	1.37×10^{-4}	100		
Chemical shift range, $\Delta\delta$ [ppm]	250	0.8	150*		

Parameter	^{129}Xe	^{3}He	^{19}F (PFP)	^{19}F (SF_6)	^{19}F (C_2F_6)
Spin-lattice relaxation time in the lungs, T_1	20 s	25 s	12.4 ms†	1.2 ms†	5.9 ms
Self-diffusion coefficient, D_0, [cm^2 s^{-1}]	0.061	2.05	0.023	0.033	0.032
Diffusion coefficient in air, D [cm^2 s^{-1}]	0.14	0.86	0.071	0.094	0.101
Ostwald solubility, L	0.17	0.0085	0.00052	0.0054	0.0015

Source: Adapted from [15, 18, 19, 58–64].

*Approximate chemical shift between PFP and SF_6 resonances. PFP also has two peaks which are separated by ~48 ppm.

†The reported T_1 values of PFP and SF_6 assume the gas is mixed with ~20% O_2.

In general, an inert fluorinated gas that is ideal for imaging would have a short T_1 and multiple magnetically equivalent ^{19}F nuclei. In fact, SF_6, C_2F_6 and PFP all contain six magnetically equivalent ^{19}F nuclei per molecule. Together, all of these properties make inert fluorinated gas imaging possible, and ^{19}F MR images of the lung can be acquired with a reasonable SNR and within a single breath-hold.

Since the presence of O_2 has a much less dramatic effect on the T_1 of fluorinated gases than it does for HP gases, the fluorinated gases can be mixed with O_2 to improve patient safety with little impact on image quality. The presence of O_2 in inert fluorinated gas mixtures can potentially allow for longer breath-holds than anoxic HP gas inhalations. Where HP gas inhalations have historically used a bolus on the order of 1 L, a mixture of inert fluorinated gas and O_2 can be breathed continuously to wash-out residual air from the lungs and to reach a steady-state concentration of the inert fluorinated gas mixture, thereby maximizing the available magnetization for imaging. Continuous breathing of inert fluorinated gases allows for dynamic imaging measurements, such as wash-in and wash-out time constants, and can potentially allow for the acquisition of more physiologically meaningful information. Likewise, multiple breath HP ^{3}He MRI techniques are currently being developed for the measurement of fractional ventilation, ADC, and P_AO_2 mapping [65, 66].

12.5 Static Breath-Hold Imaging

Edwin Heidelberger and Paul Lauterbur were the first to demonstrate gas-phase MRI in 1982, by acquiring ^{19}F images using CF_4. These initial images were acquired in phantoms and excised rabbit lungs at a field strength of 0.1 T [67]. In 1984, Rinck et al. demonstrated the first in vivo ^{19}F MR images in healthy dog lungs, also using CF_4 in a 0.1 T magnet [68, 69]. A 3D isotropic dataset was obtained during a 25 min acquisition resulting in a 65 × 65 × 65 matrix with 2–3 mm resolution. Figure 12.3 shows the whole lung projection ^{19}F MR images that were reconstructed from the 3D raw data. Unfortunately, little progress was made in the development of this technique for a number of years, likely due to the invention of HP gas MRI and the ability to produce high-quality ^{3}He MR images of the lungs.

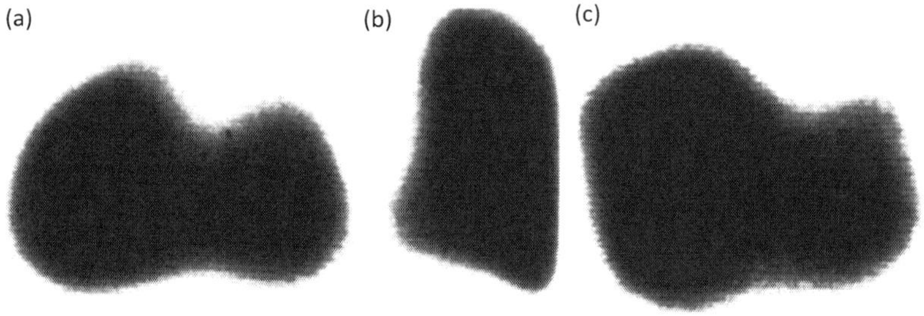

Figure 12.3 The first in vivo ^{19}F MR images acquired in dog lungs using CF_4 in the (a) coronal, (b) sagittal, and (c) axial planes. Images were reproduced with permission from Rinck et al. (1984). *Fortschr Röntgenstr.*, 140, 239–243.

In 1998, Kuethe et al. demonstrated 3D imaging of rat lungs with continuous breathing of a mixture of C_2F_6 and O_2 [15]. These images had a 700 μm resolution, and the acquisition took 4.3 h. It is interesting to note that Kuethe et al. used an image acquisition approach that was essentially the same as Paul Lauterbur's original demonstration of MRI in 1973, except a 3D inverse Fourier transform was used to reconstruct the data [21]. Figure 12.4 shows current examples of ^{19}F MR images of SF_6 in rat lungs that were acquired by our research group. Each image in Fig. 12.4 was a 2D projection in the axial or coronal plane, acquired with two separate 10 s breath-holds using the 2D x-centric pulse sequence [70, 71].

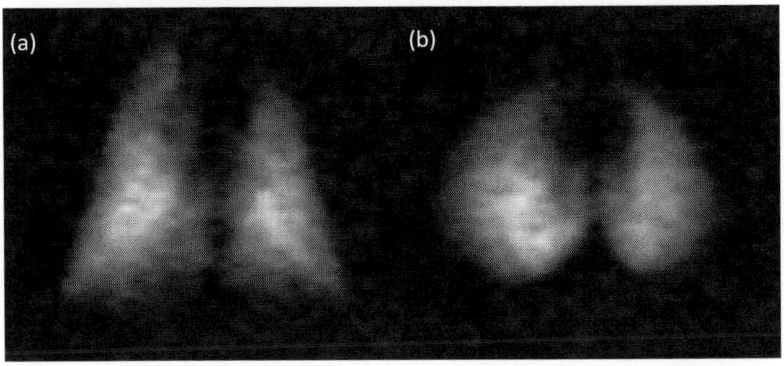

Figure 12.4 ^{19}F MR images of SF_6 in rat lungs. Each image was a 2D projection in the (a) coronal or (b) axial plane, acquired with two separate 10 s breath-holds of SF_6 using the 2D x-centric pulse sequence (Cartesian sampling, partial echo factor = 0.505, 64 × 64 matrix, 938 μm resolution). Images were reproduced with permission from Couch et al. (2014). *NMR Biomed.*, 27, 1525–1534.

Kuethe et al. hypothesized in 1998 that it would be possible to acquire inert fluorinated gas MR images in human lungs with a similar SNR efficiency as ^1H MRI. This hypothesis was based on the assumption that the short T_1 relaxation of fluorinated gases allows for more averaging, and a relatively coarse matrix size (64 × 64) would increase the number of ^{19}F nuclei per voxel [15]. Inert fluorinated gas MRI of human lungs was first demonstrated by Wolf et al. in 2008 using a mixture of SF_6 and O_2 [72]. 2D whole lung projection images were reported in a healthy volunteer, with an SNR~9. Although these initial images were very poor in quality, this work was an important benchmark as not only did safety need to be taken into consideration, but there were likely significant regulatory obstacles that needed to be overcome before human studies could begin with this new imaging modality.

More recently, inert fluorinated gas MRI has been reported in healthy volunteers by Couch et al. [19]. Figure 12.5 shows 12 slices from a ^{19}F 3D UTE dataset that were obtained in a healthy volunteer during a 15 s breath-hold of a mixture of 79% PFP and 21% O_2.

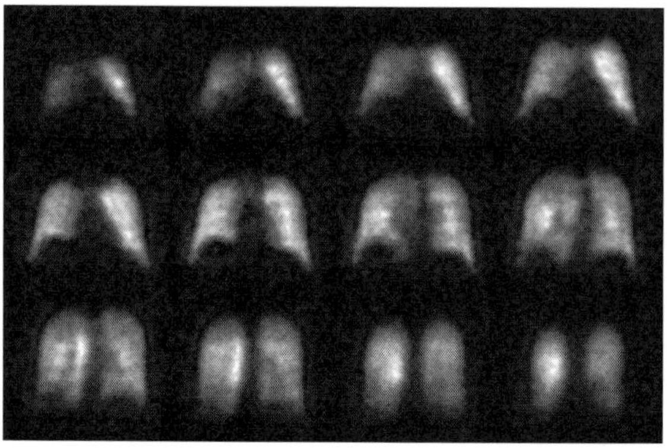

Figure 12.5 Coronal pulmonary ^{19}F 3D UTE MR images obtained in a healthy volunteer during a 15 s breath-hold after continuous breathing of a mixture of 79% PFP and 21% O_2 (64 × 64 matrix, 7 mm resolution). Images were reproduced with permission from Couch et al. (2013). *Radiology,* 269, 903–909.

As expected, the distribution of ^{19}F signal was homogeneous in this healthy individual. These images were acquired at 3 T using a flexible wrap-around quadrature transmit-receive coil (Clinical MR Solutions, Brookfield, WI). Figure 12.6 shows similar ^{19}F 3D UTE MR images that were acquired in the axial plane using the same settings as Fig. 12.5.

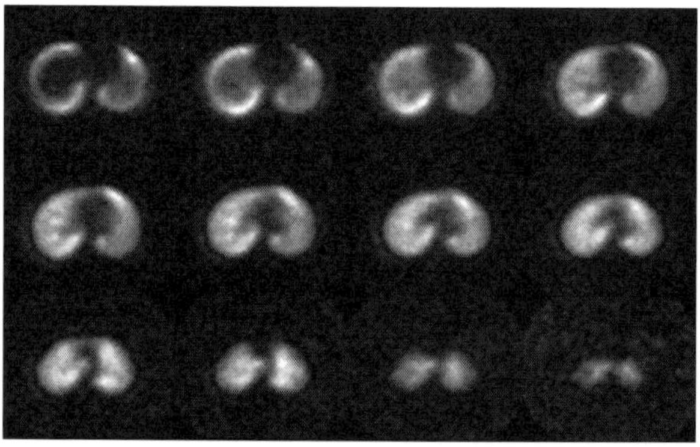

Figure 12.6 Axial pulmonary ^{19}F 3D UTE MR images obtained in a healthy volunteer during a 15 s breath-hold after continuous breathing of a mixture of 79% PFP and 21% O_2 (64 × 64 matrix, 7 mm resolution). Images were reproduced with permission from Couch et al. (2013). *Radiology,* 269, 903–909.

For the ^{19}F lung images shown in Figs. 12.5 and 12.6, images were acquired under breath-hold conditions, following continuous breathing of the PFP/O_2 gas mixture. A 5 L Tedlar bag was used to give five to seven breaths of the PFP/O_2 mixture, which effectively washed out most the residual air from the lungs, allowing the inert fluorinated gas concentration to reach an approximate steady state. In order to acquire ^{19}F MR images at a consistent lung inflation of approximately FRC + 1 L, the 5 L bag was replaced with a full 1 L bag of the PFP/O_2 mixture, which was completely inhaled before the volunteer held their breath for 15 s. For ^{19}F 3D UTE MRI using this continuous breathing protocol, Couch et al. reported a mean SNR (± standard deviation) of 32 ± 6 [19]. Due to the short T_2^* of inert fluorinated gases (as short as 2.2 ms for PFP in the lungs at 3 T), the SNR was expected to benefit from a short echo time. It should be noted, however, that the ^{19}F UTE lung images in Figs. 12.5 and 12.6 show substantial T_2^* blurring and phase artifacts around the edges of the lungs.

Halaweish et al. recently demonstrated 3D gradient echo ^{19}F MRI using the same mixture of PFP and O_2 in both healthy volunteers and in patients with COPD, asthma, and lung transplants [20]. Figure 12.7a shows an example of a ^{19}F 3D gradient echo MR image that was acquired in the coronal plane from a COPD patient with emphysema.

The ^{19}F images shown in Fig. 12.7 contain signal voids that appear to be qualitatively similar to ventilation defects in HP ^3He and ^{129}Xe MRI. Similar to Couch et al., these images were acquired at 3 T using a flexible vest coil, and imaging occurred after several breaths of the PFP/O_2 mixture were taken and the inert fluorinated gas concentration reached a steady state. The PFP/O_2 mixture was delivered to subjects using a custom-built MR-safe gas delivery system and breathing mask [73]. The computer-controlled delivery system was able to switch between air, O_2, and the fluorinated gas mixture, as well as provide additional information in real-time, such as heart rate, O_2 saturation, O_2 and CO_2 concentration, and flow rate. Ideally, a robust gas delivery system can potentially improve patient compliance, as well as provide additional information to aid in the interpretation of inert fluorinated gas images. The ^{19}F gradient echo MR images from Halaweish et al. had a mean SNR (± standard deviation) of 16 ± 6 for healthy volunteers and 12 ± 5 for patients with respiratory diseases including COPD, asthma, and lung transplants [20]. The SNR of these gradient echo images was, on average, approximately half of the UTE SNR reported by Couch et al. [19].

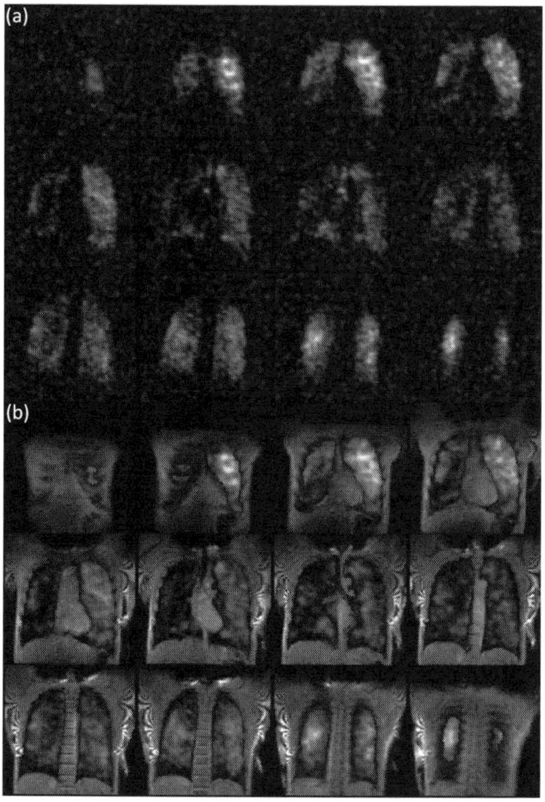

Figure 12.7 Lung images in COPD patient with emphysema. (a) Coronal ^{19}F 3D MR images obtained using a mixture of 79% PFP and 21% O_2 (64 × 64 matrix, 6.25 mm resolution). (b) Lung ^{19}F MR images overlaid in grayscale on top of the conventional ^{1}H MR images. Images were reproduced with permission from Halaweish et al. (2013). *Chest,* 144, 1300–1310.

Halaweish et al. showed that inert fluorinated gas MRI has the ability to detect ventilation defects in patients with COPD, in a manner similar to HP noble gas MRI. Future work in this area will focus on quantifying ventilation defects, and validating these measurements with a direct comparison to HP ^3He and ^{129}Xe MRI. For example, the total ventilated volume (VV) and VDV can be obtained from Fig. 12.7b, which shows the ^{19}F gradient echo image from a COPD patient with emphysema overlaid on a conventional ^1H localizer image [20, 74]. The density and diffusivity of PFP will be important in interpreting the size of ventilation defects, as a recent comparison of VDV measurements using HP ^3He and ^{129}Xe MRI in COPD patients demonstrated a significantly greater VDV as measured from HP ^{129}Xe MRI. It was suggested that the higher density and lower diffusivity of ^{129}Xe may be contributing factors that lead to slower filling in the terminal airways and larger disease-related defects [52]. On the other hand, the use of continuous breathing to reach a steady-state concentration of the inert fluorinated gas before imaging may compensate for the slow filling and make ventilation defects appear smaller. It should also be noted that an accurate measurement of VV and VDV may require corrections for B_1 inhomogeneities [75].

Overall, Couch et al. [19] and Halaweish et al. [20] have demonstrated that inert fluorinated gas MRI is a safe and well-tolerated technique and that it can be performed in volunteers and patients with severe pulmonary diseases. PFP was the gas of choice used in the recent work of both Couch et al. and Halaweish et al.; however, using SF_6 may provide some additional advantages, especially due to its shorter T_1 relaxation time and single resonance. On the other hand, the desire to use shorter repetition times with SF_6 may lead to issues with specific absorption rate (SAR) limits. Regardless of the choice of fluorinated gas, substantial improvements have clearly been made since the early animal work by Rinck et al. [68] and the early human imaging work by Wolf et al. [72], and these improvements in image quality are ongoing.

12.6 Dynamic Imaging

Since inert fluorinated gases are generally mixed with at least 21% O_2, it is possible to perform dynamic MR imaging during long periods of continuous breathing of the inert fluorinated gas mixture, without sacrificing the safety of the animal or human subject. The majority of the dynamic ^{19}F MRI literature has been performed in animals. In one study of rats continuously breathing a mixture of SF_6 and O_2, lung volumes were measured throughout the respiratory cycle by analyzing the periodic changes in whole-lung spectroscopy signals [76]. This technique could easily be translated to humans, as an SNR benefit would be expected from the differences in lung size alone. Naturally, this raises possibilities of using ^{19}F MRI to obtain information that is similar to pulmonary function tests, while using the imaging component to provide regional information that is complementary to the dynamic lung volume measurements.

A number of animal studies have used dynamic ^{19}F MRI to investigate the kinetics of pulmonary ventilation by measuring the wash-in and/or wash-out time constants of inert fluorinated gases [17, 77, 78]. For example, one study in pigs measured both the wash-in and wash-out kinetics of SF_6 by fitting mono-exponential decays to the signal enhancement curves, and the wash-out time constants were found to be longer than the wash-in time constants [17]. This result qualitatively agreed with literature expectations from xenon-enhanced CT (Xe-CT), and this difference in wash-in and wash-out time constants may be related to the density and viscosity of heavy gases [79]. Dynamic ^{19}F MRI of SF_6 wash-in measurements in pigs have been validated with a comparison to respiratory gas analysis [80].

Dynamic imaging approaches can also be used to map the fractional ventilation parameter, *r*, which is defined as the ratio of new gas entering a voxel divided by the total volume of gas inside the voxel at end inspiration [81] and is given by

$$r = \frac{V_{\text{new}}}{V_{\text{total}}} = \frac{V_{\text{new}}}{V_{\text{old}} + V_{\text{new}}} \tag{12.1}$$

In contrast to the above discussion, where bulk estimates of the time constants were determined, ventilation maps can be calculated using a pixel-by-pixel analysis of the wash-in/wash-out signal enhancement. Similar ventilation mapping techniques have been reported in animals using Xe-CT [82, 83] and HP noble gas MRI [84, 85]. Ventilation mapping using inert fluorinated gas MRI has recently been demonstrated in rats that were ventilated with PFP and SF_6 [71]. This study measured ventilation during the wash-out phase with a breathing scheme where the animal was first ventilated with a mixture of PFP/SF_6 and O_2 for at least 3 min, and then a baseline ^{19}F image was acquired during a 10 s breath-hold. The breathing mixture was then switched to 100% O_2 for a total of nine breaths, and a ^{19}F image was acquired during a 10 s breath-hold following each of the O_2 breaths. Figure 12.8 shows a series of eight axial ^{19}F 2D projection rat lung images that were obtained during SF_6 wash-out. Note that the images for the final two O_2 wash-out breaths are not shown. These images were acquired using a two-breath 2D x-centric approach, similar to the rat lung images shown in Fig. 12.4.

Figure 12.8 Series of axial ^{19}F 2D projection lung images obtained in a healthy rat during SF_6 wash-out (baseline SF_6 image followed by seven breaths of 100% O_2, final two O_2 breaths not shown). Similar to Fig. 12.4, each image was obtained with a two-breath 2D x-centric approach (Cartesian sampling, partial echo factor = 0.505).

The whole-lung ^{19}F MRI signal for the images in Fig. 12.8 is plotted as a function of O_2 wash-out breath number in Fig. 12.9, and the dashed line represents the best fit to the signal equation

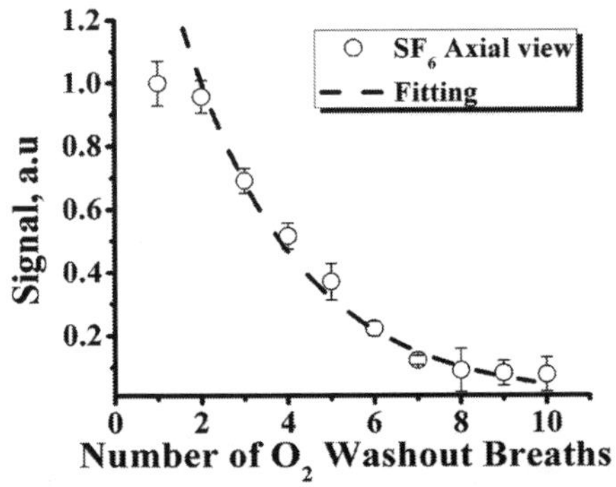

Figure 12.9 Whole-lung ^{19}F MRI signal as a function of O_2 wash-out breath number from a fractional ventilation measurement in a healthy rat that was ventilated with SF_6 at baseline. These data correspond to the images in Fig. 12.8, and the dashed line represents the best fit to Equation 12.2.

$$S(n) = \text{const.}\ (1 - r)^n \qquad (12.2)$$

where S is the signal at breath n, and r is the fractional ventilation. The mean whole lung r value in this case was 0.41 ± 0.04, which agrees with the expected r value that is based on the known ventilator settings and the expected lung size [TV = 3 mL, FRC = 8 mL, r = TV/(TV + FRC) = 0.33]. The final ventilation map is discussed below in Section 12.9.

Halaweish et al. recently investigated the wash-in and wash-out characteristics of PFP in healthy volunteers and patients with respiratory diseases, such as asthma and COPD [86]. In this study, a total of eight ^{19}F 3D gradient echo images were acquired during ~15 s breath-holds in order to capture the wash-in and wash-out of a mixture of 79% PFP and 21% O_2. Imaging was interleaved with three breaths of the PFP/O_2 mixture during the wash-in phase or with three breaths of room air during the wash-out phase. For healthy volunteers, the ^{19}F signal distribution was homogeneous, and the wash-in/wash-out curves were qualitatively similar to Xe-CT results [79]. As expected, the images from patients with respiratory diseases showed increased heterogeneity and ventilation defects; however, there was significant variability in the appearance of the wash-in/wash-out curves, which can be indicative of gas trapping. Preliminary efforts to quantify wash-in and wash-out time constants in healthy volunteers and patients with COPD are currently underway [87], and future work will focus on mapping the fractional ventilation parameter in human lungs. There has been a recent focus on ventilation mapping in humans using HP ^{3}He MRI, as it may provide complementary information to ADC and P_AO_2 mapping [65, 66]. A heterogeneous distribution of fractional ventilation has been observed in COPD patients, along with areas of high r near ventilation defects, which may be due to gas transport between well-ventilated and poorly ventilated regions of the lung [85].

12.7 Diffusion Imaging

HP noble gas ADC measurements are well known to be sensitive to the lung microstructure, and these measurements can distinguish between healthy and emphysematous tissue [44, 53]. Preliminary inert fluorinated gas ADC measurements are currently underway, and this technique may be able to provide similar information to HP noble gas MRI. The first demonstration of inert fluorinated gas ADC measurements was performed in healthy rats using SF_6, and the measured ADC was slightly less than the self-diffusion coefficient of SF_6 [16, 61]. An elastase-induced model of emphysema in rats was employed to help determine whether this technique had a potential to detect pulmonary disease, and elevated C_2F_6 ADC values were detected in emphysematous regions of rat lungs [88]. This result was expected, based on previous measurements in the literature performed using HP noble gas MRI. Significant differences in C_2F_6 ADC values were also detected in a study of healthy and emphysematous excised human lungs [89, 90], which was an important first step towards translating this technique to human imaging.

Despite the challenges of imaging gases with very short T_2^* values, it is possible to probe ADCs in humans using a UTE approach in combination with bi-polar diffusion-sensitizing gradients. As an example, Fig. 12.10a shows three slices from a ^{19}F 3D UTE image without diffusion weighting acquired in a healthy volunteer using a mixture of 79% PFP and 21% O_2 (images with diffusion-weighting not shown). The images shown in Fig. 12.10a were acquired in our lab using methods similar to the UTE images shown in Section 12.5. In order to achieve a b value of 9.59 $s \cdot cm^{-2}$ and diffusion time of 1 ms, a TE of 3.8 ms was required. In this case, these settings yielded an SNR in the center slice image (without diffusion weighting) of approximately 15. Figure 12.10b shows the corresponding ADC maps with mean ADC values of 0.034 ± 0.021 $cm^2 \cdot s^{-1}$, 0.025 ± 0.016 $cm^2 \cdot s^{-1}$, and 0.023 ± 0.011 $cm^2 \cdot s^{-1}$, respectively, where the error represents the heterogeneity in each respective ADC map.

The ADC values for the healthy volunteer shown in Fig. 12.10 were similar to previously published values for the diffusion of PFP mixed with O_2 [62]. This was to be expected for a diffusion time of 1 ms and the lightly restricted diffusion length scale that was probed. A longer diffusion time would be required to reach the restricted diffusion regime that is normally used for HP noble gas measurements of ADC. Some of these HP noble gas measurements employ a geometrical model in order to determine morphological parameters, and this method has been successfully validated against lung histology [54, 91]. Since this method requires long diffusion times that necessitate long TEs, lung morphometry measurements will not be possible with inert fluorinated gases. However, by probing multiple diffusion times, it may be possible to measure the surface-to-volume ratio (S/V) with inert fluorinated gases, which is an important biomarker of lung microstructure [90, 92]. Measuring S/V requires at least two diffusion times that are less than or on the order of T_2^*. It is also possible to mix inert fluorinated gases with other inert gases, such as ^{4}He, to increase the free diffusivity, and hence the diffusion length scale that will be probed. Although there are a number of potential difficulties, inert fluorinated gas diffusion measurements have been demonstrated in humans, and there is a possibility for probing the lung microstructure in humans with pulmonary diseases.

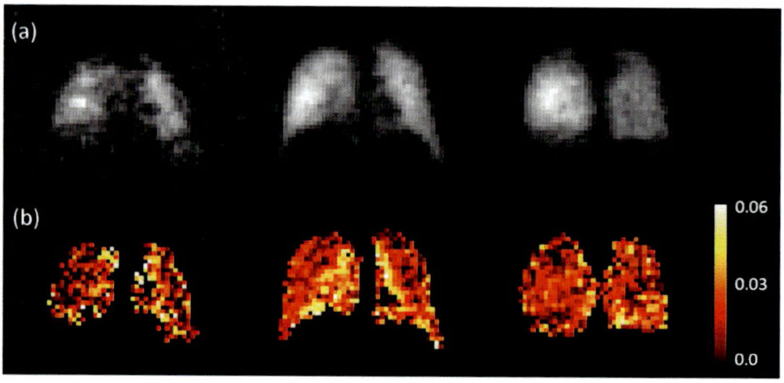

Figure 12.10 Diffusion imaging in the lung using fluorinated gas MRI. (a) Three slices from a ^{19}F 3D UTE lung image without diffusion weighting, and (b) the corresponding ADC maps. An image without diffusion-weighting and an image with b = 9.59 s·cm^{-2} were both acquired within a single 15 s breath-hold following continuous breathing of a mixture of 79% PFP and 21% O_2 in a healthy volunteer. Images were reproduced with permission from Couch et al. (2014). *NMR Biomed.*, 27, 1525–1534.

12.8 *V/Q* Measurement

The ventilation/perfusion ratio (V/Q) is known to be directly related to gas exchange efficiency, and detecting a mismatch between ventilation and perfusion can be a sensitive indicator of pulmonary disease [93]. V/Q measurements have recently been performed with ^1H MRI using a combination of three techniques: O_2-enhanced MRI to measure specific ventilation, a modified arterial spin labeling (ASL) approach to measure perfusion, and a fast gradient echo method to measure regional lung density [10]. It is also possible to measure V/Q using HP ^3He MRI, and this method uses regional P_AO_2 measurements to determine V/Q [94].

Kuethe et al. first reported inert fluorinated gas V/Q measurements using rats that continuously breathed a mixture of SF_6 and O_2 [18]. In essence, two images were acquired with different SF_6 concentrations as follows: one image was acquired using 80% SF_6 and 20% O_2, and the resulting SF_6 signal intensity was fairly homogeneous; the second image was acquired using 25% SF_6 and 75% O_2, and the resulting SF_6 signal intensity varied according to regional variations in V/Q. By taking the quotient image and numerically computing the dependency of V/Q on the fluorinated gas partial pressures, a map of V/Q was generated. In a model of obstructed ventilation, differences in V/Q distributions were detected for the left and right lungs. The same group developed another approach to V/Q mapping, which involved mapping T_1 with a modified Look-Locker technique [95]. A map of V/Q was generated by computing the dependency of T_1 on the fluorinated gas partial pressure, which in turn is related to V/Q. The T_1 method for V/Q mapping yielded similar results to the quotient method in the model of obstructed ventilation in rats [95].

Since V/Q is a meaningful biomarker for disease, it would naturally be of great interest to extend these V/Q mapping techniques to human imaging; however, it should be noted that this technique has some potential limitations. In particular, V/Q mapping using inert fluorinated gas MRI is weakly sensitive to high and low V/Q values. Although Adolphi et al. reported V/Q values ranging from about 0.01 to 10, the V/Q is weakly dependent on the fluorinated gas T_1 at both high and low V/Q values [95]. Therefore, a robust method for T_1 mapping is required, as any small errors in determining T_1 will lead to large errors in V/Q. The ongoing development and optimization of efficient image acquisition techniques will determine if there is sufficient SNR to measure T_1 and generate V/Q maps in humans within reasonable safety limits.

12.9 Gravitational Distribution

It is well known that pulmonary ventilation exhibits a gravitational gradient due to a gradient in regional compliance. In other words, ventilation is always greater in more gravitationally dependent regions of the lung (i.e. the lowest part of the lung). The intrapleural pressure surrounding the lungs is less than atmospheric pressure due to the elasticity of the lung and expansion of the chest wall during breathing. In order to support the weight of the lungs, the intrapleural pressure is less negative in dependent regions. Therefore, alveoli in dependent regions of the lung have a smaller resting volume and are easier to inflate during breathing (i.e., they are more compliant) [6].

This relationship in ventilation distribution has been previously demonstrated with a variety of imaging techniques, such as Xe-CT [82], O_2-enhanced ^1H MRI [31], and HP noble gas MRI [84, 96]. These imaging techniques typically employ a wash-in/wash-out approach to quantify regional fractional ventilation, and then gradients in ventilation are calculated in the vertical direction. Ventilation gradients are also known to be posture-dependent, as stronger gradients are expected in the supine position compared to the prone position [10, 96, 97]. It would be natural to expect that ventilation measurements obtained from inert fluorinated gas MRI will exhibit similar gravitational gradients.

Section 12.6 discussed in detail the potential for wash-in/wash-out imaging with inert fluorinated gas MRI, and efforts to perform ventilation mapping and ventilation gradient measurements are currently underway in animals and humans. Figure 12.11a shows a representative axial 2D projection fractional ventilation map that was obtained in a healthy rat in the supine position following SF_6 washout. This map was generated using the SF_6 washout images from Fig. 12.8 and a pixel-by-pixel non-linear least-squares fit to Equation 12.2. Figure 12.11b shows the corresponding fractional ventilation gradient plotted from posterior to anterior. That is, the mean *r* value from each row in Fig. 12.11a is plotted as a function of distance in the vertical direction in Fig. 12.11b. The solid line represents a linear regression, and ventilation gradient has a slope of -0.011 cm^{-1}. As expected, a negative gradient indicates that ventilation is greater in more dependent regions of the lung, and these results are qualitatively similar to Xe-CT [82] and HP gas MRI [84].

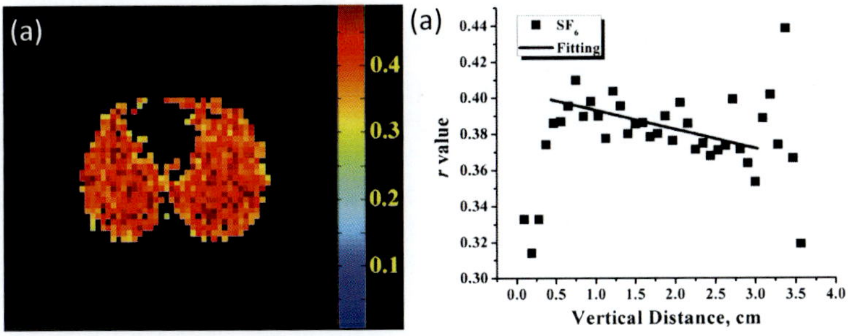

Figure 12.11 Fractional ventilation map obtained in a healthy rat. (a) Axial 2D projection fractional ventilation map using the SF_6 washout images shown in Fig. 12.8. (b) The corresponding fractional ventilation gradient plotted from posterior to anterior with a slope of -0.011 cm^{-1}.

Preliminary data demonstrating gravitational gradients of ^{19}F signal in human lungs have been reported [98]. Figure 12.12a shows one example of a whole lung axial projection image that was obtained in our lab by summing slices from a representative static breath-hold ^{19}F 3D UTE image acquired in a healthy volunteer lying in the supine position. The image shown in Fig. 12.12a was acquired using methods similar to Section 12.5, where the volunteer continuously breathed a mixture of 79% PFP and 21% O_2, and UTE images were acquired during a 15 s breath-hold. Figure 12.12b shows the normalized mean signal intensity from the axial projection image in Fig. 12.12a plotted as a function of vertical distance from posterior to anterior. The solid line represents the calculated gradient, which has a slope of -0.033 cm^{-1}, representing an 84% change in signal intensity from the anterior to posterior edges of the lung.

Although this preliminary data appears to demonstrate the expected gradient in the ^{19}F signal, it is difficult to compare these results to literature values, which generally quantify ventilation rather than analyze static breath-hold images. In the future, quantitative ventilation mapping in humans will determine if the ventilation gradients obtained from ^{19}F MRI can provide meaningful information regarding lung physiology and the gravitational distribution of ventilation in the lungs.

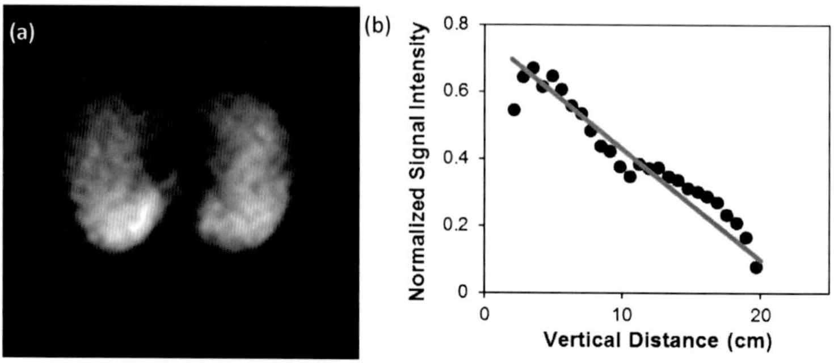

Figure 12.12 Human lung axial projection image. (a) Whole lung axial projection ^{19}F MR image acquired in a representative healthy volunteer during a 15 s breath-hold of a mixture of 79% PFP and 21% O_2. (b) Normalized signal intensity from the axial projection image plotted as a function of vertical distance in the lungs. The solid line represents the calculated gradient. Images were reproduced with permission from Couch et al. (2014). *NMR Biomed.*, 27, 1525–1534.

12.10 Conclusions

Although the first inert fluorinated gas images reported by Heidelberger and Lauterbur were extremely primitive, it was recognized in their 1982 abstract that gas-phase MRI could potentially become an important component of pulmonary imaging technology [67]. That statement still rings true today, as HP noble gas and inert fluorinated gas technologies continue to evolve. The same group that originally co-pioneered HP noble gas MRI is now helping to develop the next pulmonary imaging modality—inert fluorinated gas MRI. Once fully optimized, this technique may have significant advantages over HP noble gas MRI, since it is inexpensive, the gases are abundant, there is a sufficient thermal polarization for imaging, and the technique can be performed on any MRI scanner with broadband capability. Furthermore, inert fluorinated gas MRI is a safe technique that can be used over multiple breaths, since the gases can be mixed with O_2.

Future work will investigate the use of parallel imaging, respiratory gating, novel pulse sequences, and a wider range of inert fluorinated gases in order to further optimize this new pulmonary imaging technique. The ongoing improvements in image quality will allow for the continued investigation of static breath-hold imaging, dynamic imaging and the detection of functional biomarkers, such as *V*/*Q*, ADC and regional ventilation. All of these developments and advantages may eventually lead to regulatory approval for routine clinical imaging using inert fluorinated gas MRI. Overall, inert fluorinated gas MRI has the potential to become a viable clinical imaging modality for non-invasively imaging the lung and aiding in the management of respiratory diseases.

Acknowledgments

Portions of this chapter were reproduced with permission from Couch et al. *NMR Biomed,* 2014; DOI:10.1002/nbm.3165. This work was funded in part by the Thunder Bay Regional Research Institute (TBRRI) and Lakehead University. MJC was supported by a Natural Sciences and Engineering Research Council (NSERC) Canada Graduate Scholarship (CGS). Thanks to the Thunder Bay Regional Health Sciences Centre (TBRHSC) MR technologists for their time and assistance with MR scanning of volunteers and to Ralph Hashoian for making the human RF coils and interface electronics.

References

1. Boulet LP, Bourbeau J, Skomro R, Gupta S. Major care gaps in asthma, sleep and chronic obstructive pulmonary disease: A road map for knowledge translation. *Can Respir J,* 2013; 20: 265–269.
2. Martinez FD, Vercelli D. Asthma. *Lancet,* 2013; 382: 1360–1372.
3. Vijayan VK. Chronic obstructive pulmonary disease. *Indian J Med Res,* 2013; 137: 251–269.
4. Miller MR, Hankinson J, Brusasco V, Burgos F, Casaburi R, Coates A, Crapo R, Enright P, van der Grinten CP, Gustafsson P, Jensen R, Johnson DC, MacIntyre N, McKay R, Navajas D, Pedersen OF, Pellegrino R, Viegi G, Wanger J. Standardisation of spirometry. *Eur Respir J,* 2005; 26: 319–338.
5. Criee CP, Sorichter S, Smith HJ, Kardos P, Merget R, Heise D, Berdel D, Kohler D, Magnussen H, Marek W, Mitfessel H, Rasche K, Rolke M, Worth H, Jorres RA. Body plethysmography—its principles and clinical use. *Respir Med,* 2011; 105: 959–971.

6. West JB. Regional differences in the lung. *Chest*, 1978; 74: 426–437.
7. Mills GH, Wild JM, Eberle B, Van Beek EJ. Functional magnetic resonance imaging of the lung. *Br J Anaesth*, 2003; 91: 16–30.
8. Fazel R, Krumholz HM, Wang Y, Ross JS, Chen J, Ting HH, Shah ND, Nasir K, Einstein AJ, Nallamothu BK. Exposure to low-dose ionizing radiation from medical imaging procedures. *N Engl J Med*, 2009; 361: 849–857.
9. Johnson KM, Fain SB, Schiebler ML, Nagle S. Optimized 3D ultrashort echo time pulmonary MRI. *Magn Reson Med*, 2013; 70: 1241–1250.
10. Henderson AC, Sa RC, Theilmann RJ, Buxton RB, Prisk GK, Hopkins SR. The gravitational distribution of ventilation-perfusion ratio is more uniform in prone than supine posture in the normal human lung. *J Appl Physiol*, 2013; 115: 313–324.
11. Altes TA, Salerno M. Hyperpolarized gas MR imaging of the lung. *J Thorac Imaging*, 2004; 19: 250–258.
12. Lilburn DM, Pavlovskaya GE, Meersmann T. Perspectives of hyperpolarized noble gas MRI beyond ^{3}He. *J Magn Reson*, 2013; 229: 173–186.
13. Liu Z, Araki T, Okajima Y, Albert M, Hatabu H. Pulmonary hyperpolarized noble gas MRI: Recent advances and perspectives in clinical application. *Eur J Radiol*, 2014; 83: 1282–1291.
14. Couch MJ, Blasiak B, Tomanek B, Ouriadov AV, Fox MS, Dowhos KM, Albert MS. Hyperpolarized and inert gas MRI: The future. *Mol Imaging Biol*, 2015; 17: 149–162.
15. Kuethe DO, Caprihan A, Fukushima E, Waggoner RA. Imaging lungs using inert fluorinated gases. *Magn Reson Med*, 1998; 39: 85–88.
16. Perez-Sanchez JM, Perez de Alejo R, Rodriguez I, Cortijo M, Peces-Barba G, Ruiz-Cabello J. In vivo diffusion weighted ^{19}F MRI using SF_6. *Magn Reson Med*, 2005; 54: 460–463.
17. Schreiber WG, Eberle B, Laukemper-Ostendorf S, Markstaller K, Weiler N, Scholz A, Burger K, Heussel CP, Thelen M, Kauczor HU. Dynamic ^{19}F-MRI of pulmonary ventilation using sulfur hexafluoride (SF_6) gas. *Magn Reson Med*, 2001; 45: 605–613.
18. Kuethe DO, Caprihan A, Gach HM, Lowe IJ, Fukushima E. Imaging obstructed ventilation with NMR using inert fluorinated gases. *J Appl Physiol*, 2000; 88: 2279–2286.
19. Couch MJ, Ball IK, Li T, Fox MS, Littlefield SL, Biman B, Albert MS. Pulmonary ultrashort echo time ^{19}F MR imaging with inhaled fluorinated gas mixtures in healthy volunteers: Feasibility. *Radiology*, 2013; 269: 903–909.

20. Halaweish AF, Moon RE, Foster WM, Soher BJ, McAdams HP, MacFall JR, Ainslie MD, MacIntyre NR, Charles HC. Perfluoropropane gas as a magnetic resonance lung imaging contrast agent in humans. *Chest*, 2013; 144: 1300–1310.
21. Lauterbur PC. Image formation by induced local interactions—examples employing nuclear magnetic-resonance. *Nature*, 1973; 242: 190–191.
22. Ma W, Sheikh K, Svenningsen S, Pike D, Guo F, Etemad-Rezai R, Leipsic J, Coxson HO, McCormack DG, Parraga G. Ultra-short echo-time pulmonary MRI: Evaluation and reproducibility in COPD subjects with and without bronchiectasis. *J Magn Reson Imaging*, 2015; 41: 1465–1474.
23. Weiger M, Wu M, Wurnig MC, Kenkel D, Jungraithmayr W, Boss A, Pruessmann KP. Rapid and robust pulmonary proton ZTE imaging in the mouse. *NMR Biomed*, 2014; 27: 1129–1134.
24. Gibiino F, Sacolick L, Menini A, Landini L, Wiesinger F. Free-breathing, zero-TE MR lung imaging. *Magn Reson Mater Phys*, 2015; 28: 207–215.
25. Bieri O. Ultra-fast steady state free precession and its application to in vivo H morphological and functional lung imaging at 1.5 tesla. *Magn Reson Med*, 2013; 70: 657–663.
26. Edelman RR, Hatabu H, Tadamura E, Li W, Prasad PV. Noninvasive assessment of regional ventilation in the human lung using oxygen-enhanced magnetic resonance imaging. *Nat Med*, 1996; 2: 1236–1239.
27. Ohno Y, Chen Q, Hatabu H. Oxygen-enhanced magnetic resonance ventilation imaging of lung. *Eur J Radiol*, 2001; 37: 164–171.
28. Chen Q, Jakob PM, Griswold MA, Levin DL, Hatabu H, Edelman RR. Oxygen enhanced MR ventilation imaging of the lung. *Magn Reson Mater Phys*, 1998; 7: 153–161.
29. Togao O, Tsuji R, Ohno Y, Dimitrov I, Takahashi M. Ultrashort echo time (UTE) MRI of the lung: Assessment of tissue density in the lung parenchyma. *Magn Reson Med*, 2010; 64: 1491–1498.
30. Kruger SJ, Fain SB, Johnson KM, Cadman RV, Nagle SK. Oxygen-enhanced 3D radial ultrashort echo time magnetic resonance imaging in the healthy human lung. *NMR Biomed*, 2014; 27: 1535–1541.
31. Sa RC, Cronin MV, Henderson AC, Holverda S, Theilmann RJ, Arai TJ, Dubowitz DJ, Hopkins SR, Buxton RB, Prisk GK. Vertical distribution of specific ventilation in normal supine humans measured by oxygen-enhanced proton MRI. *J Appl Physiol*, 2010; 109: 1950–1959.

32. Bauman G, Puderbach M, Deimling M, Jellus V, Chefd'hotel C, Dinkel J, Hintze C, Kauczor HU, Schad LR. Non-contrast-enhanced perfusion and ventilation assessment of the human lung by means of Fourier decomposition in proton MRI. *Magn Reson Med*, 2009; 62: 656–664.

33. Lederlin M, Bauman G, Eichinger M, Dinkel J, Brault M, Biederer J, Puderbach M. Functional MRI using Fourier decomposition of lung signal: Reproducibility of ventilation- and perfusion-weighted imaging in healthy volunteers. *Eur J Radiol*, 2013; 82: 1015–1022.

34. Bauman G, Lutzen U, Ullrich M, Gaass T, Dinkel J, Elke G, Meybohm P, Frerichs I, Hoffmann B, Borggrefe J, Knuth HC, Schupp J, Prum H, Eichinger M, Puderbach M, Biederer J, Hintze C. Pulmonary functional imaging: Qualitative comparison of Fourier decomposition MR imaging with SPECT/CT in porcine lung. *Radiology*, 2011; 260: 551–559.

35. Bauman G, Scholz A, Rivoire J, Terekhov M, Friedrich J, de Oliveira A, Semmler W, Schreiber LM, Puderbach M. Lung ventilation- and perfusion-weighted Fourier decomposition magnetic resonance imaging: In vivo validation with hyperpolarized ^{3}He and dynamic contrast-enhanced MRI. *Magn Reson Med*, 2013; 69: 229–237.

36. Miller GW, Mugler JP, 3rd, Sa RC, Altes TA, Prisk GK, Hopkins SR. Advances in functional and structural imaging of the human lung using proton MRI. *NMR Biomed*, 2014; 27: 1542–1556.

37. Walker TG, Happer W. Spin-exchange optical pumping of noble-gas nuclei. *Rev Mod Phys*, 1997; 69: 629–642.

38. Albert MS, Cates GD, Driehuys B, Happer W, Saam B, Springer CS, Jr., Wishnia A. Biological magnetic resonance imaging using laser-polarized ^{129}Xe. *Nature*, 1994; 370: 199–201.

39. Albert MS, Balamore D. Development of hyperpolarized noble gas MRI. *Nucl Instrum Meth A*, 1998; 402: 441–453.

40. Zhou X, Sun Y, Mazzanti M, Henninger N, Mansour J, Fisher M, Albert M. MRI of stroke using hyperpolarized ^{129}Xe. *NMR Biomed*, 2010; 24: 170–175.

41. Mazzanti ML, Walvick RP, Zhou X, Sun Y, Shah N, Mansour J, Gereige J, Albert MS. Distribution of hyperpolarized xenon in the brain following sensory stimulation: Preliminary MRI findings. *PLoS One*. 2011; 6: e21607.

42. Rao MR, Stewart NJ, Norquay G, Wild JM. Spectroscopy of Dissolved ^{129}Xe in Human Brain at 1.5 T. *Proc 22nd Annual Meeting of ISMRM*, 2014; 3532.

43. Couch MJ, Ball IK, Li T, Fox MS, Ouriadov AV, Biman B, Albert MS. Inert fluorinated gas MRI: A new pulmonary imaging modality. *NMR Biomed*, 2014; 27: 1525–1534.

44. Parraga G, Ouriadov A, Evans A, McKay S, Lam WW, Fenster A, Etemad-Rezai R, McCormack D, Santyr G. Hyperpolarized ^{3}He ventilation defects and apparent diffusion coefficients in chronic obstructive pulmonary disease: Preliminary results at 3.0 Tesla. *Invest Radiol*, 2007; 42: 384–391.

45. Peterson ET, Dai J, Holmes JH, Fain SB. Measurement of lung airways in three dimensions using hyperpolarized helium-3 MRI. *Phys Med Biol*, 2011; 56: 3107–3122.

46. Diaz S, Casselbrant I, Piitulainen E, Pettersson G, Magnusson P, Peterson B, Wollmer P, Leander P, Ekberg O, Akeson P. Hyperpolarized ^{3}He apparent diffusion coefficient MRI of the lung: Reproducibility and volume dependency in healthy volunteers and patients with emphysema. *J Magn Reson Imaging*, 2008; 27: 763–770.

47. Deninger AJ, Eberle B, Ebert M, Grossmann T, Heil W, Kauczor H, Lauer L, Markstaller K, Otten E, Schmiedeskamp J, Schreiber W, Surkau R, Thelen M, Weiler N. Quantification of regional intrapulmonary oxygen partial pressure evolution during apnea by (3)He MRI. *J Magn Reson*, 1999; 141: 207–216.

48. Zhao L, Mulkern R, Tseng CH, Williamson D, Patz S, Kraft R, Walsworth RL, Jolesz FA, Albert MS. Gradient-echo imaging considerations for hyperpolarized Xe-129 MR. *J Magn Reson Ser B*, 1996; 113: 179–183.

49. Kouzes RT, Ely JH, Erikson LE, Kernan WJ, Lintereur AT, Siciliano ER, Stephens DL, Stromswold DC, Van Ginhoven RM, Woodring ML. Neutron detection alternatives to He-3 for national security applications. *Nucl Instrum Meth A*, 2010; 623: 1035–1045.

50. Hersman FW, Ruset IC, Ketel S, Muradian I, Covrig SD, Distelbrink J, Porter W, Watt D, Ketel J, Brackett J, Hope A, Patz S. Large production system for hyperpolarized ^{129}Xe for human lung imaging studies. *Acad Radiol*, 2008; 15: 683–692.

51. Nikolaou P, Coffey AM, Walkup LL, Gust BM, Whiting N, Newton H, Barcus S, Muradyan I, Dabaghyan M, Moroz GD, Rosen MS, Patz S, Barlow MJ, Chekmenev EY, Goodson BM. Near-unity nuclear polarization with an open-source ^{129}Xe hyperpolarizer for NMR and MRI. *Proc Natl Acad Sci U S A*, 2013; 110: 14150–14155.

52. Kirby M, Svenningsen S, Owrangi A, Wheatley A, Farag A, Ouriadov A, Santyr GE, Etemad-Rezai R, Coxson HO, McCormack DG,

Parraga G. Hyperpolarized ^{3}He and ^{129}Xe MR imaging in healthy volunteers and patients with chronic obstructive pulmonary disease. *Radiology*, 2012; 265: 600–610.

53. Kaushik SS, Cleveland ZI, Cofer GP, Metz G, Beaver D, Nouls J, Kraft M, Auffermann W, Wolber J, McAdams HP, Driehuys B. Diffusion-weighted hyperpolarized ^{129}Xe MRI in healthy volunteers and subjects with chronic obstructive pulmonary disease. *Magn Reson Med*, 2011; 65: 1154–1165.

54. Ouriadov A, Farag A, Kirby M, McCormack DG, Parraga G, Santyr GE. Lung morphometry using hyperpolarized (129)-Xe apparent diffusion coefficient anisotropy in chronic obstructive pulmonary disease. *Magn Reson Med*, 2013; 70: 1699–1706.

55. Qing K, Ruppert K, Jiang Y, Mata JF, Miller GW, Shim YM, Wang C, Ruset IC, Hersman FW, Altes TA, Mugler JP, 3rd. Regional mapping of gas uptake by blood and tissue in the human lung using hyperpolarized xenon-129 MRI. *J Magn Reson Imaging*, 2014; 39: 346–359.

56. Driehuys B, Martinez-Jimenez S, Cleveland ZI, Metz GM, Beaver DM, Nouls JC, Kaushik SS, Firszt R, Willis C, Kelly KT, Wolber J, Kraft M, McAdams HP. Chronic obstructive pulmonary disease: Safety and tolerability of hyperpolarized ^{129}Xe MR imaging in healthy volunteers and patients. *Radiology*, 2012; 262: 279–289.

57. Wagner PD. The multiple inert gas elimination technique (MIGET). *Intens Care Med*, 2008; 34: 994–1001.

58. Wen W-Y, Muccitelli JA. Thermodynamics of some perfluorocarbon gases in water. *J Solut Chem*, 1979; 8: 225–246.

59. Leawoods JC, Yablonskiy DA, Saam B, Gierada DS, Conradi MS. Hyperpolarized He-3 gas production and MR imaging of the lung. *Concept Magnetic Res*, 2001; 13: 277–293.

60. Moller HE, Chen XJ, Saam B, Hagspiel KD, Johnson GA, Altes TA, de Lange EE, Kauczor HU. MRI of the lungs using hyperpolarized noble gases. *Magn Reson Med*, 2002; 47: 1029–1051.

61. Ruiz-Cabello J, Perez-Sanchez JM, Perez de Alejo R, Rodriguez I, Gonzalez-Mangado N, Peces-Barba G, Cortijo M. Diffusion-weighted ^{19}F-MRI of lung periphery: Influence of pressure and air-SF_6 composition on apparent diffusion coefficients. *Respir Physiol Neurobiol*, 2005; 148: 43–56.

62. Chang YV, Conradi MS. Relaxation and diffusion of perfluorocarbon gas mixtures with oxygen for lung MRI. *J Magn Reson*, 2006; 181: 191–198.

63. Sarkar K, Katiyar A, Jain P. Growth and dissolution of an encapsulated contrast microbubble: Effects of encapsulation permeability. *Ultrasound Med Biol*, 2009; 35: 1385–1396.

64. Mugler JP, 3rd, Altes TA. Hyperpolarized ^{129}Xe MRI of the human lung. *J Magn Reson Imaging*, 2013; 37: 313–331.

65. Hamedani H, Kadlececk S, Han B, Emami K, Xin Y, Ishii M, Rossman M, Rizi R. A simultaneous multi-breath scheme for measurement of ADC, pAO_2 and fractional ventilation using ^3He MRI in human. *Proc 21st Annual Meeting of ISMRM*, 2013; 1468.

66. Horn FC, Deppe MH, Marshall H, Parra-Robles J, Wild JM. Quantification of regional fractional ventilation in human subjects by measurement of hyperpolarized ^3He washout with 2D and 3D MRI. *J Appl Physiol*, 2014; 116: 129–139.

67. Heidelberger E, Lauterbur PC. Gas Phase ^{19}F-NMR Zeugmatography: A New Approach to Lung Ventilation Imaging. *Proc 1st Annual Meeting Society of Magnetic Resonance in Medicine*, 1982; 70–71.

68. Rinck PA, Petersen SB, Lauterbur PC. NMR-Imaging von fluorhaltigen Substanzen. 19Fluor-Ventilations- und -Perfusionsdarstellungen. *Fortschr Röntgenstr*, 1984; 140: 239–243.

69. Rinck PA, Peterson SB, Heidelberger E, Acuff V, Reinders J, Bernardo ML, Hedges LK, Lauterbur PC. NMR ventilation imaging of the lungs using perfluorinated gases. *Magn Reson Med*, 1984; 1: 237–237.

70. Santyr GE, Lam WW, Parra-Robles JM, Taves TM, Ouriadov AV. Hyperpolarized noble gas magnetic resonance imaging of the animal lung: Approaches and applications. *J Appl Phys*, 2009; 105: 102004.

71. Ouriadov AV, Fox MS, Couch MJ, Li T, Ball IK, Albert MS. In vivo regional ventilation mapping using fluorinated gas MRI with an x-centric FGRE method. *Magn Reson Med*, 2015; 74: 550–557.

72. Wolf U, Scholz A, Terekhov M, Muennemann K, Kreitner K, Werner C, Dueber C, Schreiber WG. Fluorine-19 MRI of the lung: First human experiment. *Proc 16th Annual Meeting of ISMRM*, 2008; 3207.

73. Halaweish AF, Charles HC. Physiorack: An integrated MRI safe/conditional, Gas delivery, respiratory gating, and subject monitoring solution for structural and functional assessments of pulmonary function. *J Magn Reson Imaging*, 2014; 39: 735–741.

74. Kirby M, Heydarian M, Svenningsen S, Wheatley A, McCormack DG, Etemad-Rezai R, Parraga G. Hyperpolarized ^3He magnetic resonance functional imaging semiautomated segmentation. *Acad Radiol*, 2012; 19: 141–152.

75. Miller GW, Altes TA, Brookeman JR, De Lange EE, Mugler JP, 3rd. Hyperpolarized ^{3}He lung ventilation imaging with B1-inhomogeneity correction in a single breath-hold scan. *Magn Reson Mater Phys*, 2004; 16: 218–226.
76. Kuethe DO, Behr VC, Begay S. Volume of rat lungs measured throughout the respiratory cycle using ^{19}F NMR of the inert gas SF_6. *Magn Reson Med*, 2002; 48: 547–549.
77. Wolf U, Scholz A, Heussel CP, Markstaller K, Schreiber WG. Subsecond fluorine-19 MRI of the lung. *Magn Reson Med*, 2006; 55: 948–951.
78. Wolf U, Scholz A, Terekhov M, Koebrich R, David M, Schreiber LM. Visualization of inert gas wash-out during high-frequency oscillatory ventilation using fluorine-19 MRI. *Magn Reson Med*, 2010; 64: 1478–1483.
79. Chon D, Simon BA, Beck KC, Shikata H, Saba OI, Won C, Hoffman EA. Differences in regional wash-in and wash-out time constants for xenon-CT ventilation studies. *Respir Physiol Neurobiol*, 2005; 148: 65–83.
80. Scholz AW, Wolf U, Fabel M, Weiler N, Heussel CP, Eberle B, David M, Schreiber WG. Comparison of magnetic resonance imaging of inhaled SF_6 with respiratory gas analysis. *Magn Reson Imaging*, 2009; 27: 549–556.
81. Deninger AJ, Mansson S, Petersson JS, Pettersson G, Magnusson P, Svensson J, Fridlund B, Hansson G, Erjefeldt I, Wollmer P, Golman K. Quantitative measurement of regional lung ventilation using ^{3}He MRI. *Magn Reson Med*, 2002; 48: 223–232.
82. Lam WW, Holdsworth DW, Du LY, Drangova M, McCormack DG, Santyr GE. Micro-CT imaging of rat lung ventilation using continuous image acquisition during xenon gas contrast enhancement. *J Appl Physiol*, 2007; 103: 1848–1856.
83. Santyr GE, Couch MJ, Lam WW, Ouriadov A, Drangova M, McCormack DG, Holdsworth DW. Comparison of hyperpolarized (3) He MRI with Xe-enhanced computed tomography imaging for ventilation mapping of rat lung. *NMR Biomed*, 2011; 24: 1073–1080.
84. Couch MJ, Ouriadov A, Santyr GE. Regional ventilation mapping of the rat lung using hyperpolarized 129Xe magnetic resonance imaging. *Magn Reson Med*, 2012; 68: 1623–1631.
85. Emami K, Hamedani H, Kadlececk S, Han B, Xin Y, Ishii M, Rossman M, Rizi R. A pilot fractional ventilation imaging study in healthy subjects, asymptomatic smokers and obstructive pulmonary disease patients. *Proc 21st Annual Meeting of ISMRM*, 2013; 1463.

86. Halaweish AF, Foster WM, Moon RE, MacIntyre NR, MacFall JR, Charles HC. Dynamics of Pulmonary Ventilation Distribution at Steady State via ^{19}Fluorine-Enhanced MRI: Initial Experiences and Future Developments. *Proc 21st Annual Meeting of ISMRM*, 2013; 4111.

87. Charles HC, MacIntyre NR, Moon R, Foster WM, McAdams HP, Ainslie M. In-vivo visualization of slow to fast filling compartments in human lungs using ^{19}F MRI of perfluorinated gases mixed with oxygen. *Am Respir Crit Care Med*, 2014; 189: A4348.

88. Carrero-Gonzalez L, Kaulisch T, Stiller D. In vivo diffusion-weighted MRI using perfluorinated gases: ADC comparison between healthy and elastase-treated rat lungs. *Magn Reson Med*, 2013; 70: 1761–1764.

89. Jacob RE, Chang YV, Choong CK, Bierhals A, Zheng Hu D, Zheng J, Yablonskiy DA, Woods JC, Gierada DS, Conradi MS. ^{19}F MR imaging of ventilation and diffusion in excised lungs. *Magn Reson Med*, 2005; 54: 577–585.

90. Conradi MS, Saam BT, Yablonskiy DA, Woods JC. Hyperpolarized ^{3}He and perfluorocarbon gas diffusion MRI of lungs. *Progr NMR Spectr*, 2006; 48: 63–83.

91. Yablonskiy DA, Sukstanskii AL, Quirk JD, Woods JC, Conradi MS. Probing lung microstructure with hyperpolarized noble gas diffusion MRI: Theoretical models and experimental results. *Magn Reson Med*, 2013; 71: 486–505.

92. Mitra PP, Sen PN, Schwartz LM. Short-time behavior of the diffusion coefficient as a geometrical probe of porous media. *Phys Rev B Condens Matter*, 1993; 47: 8565–8574.

93. West JB. *Pulmonary Pathophysiology*. Philadelphia: Lippincott Williams & Wilkins; 2013.

94. Rizi RR, Baumgardner JE, Ishii M, Spector ZZ, Edvinsson JM, Jalali A, Yu J, Itkin M, Lipson DA, Gefter W. Determination of regional VA/Q by hyperpolarized ^{3}He MRI. *Magn Reson Med*, 2004; 52: 65–72.

95. Adolphi NL, Kuethe DO. Quantitative mapping of ventilation-perfusion ratios in lungs by ^{19}F MR imaging of T_1 of inert fluorinated gases. *Magn Reson Med*, 2008; 59: 739–746.

96. Mansson S, Deninger AJ, Magnusson P, Pettersson G, Olsson LE, Hansson G, Wollmer P, Golman K. ^{3}He MRI-based assessment of posture-dependent regional ventilation gradients in rats. *J Appl Physiol*, 2005; 98: 2259–2267.

97. Marcucci C, Nyhan D, Simon BA. Distribution of pulmonary ventilation using Xe-enhanced computed tomography in prone and supine dogs. *J Appl Physiol*, 2001; 90: 421–430.

98. Couch MJ, Ball IK, Li T, Fox MS, Biman B, Albert MS. Gravitational Distribution Gradient of Inert Fluorinated Gases in Human Lungs Using ^{19}F Ultra-Short Echo Time MRI. *Proc 22nd Annual Meeting of ISMRM*, 2014; 2304.

Chapter 13

Tracking of Capsules and Catheters in the Human Gastrointestinal Tract

Andreas Steingötter[a,b] and Tobias Hahn[a,b,c]

[a]*Division of Gastroenterology and Hepatology, University of Zurich, Zurich, Switzerland*
[b]*Institute of Biomedical Engineering, University and ETH Zurich, Zurich, Switzerland*
[c]*Madrid-MIT M+Visión Consortium, Massachusetts Institute of Technology, Cambridge, Massachusetts, USA*

steingoetter@biomed.ee.ethz.ch

13.1 ^{19}F for GI Applications

13.1.1 Gastrointestinal (GI) Function

The gastrointestinal (GI) tract is a large organ system 3–7 m long [1–3]. It is made up of various specialized organs with distinct functions (e.g., stomach, small bowel, colon) and lies within the abdominal and pelvic cavities. The primary function of the GI tract is to control the intake, processing, transport, and absorption of nutrients and the defecation of indigestible parts in food. As food is transferred to the stomach, the first intragastric digestive processes get underway by mechanical mixing of gastric acid and pepsin with ingested nutrients [4]. Controlled gastric emptying

Fluorine Magnetic Resonance Imaging
Edited by Ulrich Flögel and Eric Ahrens

ISBN 978-981-4745-31-4 (Hardcover), 978-981-4745-32-1 (eBook)
www.panstanford.com

of its content enables continuous duodenal nutrient sensing and is followed by intestinal digestion comprising the mixing of nutrients with secreted bile and pancreatic enzymes, their aboral peristaltic transport and continuous absorption. The intestinal absorption of nutrient molecules stimulates the release of GI peptide hormones that modulate visceral function and satiation feelings [5–7]. This complex mechanism allows for the delivery of food at a rate that facilitates digestion, controls postprandial glycemia and regulates energy intake [8]. In summary, GI motility, digestion and nutrient absorption are integrated functions that are regulated by complex feedback mechanisms.

13.1.2 Imaging of GI Function

Gastrointestinal function is very difficult to assess with any current technology [9]. Global measures of bowel transit are commonly achieved by monitoring the progression of ingested radiolabeled markers through the GI tract using γ-scintigraphy [10, 11]. Standard methodologies to assess gastric and intestinal motor and secretory function require the unpleasant insertion of pressure and/or pH sensing catheters via mouth or anus. The limited length and sensor number only permit locally restricted measurements of intraluminal pressure and/or pH. Catheter placement in the esophagus, stomach or rectum can mostly be controlled without additional medical imaging; however, placement in the small intestine and colon is still performed under fluoroscopic control [12].

^{1}H MRI has the ability to simultaneously assess multiple aspects of GI motor and secretory function, such as gastric emptying, secretion and peristaltic activity, gallbladder and duodenal contractions, intestinal fluid content, transit and activity, on both a local and global scale [13–19]. This represents a step improvement compared to current gold standard technology. However, some key challenges remain that make it difficult to relate these imaging-derived data to the transport, processing, and absorption of nutrients, i.e., the primary function of the GI tract. First, imaging of the morphology of the GI tract is difficult because it is continuously moving, and the complex three-dimensional anatomy is highly variable among individuals; even for a single individual, the position of the bowel loops varies over time due to feeding and physical activity [20]. Second, the GI tract exists in distinct "fasted" and "fed" functional states, characterized by fundamentally different physiologic motor and secretory activity. Third, MRI is not capable of visualizing changes in intraluminal pressure, which demands the concurrent use of MR compatible catheters for method validation and clinical acceptance. Fourth, even with the use of modern MR contrast agents, MRI has difficulties in detecting and differentiating macronutrients within the GI tract. This is partly due to the fact that structural properties of macroscopic luminal contents change during their passage from the stomach to the rectum, and partly due to the enormous flux within the bowel.

Current intestinal MRI methods apply bowel preparation using paramagnetic MR contrast agents together with osmotic laxatives to artificially extend the bowel lumen and improve luminal contrast [21, 22]. However, paramagnetic contrast agents provide only a relatively small contrast enhancement in comparison to the range of signal intensities of the adjacent tissues [23]. The use of super-paramagnetic contrast agents is limited by inherent susceptibility artifacts [24]. Higher intraluminal contrast, without the need of special bowel preparations, can be achieved if MR active compounds with no similar in vivo counterparts are exogenously administered, effectively making these compounds the only signal sources. ^{19}F, as the MR detectable element with the highest MR sensitivity of the stable isotopes after ^{1}H, is very well suited for this purpose, since nearly no ^{19}F is found in the human body. Therefore, combined ^{19}F and ^{1}H MRI has the potential to help overcome some of the above-mentioned limitations and to provide improved non-invasive and simultaneous imaging of GI morphology and function. Moreover, ^{19}F labeling and tracking of GI catheters would allow for catheter-guided $^{1}H/^{19}F$ imaging of local and global GI motor function and morphology and vice versa allow for MR image-guided GI catheter placement.

13.1.3 Monitoring of GI Drug Delivery

Besides the processing and absorption of nutrients, the GI tract is also the most important route for the delivery of drugs to the systemic circulation. Alike nutrient digestion, drug bioavailability depends on gastric emptying, intestinal transport, and the intraluminal (mechanical) processing of the ingested dosage form [25, 26]. Depending on the absorption window of the drug in the intestine, various oral dosage forms have been developed exhibiting dedicated controlled release characteristics and/or defined GI transit times. γ-scintigraphy is the current gold standard for monitoring the GI transit of radiolabeled drug delivery systems [10, 27]. Magnetic marker monitoring (MMM) [28] and magnetic pill tracking [29] have been proposed as alternative radiation-free imaging modalities for biopharmaceutical studies. ^{19}F MRI shows potential as another radiation free alternative for monitoring the GI transit of small fluorine-containing markers in the GI tract [30]. Combined ^{19}F and ^{1}H MRI would then complement the monitoring of labeled oral dosage forms by integrating information on organ morphology and function.

13.1.4 Requirements for Combined ^{19}F/^{1}H MRI of the GI Tract

In standard MRI, only a few part per million (ppm) of excited nuclear spins contribute to the net magnetization forming the detected MR signal. In addition to this inherent low sensitivity of MRI in general, several unresolved technological issues must be addressed before combined ^{19}F and ^{1}H MRI can be fully exploited in the human GI research setting. Multi-channel, multi-nuclear MRI systems with efficient abdominal coils must become readily available. A field of view minimum of 35 cm (feet-head-direction), by 35 cm (left-right-direction), by 20 cm (anterior-posterior-direction) must be covered by a preferably homogenous radio frequency (RF) excitation (B1) field. Ideally, these must be actively decoupled to allow for unperturbed simultaneous ^{1}H and ^{19}F imaging. Moreover, a range of sequential and/or simultaneous abdominal scanning sequences providing different tissue and luminal contrasts is required for GI anatomy to visualize transport and motor activity, along with the ^{19}F data.

The low inherent sensitivity of ^{19}F MRI also demands the use of very signal-efficient exogenous ^{19}F markers. Perfluorocarbons (PFCs) exhibit high ^{19}F atom density and consists mostly of carbon and fluorine atoms. For imaging, also the degree of symmetry of the carbon-fluorine bonds determines the MR spectrum and useful signal intensity. Perfluoro-[15]-crown-5 ether (PFCE) and hexafluorobenzene (HFB) with their symmetric structure exhibit only a single peak in their MR spectra and provide high signal intensity for imaging. Like most perfluorocarbons, these two compounds are biologically inert and lack acute toxicity and are therefore promising for human applications [31]. Due to its high volatility, HFB is rapidly lost through the lungs by respiration [32, 33]. In contrast, the low vapor pressure of PFCE results in very slow clearance (up to several months) from the reticuloendothelial system [34, 35]. None of the two compounds has been approved for human use, and accordingly they cannot be applied through direct oral intake. Nevertheless, oral application of ^{19}F compounds in humans can be realized by safe encapsulation.

13.2 ^{19}F Labeling of Capsules and Catheters*

To monitor GI function non-invasively, including local and global transit times, as well as intestinal motor activity, our laboratory aimed at developing a background-free ^{19}F MRI approach making use of encapsulated ^{19}F compounds. A proof-of-principle demonstration of human ^{19}F/^{1}H GI MRI has been achieved by Schwarz et al. [37], who successfully visualized the intestinal movements and transit of a large-sized capsule (length 22 mm, diameter 7 mm) of polychlorotrifluoroethylene (PCTFE) shell material filled with 350 μl perfluorononane (C9F20) in a healthy subject. However, to better emulate nutrient-like luminal passage and improve patient comfort, the capsule containing the ^{19}F compound should be smaller in size within the limits of known sensitivities and without compromising safety. Various methods for encapsulation of liquid ^{19}F-based core materials into nano- and micro-capsules have been proposed [38–40]. However, capsules with diameters ranging from micrometers to nanometers in size have been reported to be absorbed into the bloodstream [41]. Considering this, and taking into account that possible shell material degradation or liquid leakage is possible [40], micro-and nanometer-sized capsules should currently not be considered for human ^{19}F MRI GI applications, unless full toxicological tests are performed.

* This section is based on work presented in Chapter 4 of [36].

13.2.1 Dual-Shell ^{19}F Capsule

A capsule-in-capsule prototype was developed that has smaller dimensions and an advanced leakage protection system compared to the initially proposed, large-sized capsules. The capsule design is presented in Fig. 13.1. The dual-shell capsule design comprises an inner capsule made from polyether ether ketone (PEEK) (cf. Fig. 13.3a), an outer capsule made from PCTFE (density 2.12 g/cm^3), an o-ring made from ethylene propylene diene monomer rubber (EPDM), and a coating of SILASTIC (Dow Corning Corporation, Midland, MI, USA) (Fig. 13.3b). The inner capsule consisted of two halves that were fitted together in liquid for reduced air inclusion during filling. The filling volume of this inner capsule was 65 μl. The two halves of the outer capsule were screwed together for increased mechanical stability. Capsule shells were generated using a high-precision turning machine (Super-Precision Conquest SP, Hardinge, Elmira, NY, USA). Prior to oral delivery, each capsule was kept in a closed laboratory environment and systematically tested for leakage.

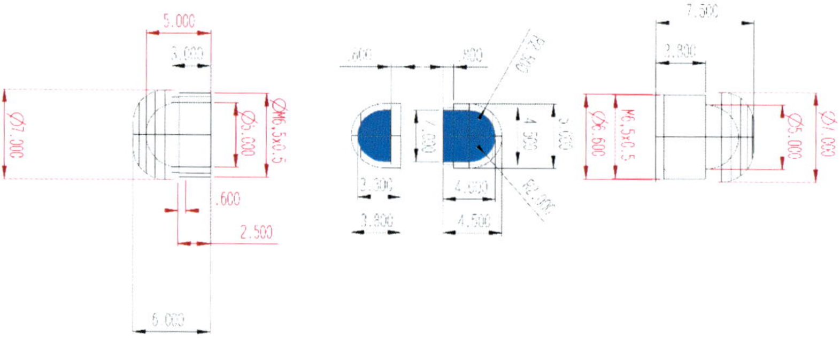

Figure 13.1 Technical drawing of the capsule-in-capsule design. Diameters (Ø) and lengths are given in mm. For a picture, see Fig. 13.3 a,b. The filling volume of 65 μl is depicted in blue.

Capsule leakage was tested by 24–72 h exposure to pH 1 and constant mechanical stress at ~37°C using a magnetic stirrer or shaker. A safety control scheme was defined to provide a decision criterion on the capsule's in vivo safety. A leakage tolerance threshold was established that allowed determination of capsules suitability for oral use. The dual-shell capsule design, together with the safety control scheme, was reported as a medical device to the authorities and approved by the local ethics committee. Due to the above-described advantages with regard to signal efficiency and toxicity, PFCE (99%, Exfluor Research Corp., Round Rock, TX, USA) and HFB (99%, Apollo Scientific, Cheshire, UK) were used as capsule core liquids. Respective safety margins were defined according to available toxicological data. For PFCE, the acceptance criterion was defined as a maximum weight loss of 0.03 mg/day. For HFB, the acceptance criterion was defined as 0.1 mg/day. The ^{19}F capsules were expected to be excreted from the body within 72 h [42, 43]. The expected maximum exposition, therefore, was 0.3 mg (0.2 μl) HFB and 0.09 mg (0.05 μl) PFCE. Leakage tests were performed using a high-precision scale with standard deviation ± 0.015 mg (XP205DR, Mettler Toledo, Columbus, Ohio). The detected weight change over time of all produced dual-shell capsules is shown in Fig. 13.2.

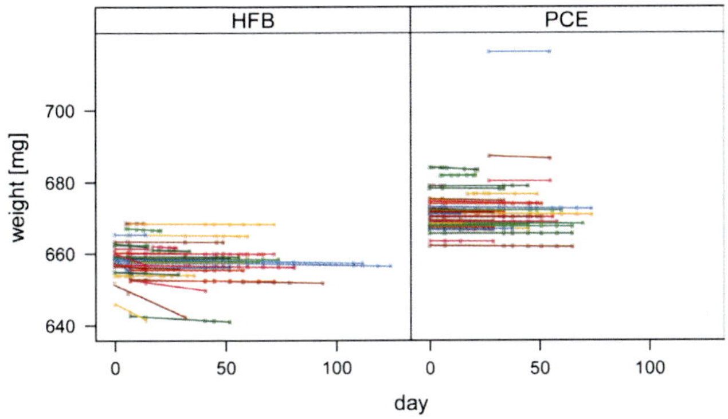

Figure 13.2 Measured capsule weight over time (dots) together with fitted leakage curves (lines). The standard deviation of the used scale was ± 0.015 mg.

Data were analyzed using a linear mixed-effect model. The analysis suggested a less predictable behavior of the HFB filled capsules. Table 13.1 summarizes the leakage estimates and 95% confidence intervals for all analyzed capsules. Capsules outside the safety margins had leakages greater than 0.1 mg/day for HFB and 0.03 mg/day for PFCE. The group estimate of the dual-shell capsule, including all capsules that were not accepted for oral administration, is well below the defined leakage tolerance criterion, both for HFB and PFCE.

Table 13.1 Summary of leakage test results reported as mean (95% confidence interval)

^{19}F marker	Leakage in mg/day	No. of analyzed capsules	Outside safety margin
HFB	0.049 (0.026, 0.073)	38	1
PFCE	0.007 (−0.015, 0.028)	45	3

Two dual-shell ^{19}F capsules were administered to 15 healthy volunteers and their intestinal movement and transit were monitored by $^{19}F/^{1}H$ MRI for 4 h after gastric transit. Capsules were retrieved after excretion, and orocecal transit time as well as possible defects and adverse events were noted. No adverse events were reported in the in vivo study. One out of the total 60 ^{19}F capsules was not retrieved after excretion; the volunteer stated that they did not have the opportunity to retrieve the capsule later than 48 h after capsule intake. The other 59 capsules were retrieved and no defects were detected. Mean and standard deviation of the time until excretion of the 59 retrieved capsules were 31 h : 18 min ± 15 h : 37 min.

13.2.2 Single-Shell ^{19}F Capsule

To further reduce capsule size and handling, a second type of ^{19}F capsule was developed allowing for more efficient procedures with regard to production and filling (Fig. 13.3c). Ultrasonic welding was used for bonding of the halves of the new generation capsule. Different shell materials were tested, revealing different mechanical and MR properties. The shell materials tested were PEEK (density 1.40 g/cm^3), acrylonitrile butadiene styrene (ABS) (density 1.04 g/cm^3), polymethyl methacrylate (PMMA) (density 1.18 g/cm^3) and ABS HD-15 (density 1.05 g/cm^3). Single-shell capsules were produced, stored and analyzed according to previously developed procedures, but were not tested in vivo. ABS capsules showed the smallest leakage and narrowest (95%) confidence interval. Leakage from ABS capsules was estimated to 0.003 mg/day (HFB) and 0.002 mg/day (PFCE), respectively. Figure 13.3d depicts a standard Diclofenac pill for size comparison.

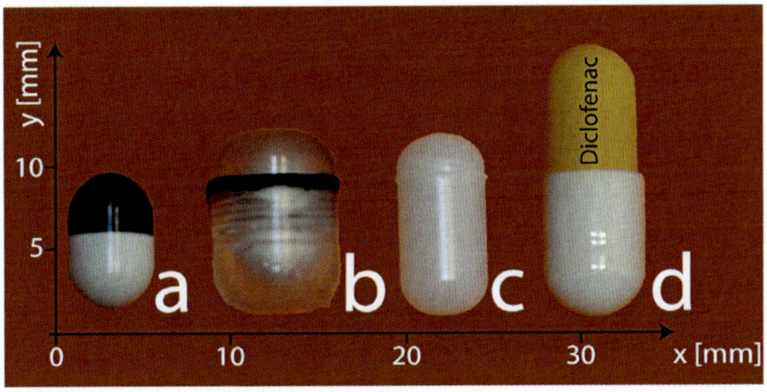

Figure 13.3 Size comparison of the developed capsule designs and a standard Diclofenac pill. (a) Inner capsule of the dual-shell ^{19}F capsule design. (b) Dual-shell ^{19}F capsule design. (c) Single-shell ^{19}F capsule design after ultrasonic welding. (d) Standard Diclofenac pill.

13.2.3 ^{19}F-Labeled GI Catheter

Four capsules (Fig. 13.3a) made of PEEK and filled with 65 μl PFCE were inserted into a standard gastroduodenal Levin tube (B. Braun Medical AG, Melsungen, Germany) with a 6 mm outer diameter (Fr 18). An additional capsule containing 65 μl HFB was inserted at the catheter tip (Fig. 13.4). In vitro and in vivo experiments using this ^{19}F catheter were performed to demonstrate tracking reliability (see Sections 13.3.3 and 13.4).

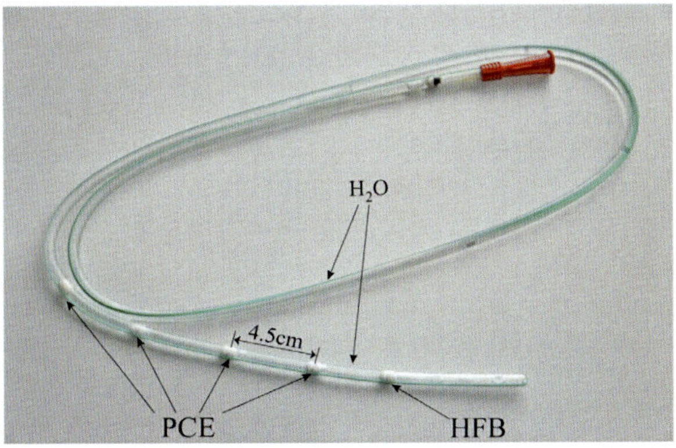

Figure 13.4 Modified ^{19}F labeled gastroduodenal Levin tube (B. Braun Medical AG, Melsungen, Germany) for ^{19}F catheter tracking.

13.3 In vivo ^{19}F Tracking: Methodology and Application†

The defined geometry of the capsules and the sparsity of the ^{19}F signal within the GI tract enable important improvements in the data acquisition and data processing strategy, such that even real-time ^{19}F MRI tracking of capsules is feasible. In the following, we describe different tracking methodologies and their respective applications in the human GI tract.

13.3.1 Tracking by Cartesian Projection

Methodologically, the goal of tracking is to determine the location of an object over time in 3D, i.e., to determine its x, y and z coordinates on a Cartesian grid. Applying ^{19}F MRI in humans implies that, except for image noise, the exogenously introduced ^{19}F-labeled object is the only sparse signal source in the image field of view. Based on this, a straightforward approach for determining the x, y and z coordinates of the ^{19}F-labeled signal source using ^{19}F MRI is given by the Cartesian projection scheme. This scheme is based on the projection slice theorem, which states that the following two procedures yield the same result:

(1) Acquisition of a single k-space line that crosses or "slices" through the center of 2D k-space,
(2) Fourier transformation of a projection of 2D image space onto a 1D line that is parallel to the k-space line in 1.

If the k-space line is selected along one of the Cartesian directions, a peak detection algorithm applied to its Fourier Transform will yield the corresponding Cartesian coordinate. Accordingly, acquiring three projections along x, y, and z will yield the 3D location of the sparse signal source. The Cartesian projection approach is advantageous with regard to reconstruction and sequence programming simplicity. Minor adaptations in standard MR sequences allow for a real-time monitoring of the

†This section is based on work presented in Chapter 3 of [36] and work published in [44].

three Cartesian coordinates of the tracked signal sources. In Fig. 13.5, an exemplary reconstructed image is shown for a modified balanced FFE sequence (TR/TE = 4.43/1.98 ms, 320 mm field of view, 4 mm resolution in measurement direction) with switched-off phase encode gradients and a 90° rotation of the frequency encode direction after 10 projections, such that the top 10 lines represent 10 projections along *x*, and the following lines projections in *y* and *z* direction, respectively. The temporal resolution for the 30 profiles was 133 ms. A tracking scheme for three orthogonal Cartesian projections is presented in Fig. 13.6.

Figure 13.5 5 Stacked ^{19}F projections along *x*-, *y*-, and *z*-directions of the laboratory reference frame, respectively, with each stack consisting of 10 projections.

In this scheme, the tracking problem is rephrased as a curve detection problem after stacking the 1D projections over time. This temporal stacking yields a 2D image where the *y*-direction/ordinate represents one Cartesian direction and the abscissa represents time (Fig. 13.7). The filters that are applied in the curve detection process, such as low-pass filter and spline fitting, are possible ways of incorporating boundary conditions on the maximum possible velocity and acceleration of the signal source.

Without any prior knowledge, ^{19}F signal sources with identical MR properties cannot be uniquely identified over time or tracked. In the case that the ^{19}F signal sources are labeled by ^{19}F compounds with different ^{19}F resonance frequencies, multiple signal sources can be tracked using Cartesian projections. The required ^{19}F compounds, such as different perfluorocarbons, must provide sufficient MR signal strength, must have non-overlapping resonance frequencies and must be excited in an interleaved manner. This way of tracking is unambiguous because the position of each ^{19}F signal source can be uniquely identified at all times and without additional incorporation of a priori knowledge.

13.3.2 In vivo Dual Compound Tracking by Cartesian Projection

The spectral separation of PFCE and HFB is 80 ppm, i.e., approximately 10 kHz using a 3 T MRI system. Because of this relatively large spectral separation, frequency selective excitation pulses can be applied to independently excite either HFB or PFCE.

In the following, results of a study are shown that demonstrate that HFB and PFCE-labeled capsules can be independently and unambiguously tracked in the human abdomen by frequency selective excitation. A tracking algorithm incorporating a posteriori knowledge of the overall capsule trajectory was applied. Experiments were performed using a whole-body 3 T MRI system (Achieva, Philips Healthcare, Best, Netherlands). A dual-channel ^{19}F transmit-receive surface coil (PulseTeq Ltd, UK, 20 cm diameter) was used in quadrature mode for ^{19}F imaging. Cartesian projections of the ^{19}F MR signal of two dual-shell capsules filled with 65 μl of either PFCE or HFB (2.39 × 10^{21} ^{19}F atoms and 2.03 × 10^{21} ^{19}F atoms, respectively) were acquired using a dynamic balanced steady-state free precession (SSFP) sequence. Scan parameters were: 32 × 32 × 32 cm^3 field of view, 4 mm resolution in measurement direction. TR and TE were 3.79 ms and 1.82 ms, respectively. The flip angle was 39°. A sinc-Gaussian pulse with 2,490 Hz bandwidth was applied for frequency selective excitation. Excitation pulses exceeding 3 kHz bandwidth, as may be advantageous for compensating shimming deficiencies, are not advised since such pulses might lead to unwanted signal smearing. The dynamic scans consisted of 10 projections in each *x*-, *y*-, and *z*-direction. The dynamic scan time was 2 × 114 ms = 228 ms for each ^{19}F compound.

Two dual-shell capsules (see Section 13.2.1) were filled with HFB and PFCE, respectively, and orally administered to 15 volunteers on two study days 1 h and 15 min prior to each measurement session. The two study days were separated by one week during which all participants received a dietary intervention, i.e., additional intake of a laxative (Normacol®), to test its effect on intestinal transit and gas. Each session lasted for 4 h and consisted of three 1 h scan intervals that were separated by short pauses where the volunteers were allowed to leave the scanner. These scan intervals were further subdivided into 3 to 4 continuous ^{19}F tracking periods. ^{1}H abdominal survey scans were performed in between for registration of the ^{19}F tracking data. The data processing part of this dual-compound tracking approach started with automatically co-registering detected capsule coordinates by cross-correlation of the corresponding ^{1}H survey scans acquired immediately before or after each tracking period. The 3D geometry of the intestine was reconstructed by performing a local regression and a polynomial fitting of "trusted" capsule coordinates. Intestinal motility reflected in the capsule velocity was analyzed based on breathing compensated capsule coordinates. Anatomic and capsule velocity data were visualized using The Visualization Toolkit (VTK, Kitware Inc., Clifton Park, NY). Intestinal 3D geometries based on the PFCE and HFB capsules were overlaid for qualitative comparison (see below). Capsule spacing over time was determined for assessment of the active segmentation of intraluminal content by intestinal contractions.

The tracking of the PFCE- and HFB-labeled capsules in the human intestine was successfully performed. No signal smearing between the PFCE and HFB signal was detected, and the proposed tracking scheme allowed for a robust detection of capsule positions (Fig. 13.7).

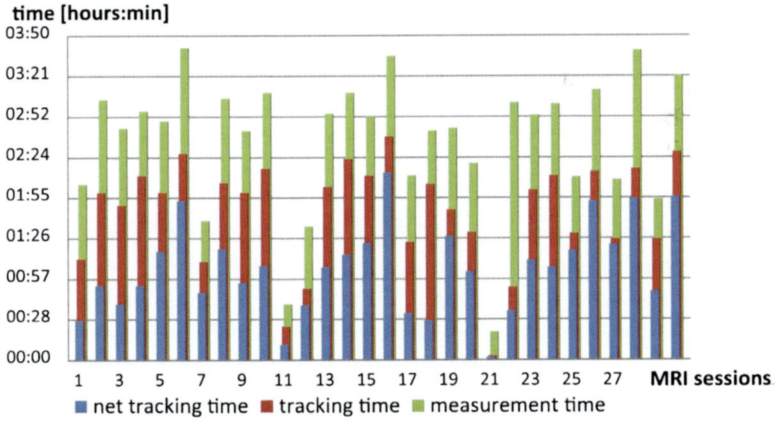

Figure 13.8 Performance of the tracking algorithm for > 30 MRI sessions. The time from the start of the first tracking scan to the end of the last tracking scan, including all pauses, is denoted as "measurement time" (green). "Tracking time" (red) denotes the total time spent for tracking, excluding tracking pauses in between. "Net tracking time" (blue) denotes the total time where capsule positions could reliably be detected. Time includes both the times for PFCE and HFB.

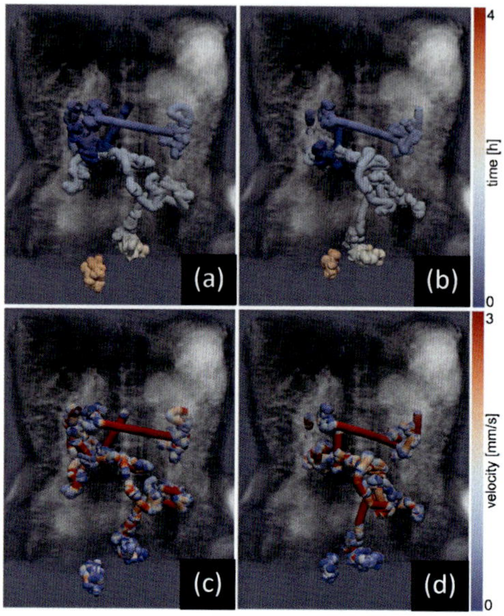

Figure 13.9 In vivo dual compound tracking in a healthy volunteer. (a, b) 3D visualization of PFCE (a) and HFB (b) capsule trajectories. The color of the reconstructed intestinal tubes continuously changes with tracking time. (c, d) Reconstructed capsule trajectories with color-coding according to local capsule velocities for PFCE (c) and HFB (d) capsule trajectories. In areas with low SNR and/or tracking pauses, no capsule trajectory is reconstructed. Contours of stomach (S, solid line), colon ascendens and descendens (C_a and C_d, dashed line) and liver (L, dotted line) are indicated in (b).

The tracking of the PFCE- and HFB-labeled capsules in the human intestine was successfully performed. No signal smearing between the PFCE and HFB signal was detected, and the proposed tracking scheme allowed for a robust detection of capsule positions (Fig. 13.7). Tracking performance was different between the HFB and PFCE filled capsules: The HFB-filled capsules exhibited lower SNR values resulting in lower tracking reliability compared to the PFCE-filled capsules. The tracking performance of the PFCE filled capsules was approximately 25% better. The lower SNR and corresponding tracking reliability of HFB compared to PFCE can be explained by the 15% lower ^{19}F content as well as by its higher volatility, leading to expected higher losses during the capsule filling process. In Fig. 13.8 the tracking performance for all 30 MRI sessions is shown. The visualized intestinal 3D geometries based on the PFCE and the HFB capsule trajectory, respectively, showed good conformity (Fig. 13.9a,b). As expected, intestinal passage showed a strongly varying capsule velocity with a high spatio-temporal dependence [45, 46] (Fig. 13.9c,d). The spacing between the PFCE and the HFB filled capsules, indicating segmental contractile activity of the intestine, varied over time. The maximum detected capsule spacing was 9.4 cm. Targeted experiments still need to be performed in order to quantify the spatio-temporal variance of intestinal motility in more detail.

This study showed the feasibility of fast, unambiguous, "dual-color" 3D tracking of two small ^{19}F-labeled capsules in the human GI tract. This enabled the analysis of functional intestinal parameters such as local velocity or luminal content segmentation, in addition to global anatomic analysis of intestinal 3D geometry. The sensitivity of the tracking approach is still below 100%, indicating the need for methods that can further optimize SNR and improve tracking reliability.

13.3.3 Tracking Multiple ^{19}F Signal Sources by 3D Golden Angle Radial Imaging

A solution for the tracking of multiple signal sources having the same resonance frequency is provided by the 3D Golden Angle radial projection scheme [47] (Fig. 13.10a–c). This tracking scheme modifies the projection directions from the three orthogonal Cartesian projections, such that *k*-space coverage is relatively uniform and quasi-random at any time and for any chosen reconstruction window size (i.e., numbers of consecutive *k*-space profiles used for reconstruction). This uniform nature of *k*-space acquisition allows for the reconstruction of 3D images at any time and for an arbitrary number of acquired consecutive *k*-space profiles. The 3D images can then be used to detect and identify the sparse ^{19}F signal sources. Given the high sparsity of ^{19}F tracking, high undersampling factors can be used for the delineation of the signal source locations. Figure 13.10d–f illustrates that in the case of 3D Golden Angle, already 3 projections suffice for locating a single signal source–as one would expect, given that 3 projections were required for locating a single signal source using Cartesian projections. The quasi-random nature of the *k*-space profile distribution leads to a high incoherence of undersampling artifacts. This prevents the occurrences of intense local signal artifacts that impair the tracking of the object of interest. Furthermore, it offers the feasibility for Compressive Sensing reconstruction methods [48] for optimized SNR.

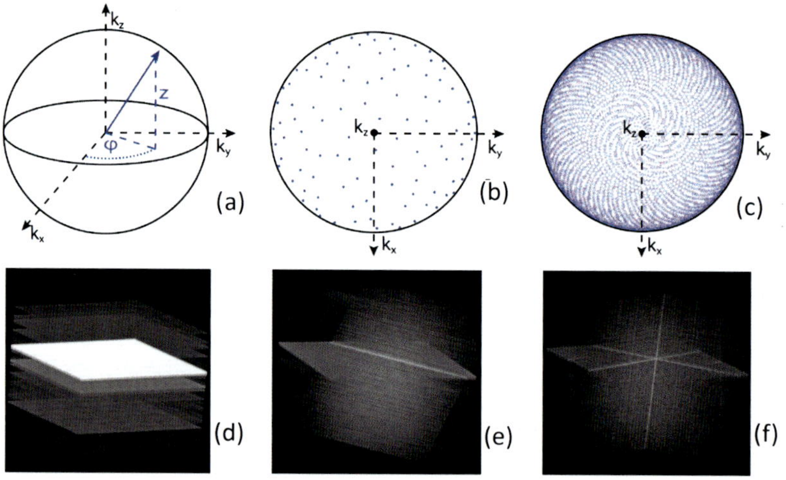

Figure 13.10 Principle of 3D Golden Angle radial imaging. (a) The 3D Golden Angle profile with index *m* is defined with the 2D golden means A = 0.6823 and B = 0.4656, the azimuth angle $\varphi = 2\pi \cdot \mathrm{m} \cdot \mathrm{A}$ and the projection length *z* in *kz* direction being the modulus 1 of m·B. (b-c) Positions of the *k*-space profile tips for a 3D Golden Angle radial imaging using 100 (b) and 6400 (c) projections, illustrating a relatively uniform *k*-space coverage. (d-f) 3D Golden Angle image reconstruction of a ^{19}F labeled capsule with different reconstruction window sizes Ω. (d) Ω = 1, (e) Ω = 2, (f) Ω = 3. From a minimum of Ω = 3, the 3D position of a ^{19}F labeled capsule can be determined. The use of larger reconstruction window sizes leads to increased SNR values.

Like Cartesian projection, robust identification of one or multiple signal sources usually requires the definition of boundary conditions. In the following, three different tracking schemes with different types of boundary conditions (a priori or a posteriori knowledge) are presented that have been tested for in vitro and in vivo application:

(1) Knowledge of size of the signal source. For each time point, a 3D image is reconstructed from the *k*-space profiles of the respective reconstruction window. Then, data is mapped onto a Cartesian grid and Fourier transformed. Next, the 3D position corresponding to the signal intensity maximum in the reconstructed 3D image is determined, defining the position of the first signal source. Subsequently, this detected signal source is cut out from the 3D image with a cut-off diameter defined by the physical size of the signal source. This process is then repeated to find the locations of the remaining signal sources. The SNR corresponding to each signal source should be computed and locations should be discarded if the corresponding SNR falls below a certain SNR threshold.

(2) Knowledge of temporal trajectory. The determination of the position of the signal sources can be constrained to the overall trajectories determined over longer periods of time. This can be helpful if the chosen temporal resolution yields a high rate of rroneously tracked coordinates. By using the described method, outliers can be effectively excluded.

(3) Knowledge of object configuration. If a priori knowledge about the physical separation of the signal sources exists, such as in the case of ^{19}F catheters [44, 49], larger cut-off diameters can be chosen, thereby effectively narrowing down the volume of signal peak detection.

As shown in Fig. 13.11, the tracking reliability increases with increasing reconstruction window size, i.e., decreasing temporal resolution. In Fig. 13.12 [44] the tracking of a ^{19}F-labeled gastroduodenal catheter in vivo is visualized; the catheter described in Section 13.2.3 (Fig. 13.4) was nasally introduced into the stomach and subsequently tracked while being pulled back through the stomach and the esophagus. Tracking with a temporal resolution of 252 ms was performed using knowledge of the size of the signal source and the temporal trajectory.

3D Golden Angle radial imaging does not only allow for tracking of multiple signal sources, but also provides the means for simultaneous $^{1}H/^{19}F$ imaging. The unique properties of freely selecting any reconstruction window size fulfills the demand for the relatively small reconstruction window needed for ^{19}F tracking and the large reconstruction window size required for ^{1}H anatomical imaging. As described in Chapter 2 of this book, dual-frequency combined $^{19}F/^{1}H$ coils are under development for interleaved or simultaneous $^{19}F/^{1}H$ imaging [50, 51].

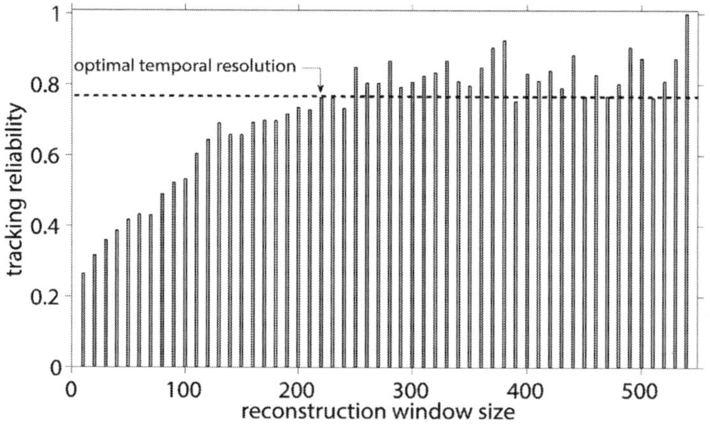

Figure 13.11 Tracking reliability as a function of reconstruction window size, based on in vitro tracking data [44]. Tracking reliability increases with reconstruction window size and remains stable after a specific reconstruction window size, here, after approximately 220 profiles.

13.4 Real-Time ^{19}F Tracking System

The methodologies described above show potential for application in two major fields, including, quantification of gastrointestinal motility and MR guidance of interventions. In particular for the latter, it is critical that the tracking methodology is applicable in real time. In this context, "real time" is understood as allowing for instant usage, i.e., instant reconstruction of the acquired data with a maximal lag time and temporal resolution of 0.5 s each. The proposed real-time tracking system allows for (1) real-time data transfer from the MR scanner's database to an external reconstruction computer and for (2) online reconstruction of ^{19}F images and fast localization of the ^{19}F signal sources. Furthermore, it allows for (3) real-time visual feedback projected inside the MR scanner room for precise control of the velocity and position of the catheter, as well as for (4) catheter-based selection of image plane geometry, which in return enables immediate ^{1}H imaging of the catheter position.

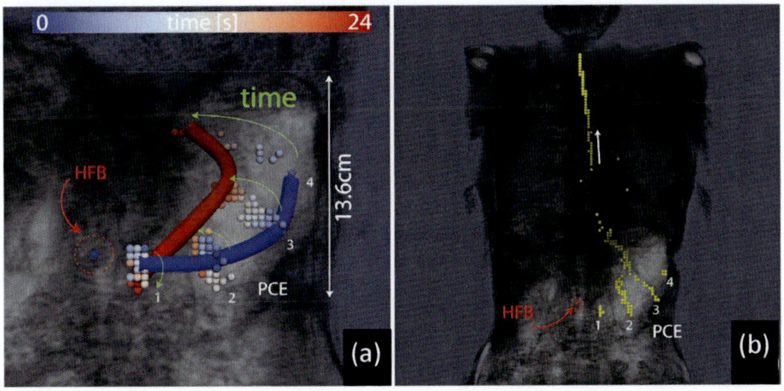

Figure 13.12 In vivo catheter tracking results as presented in [44]. (a) Intragastric ^{19}F catheter tracking with 252 ms temporal resolution by incorporation of *a priori* knowledge on the size of the signal source. The tracked capsule locations are color coded from blue (tracking beginning) to red (tracking end). The initial and final positions of the catheter are reconstructed (blue/red), as well as the initial position of the catheter tip determined by HFB tip tracking. (b) In vivo catheter tracking results overlaid onto an anatomical reference scan; catheter tracking was performed during pulling back of the ^{19}F catheter from the stomach through the esophagus at a temporal resolution of 252 ms by incorporating knowledge of the overall capsule trajectory. Tracked capsule positions are given in yellow. Gaps in the capsule trajectories reveal low SNR areas at the level of the heart and in the gastric fundus indicating increased artifacts in these areas due to cardiac and respiratory motion. The data shown consists of three individual tracking periods (3 × 6400 *k*-space lines) with short pauses between periods. The initial position of the HFB labeled catheter tip is shown in red. The stomach is outlined by the dashed white line.

Real-time ^{19}F tracking was performed in a phantom study using a 3 T whole-body clinical MRI system and the ^{19}F transmit-receive coil described in Section 13.3.2. Four capsules made of PEEK (Fig. 13.3a) were filled with 65 μl Perfluoro-15-crown-5-ether (PFCE). These ^{19}F capsules were placed inside a gastroduodenal Levin tube that was modified as described above (Fig. 13.4, Section 13.2.3), with the exception that no additional HFB-filled capsule was added at the catheter tip. The tube was pulled with different velocities through a guidance tube that had been immersed inside a gel phantom. 3D Golden Angle radial imaging [47] as described above was used for ^{19}F tracking, and radial *k*-space profiles were immediately transferred to the external reconstruction computer.

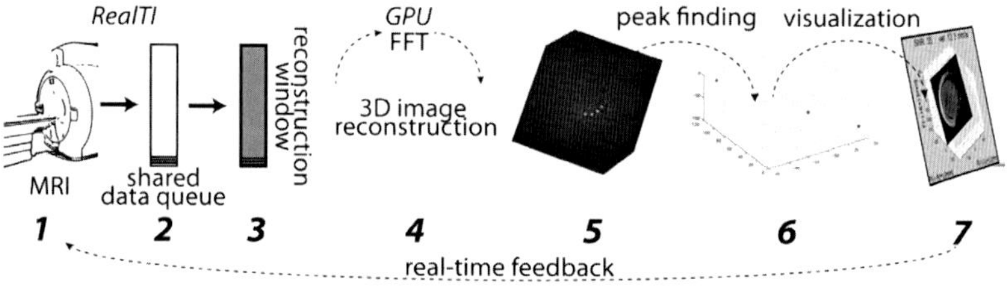

Figure 13.13 Processing steps of the real-time tracking system [49]. Acquired data is transferred from the MR scanner to a data queue shared between MRI server and Matlab client (1–2). Profiles are combined as defined by the reconstruction window size (3) for 3D image reconstruction in Matlab (4). Capsules within the catheter are detected through 3D image signal peak finding (5). SNR and object velocity as well as ^{1}H image geometries are computed for providing a visual feedback (6). The visual feedback is projected in real-time inside the MR room (7). Figure taken with permission from Ref. [49].

The real-time data transfer of each acquired k-space profile was realized using a modification of the RealTI toolkit (RTTech/ADERA, Pessac, France). The software package MRecon 3.0 (Gyrotools, Zurich, Switzerland) and a purpose-built 3D Golden Angle reconstruction algorithm were used to combine and reconstruct the acquired profiles. ^{19}F markers were tracked by signal peak finding on the gridded and reconstructed 3D images as follows. First, the 3D coordinate of the signal source with corresponding highest signal intensity was determined. Then, all voxels inside a sphere of 6 cm diameter centered on this position were then set to zero. This process was repeated for all other potential signal sources. Coordinates of detected signal sources with SNR < 15 were discarded and a real-time visual feedback was provided on the video screen indicating low SNR. A fit through the detected 19F marker positions determined slice angulation and offset parameters for the immediate planning of a 2D 1H image plane. In this way, guidance for 1H anatomic imaging [52] was implemented. The technical setup did not allow for concurrent ^{19}F and ^{1}H acquisition. Accordingly, computed imaging plane geometries were evaluated by ^{1}H scans at the end of ^{19}F tracking phases.

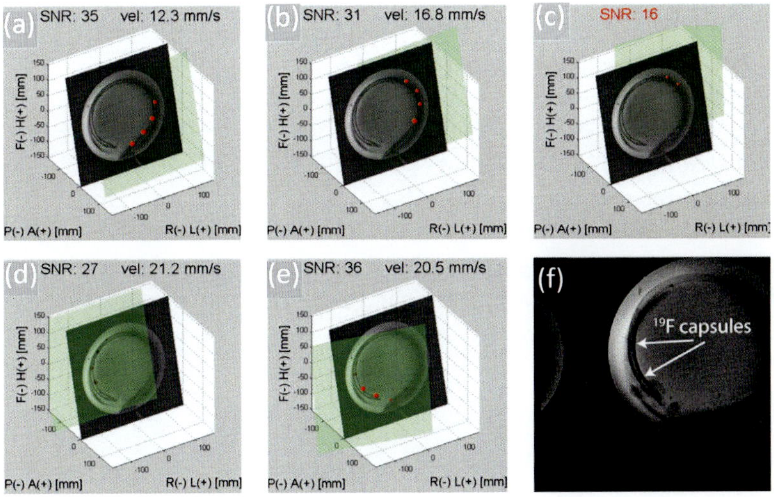

Figure 13.14 (a-e) Visual feedback during ^{19}F catheter tracking [49]. Screenshots of the feedback were acquired during real-time catheter tracking with mean catheter velocity v = 19.3 mm/s. All positions of capsules with corresponding SNR > 15 are shown as red dots. Dot size linearly scales with SNR. If all capsule positions are available, catheter velocity is additionally computed. Angulations and offsets are calculated for the ^{1}H 2D imaging plane that is fitted through high-SNR capsule positions. The computed ^{1}H imaging plane is visualized in semi-transparent green. (f) Corresponding ^{1}H image as guided by ^{19}F catheter position in panel (e). Figure taken with permission from Ref. [49].

In Fig. 13.13, the processing steps of the real-time framework are delineated. Figure 13.14 shows the results of a real-time tracking experiment with a temporal resolution of 370 ms. The tracking reliability for this experiment was shown to be 97.3%. With increasing temporal resolution and/or catheter velocity, the tracking reliability was found to generally decrease, as expected. Accordingly, the temporal resolution has to be wisely selected, depending on the precise requirements for accuracy and the expected maximum catheter velocity. The SNR values detected in this phantom study were similar to previously published in vivo SNR values [44], which indicates that the proposed method can be applied to real-time tracking of ^{19}F-labeled gastroduodenal tubes, endoscopes and pressure/pH meters in humans.

13.5 Conclusion and Outlook

The described in vitro and in vivo validation of the proposed methodology indicates that ^{19}F MRI is feasible for real-time tracking of ^{19}F-labeled capsules and catheters in humans. In combination with ^{1}H imaging, it shows potential to be applied for physiological research aiming to understand the human GI function and for pharmaceutical research studying the in vivo behavior of drug delivery systems.

References

1. Ahrens, E. H., Jr. (1956). Measurement of the human intestinal length in vivo and some causes of variation, *Gastroenterology,* **31**, pp. 274–284.

2. Fanucci, A. (1984). Small bowel length measured by radiography, *Gastrointest. Radiol.,* **9**, pp. 349–351.

3. Underhill, B. M. (1955). Intestinal length in man, *Br. Med. J.,* **2**, pp. 1243–1246.

4. Camilleri, M. (2006). Integrated upper gastrointestinal response to food intake, *Gastroenterology,* **131**, pp. 640–658.

5. Beglinger, C. (2004). Fat in the intestine as a regulator of appetite—role of CCK, *Physiol. Behav.,* **83**, pp. 617–621.

6. Beglinger, C. (2006). Gastrointestinal satiety signals in humans–physiologic roles for GLP-1 and PYY?, *Physiol. Behav.*, **89**, pp. 460–464.
7. Langhans, W. (2011). Introduction to the intestinal wall-THE regulatory interface in energy homeostasis, *Physiol. Behav.*, **105**, pp. 1–3.
8. Strader, A. D. (2005). Gastrointestinal hormones and food intake, *Gastroenterology*, **128**, pp. 175–191.
9. Schwizer, W. (2003). Non-invasive investigation of gastrointestinal functions with magnetic resonance imaging: Towards an "ideal" investigation of gastrointestinal function, *Gut*, **52**(Suppl 4), pp. iv34–39.
10. Odunsi, S. T. (2009). Selected interventions in nuclear medicine: Gastrointestinal motor functions, *Semin. Nucl. Med.*, **39**, pp. 186–194.
11. Yeong, C. H. (2013). Fusion of gamma scintigraphic and magnetic resonance images improves the anatomical delineation of radiotracer for the assessment of gastrointestinal transit, *Nucl. Med. Commun.*, **34**, pp. 645–651.
12. Lindberg, G. (2013). High-resolution manometry changes our views of gastrointestinal motility, *Neurogastroenterol. Motil.*, **25**, pp. 780–782.
13. Chaddock, G. (2014). Novel MRI tests of orocecal transit time and whole gut transit time: Studies in normal subjects, *Neurogastroenterol. Motil.*, **26**, pp. 205–214.
14. Curcic, J. (2010). Gastroesophageal junction: Structure and function as assessed by using MR imaging, *Radiology*, **257**, pp. 115–124.
15. Marciani, L. (2011). Assessment of gastrointestinal motor functions by MRI: A comprehensive review, *Neurogastroenterol. Motil.*, **23**, pp. 399–407.
16. Marciani, L. (2013). Effects of various food ingredients on gall bladder emptying, *Eur. J. Clin. Nutr.*, **67**, pp. 1182–1187.
17. Sauter, M. (2012). Measuring the interaction of meal and gastric secretion: A combined quantitative magnetic resonance imaging and pharmacokinetic modeling approach, *Neurogastroenterol. Motil.*, **24**, pp. 632–638, e272–e633.
18. Schwizer, W. (2006). Magnetic resonance imaging for the assessment of gastrointestinal function, *Scand. J. Gastroenterol.*, **41**, pp. 1245–1260.

19. Treier, R. (2006). Gastric motor function and emptying in the right decubitus and seated body position as assessed by magnetic resonance imaging, *J. Magn. Reson. Imaging,* **23**, pp. 331–338.

20. Meyers, M. A. (2000). The small bowel: Normal and pathologic anatomy. In *Dynamic Radiology of the Abdomen*, 5th ed. (Springer).

21. Menys, A. (2013). Global small bowel motility: Assessment with dynamic MR imaging, *Radiology,* **269**, pp. 443–450.

22. Menys, A. (2014). The challenge of segmental small bowel motility quantitation using MR enterography, *Br. J. Radiol.,* **87**, pp. 20140330.

23. Schwarz, R. (1999). ^{19}F-MRI of perfluorononane as a novel contrast modality for gastrointestinal imaging, *Magn. Reson. Med.,* **41**, pp. 80–86.

24. Mattrey, R. F. (1994). Perflubron as an oral contrast agent for MR imaging: Results of a phase III clinical trial, *Radiology,* **191**, pp. 841–848.

25. Kostewicz, E. S. (2014). PBPK models for the prediction of in vivo performance of oral dosage forms, *Eur. J. Pharm. Sci.,* **57**, pp. 300–321.

26. Sjogren, E. (2014). In vivo methods for drug absorption—comparative physiologies, model selection, correlations with in vitro methods (IVIVC), and applications for formulation/API/excipient characterization including food effects, *Eur. J. Pharm. Sci.,* **57**, pp. 99–151.

27. Perkins, A. C. (2004). Radionuclide imaging in drug development, *Curr. Pharm. Des.,* **10**, pp. 2907–2921.

28. Weitschies, W. (2010). Magnetic marker monitoring: High resolution real-time tracking of oral solid dosage forms in the gastrointestinal tract, *Eur. J. Pharm. Biopharm.,* **74**, pp. 93–101.

29. Worsoe, J. (2011). Gastric transit and small intestinal transit time and motility assessed by a magnet tracking system, *BMC Gastroenterol.,* **11**, p. 145.

30. Hahn, T. (2011). Visualization and quantification of intestinal transit and motor function by real-time tracking of ^{19}F labeled capsules in humans, *Magn. Reson. Med.,* **66**, pp. 812–820.

31. Yu, J. (2005). ^{19}F: A versatile reporter for non-invasive physiology and pharmacology using magnetic resonance, *Curr. Meicinal Chem.,* **12**, pp. 819–848.

32. Grosman, Y. S. (1973). Pharmacology and toxicology of hexafluorobenzene, *Izv. Estestvennonauchu. Inst. Pevinsk,* **15**, pp. 155–163.

33. Oakwood Products, Inc. (2013, November 12). Hexafluorobenzene [Material Safety Data Sheet]. Retrieved from: http://www.oakwoodchemical.com/ShowMSDS.aspx?Number=001356.

34. McGoron, A. J. (1994). Perfluorocarbon distribution to liver, lung and spleen of emulsions of perfluorotributylamine (FTBA) in pigs and rats and perfluorooctyl bromide (PFOB) in rats and dogs by 19F NMR spectroscopy, *Artif. Cells Blood Substit. Immobil. Biotechnol.,* **22**, pp. 1243–1250.

35. Oakwood Products, Inc. (2011, October 31). Perfluoro-15-crown-5-ether [Material Safety Sheet]. Retrieved from: http://www.oakwoodchemical.com/MSDS/MSD_PDF_009/N009312.pdf.

36. Hahn, T. (2012). ^{19}F MRI for the assessment of intestinal transport and tracking of gastroduodenal catheters. PhD (ETH Zurich, Zurich).

37. Schwarz, R. (2002). Gastrointestinal transit times in mice and humans measured with ^{27}Al and ^{19}F nuclear magnetic resonance, *Magn. Reson. Med.,* **48**, pp. 255–261.

38. Flögel, U. (2008). In vivo monitoring of inflammation after cardiac and cerebral ischemia by fluorine magnetic resonance imaging, *Circulation,* **118**, pp. 140–148.

39. Lanza, G. M. (2002). Targeted antiproliferative drug delivery to vascular smooth muscle cells with a magnetic resonance imaging nanoparticle contrast agent: Implications for rational therapy of restenosis, *Circulation,* **106**, pp. 2842–2847.

40. Pisani, E. (2008). Surfactant dependent morphology of polymeric capsules of perfluorooctyl bromide: Influence of polymer adsorption at the dichloromethane-water interface, *J. Colloid Interface Sci.,* **326**, pp. 66–71.

41. Fincher, J. H. (1968). Particle size of drugs and its relationship to absorption and activity, *J. Pharm. Sci.,* **57**, pp. 1825–1835.

42. Bergstrand, M. (2009). Mechanistic modeling of a magnetic marker monitoring study linking gastrointestinal tablet transit, in vivo drug release, and pharmacokinetics, *Clin. Pharmacol. Ther.,* **86**, pp. 77–83.

43. Maqbool, S. (2009). Wireless capsule motility: Comparison of the SmartPill GI monitoring system with scintigraphy for measuring whole gut transit, *Dig. Dis. Sci.,* **54**, pp. 2167–2174.

44. Hahn, T. (2012). 19F MR imaging golden angle–based capsule tracking for intestinal transit and catheter tracking: Initial in vivo experience, *Radiology,* **265**, pp. 917–925.

45. Hedsund, C. (2013). Magnet tracking allows assessment of regional gastrointestinal transit times in children, *Clin. Exp. Gastroenterol.*, **6**, pp. 201–208.
46. Weitschies, W. (1997). High-resolution monitoring of the gastrointestinal transit of a magnetically marked capsule, *J. Pharm. Sci.*, **86**, pp. 1218–1222.
47. Chan, R. W. (2009). Temporal stability of adaptive 3D radial MRI using multidimensional golden means, *Magn. Reson. Med.*, **61**, pp. 354–363.
48. Lustig, M. (2007). Sparse MRI: The application of compressed sensing for rapid MR imaging, *Magn. Reson. Med.*, **58**, pp. 1182–1195.
49. Hahn, T. (2014). Real-time multipoint gastrointestinal 19-fluorine catheter tracking, *Magn. Reson. Med.*, **71**, pp. 302–307.
50. Hockett, F. D. (2011). Simultaneous dual frequency ^{1}H and ^{19}F open coil imaging of arthritic rabbit knee at 3T, *IEEE Trans. Med. Imaging*, **30**, pp. 22–27.
51. Keupp, J. (2011). Simultaneous dual-nuclei imaging for motion corrected detection and quantification of ^{19}F imaging agents, *Magn. Reson. Med.*, **66**, pp. 1116–1122.
52. Wacker, F. K. (2004). The catheter-driven MRI scanner: A new approach to intravascular catheter tracking and imaging-parameter adjustment for interventional MRI, *Am. J. Roentgenol.*, **183**, pp. 391–395.

Part 7
Perspectives

Chapter 14

Perfluorocarbon Theranostic Nanomedicines: Pharmaceutical Scientist's Perspective

Jelena M. Janjic and Sravan K. Patel

Graduate School of Pharmaceutical Sciences,
Mylan School of Pharmacy, Duquesne University, Pittsburgh, Pennsylvania, USA

janjicj@duq.edu

Abbreviations

PFPE: Perfluoropolyether; PFTE: Perfluoro-tert-butyl ether; PFC: Perfluorocarbon; PFC-NE: Perfluorocarbon-Nanoemulsion; PFCE: Perfluoro-15-crown-5-ether; PFOB: Perfluorooctylbromide; PFDB: Perfluorodecylbromide; HC-Hydrocarbon; DFNE: Drug-free nanoemulsions; CXBNE: Celecoxib nanoemulsion

14.1 Theranostic Nanomedicines as Future Medicines

Theranostic nanomedicine aims to personalize treatment by combining therapeutics with diagnostics into one nanosystem. Variability is expected during an individual patient's course of disease progression and treatment because the human body

---

Fluorine Magnetic Resonance Imaging
Edited by Ulrich Flögel and Eric Ahrens

ISBN 978-981-4745-31-4 (Hardcover), 978-981-4745-32-1 (eBook)
www.panstanford.com

presents dynamic pathophysiology. The ability to move treatment along the path of these changes would allow a patient the opportunity to recieve exactly the treatment needed, no more and no less. Therefore, this would contribute to better understanding of the disease and the effect of the drug, as well as the ability to adapt the treatment to follow these changes in every patient. Ideally, a diagnostic agent in a theranostic nanomedicine works in synergy with the therapeutic moiety (drug) by providing input on drug reaching its anatomical, cellular and/or molecular target, feedback on drug effects on the underlying pathology and therapeutic efficacy and/or side effects. Further, by seeing the changes of pathophysiology as a drug is administered in real time, the drug treatment can be fine-tuned to match these changes. The diagnostic should also provide clues on the mechanisms involving the drug target and indirectly help fine-tune both the dose and dosing regimen. Therefore, an ideal theranostic nanomedicine provides an opportunity to personalize treatment during the entire course of therapy in an individual patient. To achieve these goals, theranostic nanomedicine must grow beyond smart engineering and into pharmaceutical development. Adding an imaging agent or drug to an existing drug delivery system or imaging platform is only a partial solution and does not move the system closer to personalized treatment. The two entities have to work in synergy, which can be achieved if the whole theranostic nanosystem is viewed from a lense of drug design, and that vision must be maintained from the inception of the nanosystem through all steps of development until reaching a patient. For a pharmaceutical scientist, theranostic nanomedicines are future medicines, meeting the same standards of effectiveness and safety as all other clinically approved treatments. Figure 14.1 summarizes the key goals for theranostic nanomedicines developed today.

This chapter will discuss perfluorocarbon (PFC) based materials that offer unique opportunities for advancing theranostic nanomedicines to clinically viable therapeutic and diagnostic solutions for varied diseases. One of the key advantages of PFCs for theranostic development comes from their MRI properties, and hence, the applicability of ^{19}F MRI in the context of theranostic design will be examined. Specifically, this chapter will focus on the challenges theranostic nanomedicine development presents with PFC formulations. PFCs are chemically and biologically inert ^{19}F MR agents that allow for quantitative and qualitative assessment of inflammation in vivo without apparent background [3–5]. As inflammation is one of the key pathological processes behind many chronic diseases and the body's response to injury and infection, discussion will focus on inflammation as a key target for theranostic nanomedicine development.

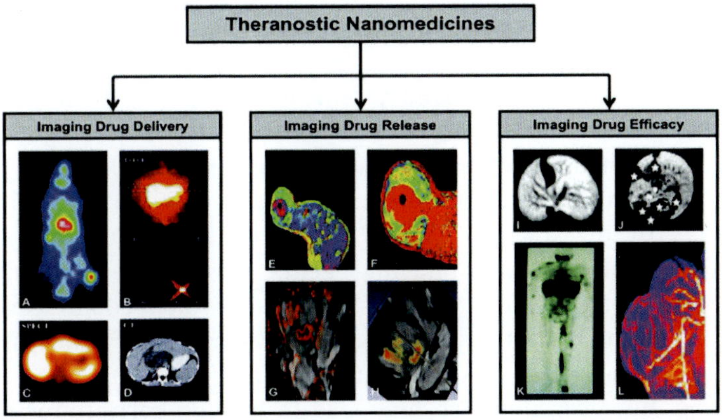

Figure 14.1 The goals of theranostic nanomedicines design. Reproduced with permission from Ref. [2].

The first issue in theranostic nanomedicines development is the inherent complexity of combining therapeutic entity with diagnostic into one system. Typically nanosystems, including theranostics, are targeted to diseased organs and tissues either passively or actively for both imaging and drug delivery. Pharmacokinetic and pharmacodynamic profiles of each component, therapeutic and diagnostic, contribute to the overall theranostic nanosystem individually and in combination. For example, if the aim is to achieve high tumor targeting with an antiproliferative agent using a nanosystem, consideration must be made for drug loading and release from the nanosystem as well as the biodistribution and targeting of the nanosystem alone. If the goal is to simply image biodistribution of the nanosystem, it may not be desirable for the imaging agent to be released along with the drug, or a completely different release profile may be needed. In simplistic terms, we introduced an imaging agent to a drug delivery nanosystem because the drug carrier biodistribution is what we are assessing. This approach to theranostics development [2] provides very useful information on the drug delivery system performance with respect to stability, targeting and retention. Though it is very useful to image biodistribution of the drug carrier in the body, does this truly define a theranostic? Biodistribution of a nanosystem alone still does not provide clues on how well the drug is working or what the changes are in the tissue or an organ in a particular pathophysiological state of the disease in response to drug delivery. Further, dosing balance between imaging agent and the drug in a theranostic nanosystem is crucial. In their recent review, Janib et al. [6] point that for some types of diagnostics to achieve their full imaging efficacy, the needed dose may lead to toxicity and confound understanding of the very pathology it aims to investigate. On the other hand, to counterbalance toxicity associated with imaging agents, we may need to adjust the overall theranostic nanosystem dose, which could lead to subtherapeutic levels of the drug delivered with the nanosystem. One way to solve this problem would be to increase drug loading. However, doing so may lead to loading barriers for both imaging and therapeutic moiety. These are some of the key challenges faced by current developers of theranostic nanosystems. Approaching the design of theranostic nanosystems with the full appreciation of these challenges, and making smart material choices to overcome these challenges, one can potentially reach the goal of truly personalized nanomedicine.

In the drug discovery field, fragment-based drug design has been used to overcome barriers to balance efficacy and safety of new drugs. Fragment-based drug design considers optimization of fragments of a drug molecule and subsequent incorporation of these fragments into a functional whole [7]. If we consider the same principles for rational design of theranostic nanomedicines, it is possible to see solutions for balancing imaging with drug delivery. Furthermore, we can consider viewing each component of the nanosystem (e.g., surfactants, shells, etc.) as fullfilling more than one role. For example, the drug carrier could also serve as an imaging agent. The imaging carrier in a nanosystem can potentially control drug release. Materials that can potentially fulfill these roles are PFCs.

14.2 Perfluorocarbons as Building Blocks for Theranostic Nanomedicines

PFCs as magnetic resonance imaging (MRI) agents are discussed in significant detail throughout this text, with focus here on their imaging properties in context of theranostic nanomedicine development. Perfluorocarbons designed for MRI can be divided into three general groups: (1) perfluorinated small molecules (MW < 1000 Da) [8], (2) perfluorinated polymers [9–13] and (3) perfluorocarbon-hydrocarbon conjugates (e.g., perfluorinated surfactants, PEG conjugates, etc.) [14–16].

One of the advantages of theranostic nanosystems built with these molecules includes background-free monitoring of in vivo biodistribution of the theranostic, which can indirectly provide insight into therapeutic effect of the drug delivered. For example, changes in inflammation in response to anti-inflammatory drug delivery can be assessed [17], as discussed later in this text. PFCs are biologically and metabolically inert, chemically stable in vastly changing environments and do not interact with biological fluids. Hence, PFCs can be used to build theranostic nanomedicines with increased safety profile as compared to other materials, such as radioactive tracers and heavy metals. As ^{19}F MRI lacks tissue depth-related detection limitations, this imaging technique allows for monitoring of theranostic nanosystem biodistribution throughout the body non-invasively, quantitatively and repeatedly over time. One concern that arises from using PFCs is their slow excretion. However, in preclinical studies, this can be used to an advantage, allowing long term follow-up of the theranostic in vivo using ^{19}F MRI. Further, PFCs can be conjugated to other molecules, which opens up ways to combine their MRI properties with other modalities such as optical imaging [13] and PET [18]. This further expands their utility as building blocks of theranostics.

Ideal theranostic nanosystems should be able to provide insight into the therapeutic effect of the drug delivered. This goal can be achieved using PFCs. Kadayyakara et al. [19] demonstrated the utility of PFCs as oxygen sensors using ^{19}F MRI in rats. The changes in ^{19}F MR signal were directly correlated to the effects of chemotherapeutic drug bischloroethylnitrosourea (BCNU) delivered to rats with growing tumors in the brain. Though drug was delivered as a solution and not with the PFC as a unique nanosystem, the oxygen sensing properties of PFCs demonstrated their theranostic potential. Biosensing properties of PFCs can also be achieved by conjugation to pH-sensing fluorescent probes [20]. Theranostic nanosystems built with these conjugates could provide insight into drug delivery to specific cellular compartments. Another interesting application discussed in this chapter is cell tracking with PFCs which is applied to monitor anti-inflammatory action of delivered drugs. Taken together, PFCs are highly advantageous materials for developing theranostic nanosystems as they can fulfill multiple roles. PFCs seem to be best utilized as theranostic building blocks when formulated into nanoemulsions.

14.3 Triphasic Perfluorocarbon Nanoemulsions as a Theranostic Platform

Nanoemulsions are kinetically stable emulsions with a droplet size typically between 50 and 500 nm, with high oil content and low amount of surfactants [21]. Nanoemulsions have wide applications in pharmaceutical industry where they are typically used to increase tissue penetration of poorly soluble drugs (transdermal delivery), as well as to improve drug solubility and bioavailability (parenteral and oral) [22–26]. They are also easily incorporated into other dosage forms, such as capsules and gels, and can be produced on an industrial scale [26–28]. A variety of processes are used to produce pharmaceutical nanoemulsions that can be divided in two main groups, that include low-energy and high-energy methods. Phase inversion-based low-energy methods were developed for nanoemulsion production on scales amenable to further industrial production [29–31]. However, consistency of size, low polydispersity, and drug loading ability give high-energy processes an advantage over low-energy emulsification. High-energy processes include high-pressure homogenization, high-shear processing (e.g., microfluidization), rotor-stator high-speed stirring [32], and acoustic cavitation processing [33]. High-energy processing methods seem to be less sensitive to composition changes and varying drug type, so that oils and surfactants can be better tolerated. Nanoemulsification methodologies are continuously being developed for both high- and low-energy processes [34–36]. In our work, microfluidization is the method of choice for theranostic nanoemulsion preparation due to ease of operation, predictable nanoemulsion properties and high colloidal stability of produced nanoemulsions with varied payload and composition [1, 10, 37–40]. What distinguishes most nanoemulsions in the pharmaceutical literature from theranostic nanoemulsions prepared with PFCs is the nature of the internal (oil) phase. Most nanoemulsions for drug delivery purposes are made with hydrocarbon (HC) oils and surfactants, whereas PFC is the major component of the internal phase in nanoemulsions developed for imaging and theranostic applications. As imaging agents, PFC nanoemulsions are easily combined with multiple types of imaging modalities. Janjic et al. [13] reported perfluoropolyether (PFPE) conjugates to several fluorescent dyes. PFC nanoemulsions can also be labeled with NIR dyes by entrapment during nanoemulsification in the lipid or surfactant layer [10, 39, 41]. Additionally, PFC nanoemulsions designed as dual-mode imaging agents for optical and photoacoustic imaging were reported by Akers et al. [42] and were also investigated as ultrasound imaging contrast agents in preclinical models [43].

Classical nanoemulsions are composed of two phases, oil and water, where oil droplets are dispersed in water and stabilized by surfactants. Triphasic PFC nanoemulsions are composed of three distinct phases, where both fluorous phase (PFC) and organic phase (HC oil) are combined and the droplets are stabilized in water (aqueous phase) by surfactants. Early development of PFC nanoemulsions followed the existing HC oil emulsification strategies. PFCs were dispersed in aqueous phase and stabilized by a variety of surfactants, from lipids to non-ionic polymers [44, 45]. These nanoemulsions were first developed as blood substitutes and quickly their potential was realized as ^{19}F MRI agents. Only recently, their use in drug delivery was investigated with highly promising results in preclinical models of cancer [46–48] and in reducing angiogenesis [49]. All of these examples are utilizing what would be called biphasic systems, where the internal phase is PFC (oil phase), which is stabilized with surfactants in water (external phase). The drug-carrying capacity was mostly realized by entrapment or anchoring to the surfactant layer of the nanoemulsion droplet. In one such example, Wickline et al. incorporated a cytolytic peptide (melittin) as anticancer agent into the lipid surfactant layer of the PFC nanoemulsion droplet [50]. They showed that tumor targeting of incorporated melittin leads to dramatic reduction of tumor size [51]. At the same time, the nanoemulsion provided tumor specific ^{19}F MR imaging in a mouse model. In another example, paclitaxel was incorporated into a block-polymer non-ionic surfactant stabilized perfluoropentane nanoemulsion designed for image-guided delivery using ultrasound and MR imaging in a mouse tumor model [43]. Lanza et al. also reported targeted delivery of anti-proliferative drugs to smooth muscle cells using perfluorooctyl bromide (PFOB) nanoemulsions [52]. PFC nanoemulsions were also reported to deliver antigen to dendritic cells (DC) with the goal of boosting the immune response in DC-based vaccines [53]. All together, these reports strongly suggest PFC nanoemulsions are highly versatile theranostic development platform. In all of these examples, the delivery of relatively low doses of the therapeutic entity to the targeted organs and tissues was required. Hence the loading of PFC nanoemulsion with drugs via the surfactant layer worked very well. However, this strategy limits loading of other drugs and limits control of drug release. To improve drug loading of poorly soluble small molecule (<1000 Da) drugs into PFC-based nanoemulsions, we redesigned them into triphasic systems. Figure 14.2 shows side-by-side comparison demonstrating key differences between two-phase (PFC/water) and triphasic (PFC/HC/water) nanoemulsions.

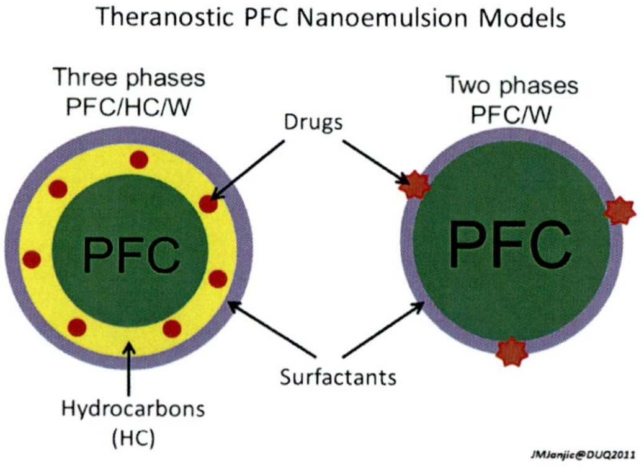

Figure 14.2 Two models of PFC nanoemulsion-based theranostic designs.

Increasing the HC oil fraction to as high as 1:1 v/v ratio of HC to PFC oil provided increased capacity for carrying poorly water soluble drugs [10, 38–40] and water insoluble NIR dyes, such as quaterrylene dye [1]. Table 14.1 summarizes a few representative examples of poorly soluble drug and/or dye loaded triphasic nanoemulsions. Furthermore, this high oil content led to increased stabilization of PFC nanoemulsions towards Ostwald ripening. Nanoemulsions of perfluoro-15-crown-5 ether (PFCE) are used for cell tracking by MR imaging [5, 9, 47, 54–57] and drug delivery [43, 46, 47, 58]. Its symmetrical structure offers 20 magnetically equivalent ^{19}F atoms per molecule showing only one resonance peak in the NMR spectrum. This attractive feature for MR imaging is hindered by nanoemulsion stability issues that arise from Ostwald ripening. It was shown that the smaller the molecular weight of HC or PFC molecule, the faster the Ostwald ripening becomes [59]. When PFCE nanoemulsion was formulated as a triphasic system by combining with HC oil Miglyol 812, in 1:1 v/v ratio, the resultant nanoemulsion remained stable for 182 days, showing no change in either droplet size or polydispersity [1]. Whereas, nanoemulsion with the same surfactant to oil ratio prepared only with PFCE showed steady increase in droplet size over the 100 days of follow-up by dynamic light scattering (DLS) measurements indicating the superior stability of triphasic system.

Triphasic PFC/HC/W nanoemulsions offer an additional advantage of dramatic improvement in effectiveness of lipid soluble NIR dyes. These dyes are easily incorporated into a high volume fraction of HC oil of triphasic nanoemulsions as opposed to surfactant layer of two-phase (PFC/W) nanoemulsions. One such example pertains to novel and exceptionally stable quaterrylene dyes. These dyes showed dramatic improvement in fluorescence signal when incorporated into lipid-rich triphasic PFC nanoemulsions over other media. Figure 14.3 summarizes key features of triphasic nanoemulsions such as stability in biologically representative media, photostability, and lack of cellular toxicity.

Triphasic nanoemulsions can be easily prepared by microfluidization with varied lipophilic payload (dye and drug), as shown in Table 14.1. Our data indicates that changes in composition can be tolerated while maintaining small droplet size

(100–200 nm), narrow size distribution, and low polydispersity index (PDI < 0.2). This is advantageous for theranostic platform development because it allows for adaptations to imaging and drug-carrying properties without significant changes to the formulation process.

Table 14.1 Triphasic PFC nanoemulsions produced by microfluidization with varied PFCs and hydrocarbon oils, and loaded with diverse lipophilic payload

PFC	HC oil	Payload	Size (nm)	PDI
PFPE	Miglyol 812N	Celecoxib*	160	0.15
PFPE	Olive oil	—	145	0.16
PFPE	Capmul® PG-8 NF	Curcumin*	103	0.19
PFPE	Olive oil	Resveratrol*	138	0.15
PFCE	Miglyol 812N	QR-4Py	135	0.15
PFCE	Miglyol 812N	DiR	140	0.18

PFCE: Perfluoro-15-Crown-5 ehter, PFPE: perfluoropolyether; PDI: polydispersity index; HC: hydrocarbon oil; DiR-cyanine NIR dye; QR-4Py: quaterrelene NIR dye [1]; *lipophilic drug.

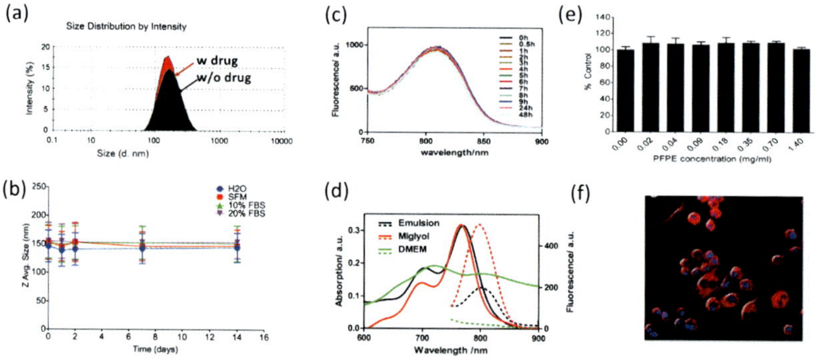

Figure 14.3 NIR-labeled triphasic PFC nanoemulsion characterization. (a) Size and polydispersity with/without drug; (b) Fluorescence spectra of quaterrylene NIR dye (QR-4Py) in emulsion, oil and media; (c) Stability in serum-containing culture media for 2 weeks; (d) no change in cell viability upon 24 h exposure to increasing drug-free nanoemulsion concentrations; (e) Photostability upon prolonged light exposure; (f) nanoemulsion droplets (red) in cells. Reproduced with permission from Refs. [1, 17].

The basic nanoemulsion preparation process is as follows. The lipophilic payload, NIR dye (e.g., DiR) and a drug (Celecoxib), is dissolved at room temperature at desired concentration (e.g., 5 mg/mL) in Miglyol 812N® (HC oil). This oil solution is further combined with PFC and surfactant mixtures in a pre-emulsification step using high-speed mixing on a Planetary Centrifugal Mixer "THINKY MIXER" ARE-310 at room temperature under ambient pressure. The pre-emulsion then processed on Microfluidizer M110S (Microfluidics) [39, 60]. When produced by microfluidization, the triphasic nanoemulsions show a narrow size distribution (Fig. 14.3a), which is not affected by the presence of the lipophilic drug. Intravenously administered nanoemulsions are subjected to complex biological environment, which we modeled in vitro by exposing nanoemulsions to serum and cell culture media. Any colloidal destabilization of the nanoemulsion droplets would lead to changes in biodistribution negatively affecting both imaging and drug delivery.

14.4 Macrophage-Targeted Perfluorocarbon Theranostic Nanoemulsions

Passive targeting to circulating inflammatory cells by PFC nanoemulsions has been extensively investigated for the purpose of detecting inflammation using ^{19}F MRI [41, 61–63]. Macrophages have emerged as the most accessible cellular targets due to their ability to take up nanoparticles injected in the blood stream. Multiple studies show preferred labeling of macrophages with PFC nanoemulsions, which is used to quantify inflammation by ^{19}F MRI in transplantation [5], abscess [64], inflammatory bowel disease [65], cancer [66], RA [67] and pulmonary inflammation [68]. Balducci et al. [67] utilized this approach to quantify inflammation and assess treatment in a rat model of RA. Excellent correlation was observed between clinical score of arthritis and the number of ^{19}F atoms detected in the arthritic joints. Furthermore, the effect of standard corticosteroid treatment showed reduction in ^{19}F content indicating reduced inflammation. ^{19}F MRI has been used in other inflammation models (e.g., IBD and graft rejection), where the macrophage burden is reduced with therapeutic intervention concomitant with reduced ^{19}F content [65, 69]. Recently, multimodal fluorescent PFC nanoemulsions have been investigated, where the introduction of fluorescent dye enabled characterization by commonly employed fluorescent methods [13]. Balducci et al. [70] reported a NIRF fluorescent PFC nanoemulsion to detect tumor-associated inflammation by ^{19}F MRI and NIRF. A strong correlation between these two imaging signals was observed in different organs. The presence of fluorescent dye facilitated assessment of nanoemulsion accumulation in immune cells using flow cytometry and histology.

Macrophages are an attractive target for theranostic nanoemulsions. In a recent review [71], Patel and Janjic summarized current state of the art developments in macrophage-targeted theranostics (see Fig. 14.4).

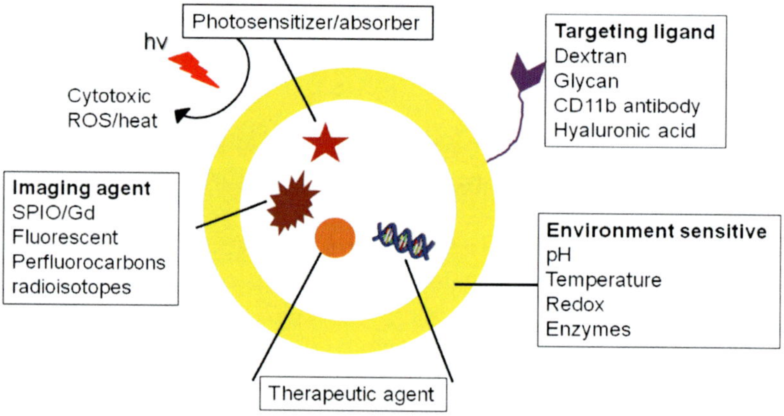

Figure 14.4 Functionalities of macrophage-targeted theranostic nanosystems. Reproduced from Ref. [71].

To demonstrate PFC nanoemulsion potential as macrophage-targeted theranostics, we prepared triphasic nanoemulsions with COX-2 inhibitor, celecoxib and a lipid soluble NIR dye. These nanoemulsions had dual mode (NIR/^{19}F MRI) properties, and when introduced to activated macrophages, reduced their production of pro-inflammatory mediator PgE2 [38–40]. The optimized nanoemulsion, with and without celecoxib, was tested at different doses in macrophages and viability was unaffected after 24 h exposure (Fig. 14.5a).

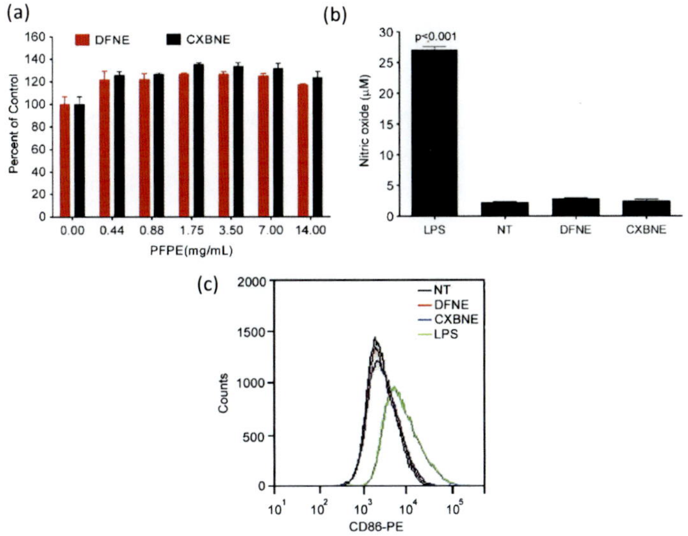

Figure 14.5 Effect of nanoemulsion on macrophage viability and activation. (a) Viability of RAW 264.7 macrophages exposed to DFNE and CXBNE for 24 h. Data is shown as mean of percent unexposed cells ± SD (n = 4). (b) Nitric oxide release from macrophages exposed to nanoemulsions for 24 h. (c) CD86 expression in macrophages exposed to nanoemulsions for 24 h. Cells not exposed to any treatment are designated as NT (no treatment). Reproduced with permission from Ref. [17].

In order to ascertain that these nanoemulsions do not produce any pro-inflammatory effects, NO release and CD86 expression changes were monitored after 24 h of exposure. A non-toxic concentration of LPS was used and cells were activated for 18 h. We found that CD86 expression and NO release were not altered by the nanoemulsions (Fig. 14.5b,c). CD86 and NO levels were comparable to untreated cells, and significantly lower than LPS-activated cells, which clearly indicate that nanoemulsions do not stimulate a pro-inflammatory phenotype in macrophages.

To understand the uptake and intracellular fate of nanoemulsions, cellular uptake studies were conducted in RAW 264.7 macrophages. Fluorescence analysis of cells exposed to nanoemulsion showed an identical emission profile (Fig. 14.6a) to that of dye in solution and nanoemulsion whereas untreated cells did not display fluorescence (Fig. 14.6). This result confirms that the fluorescent properties of the theranostics are not altered and can be detected in cells using fluorescence microscopy. Confocal microscopy showed that the nanoemulsions were readily internalized by macrophages (Fig. 14.6b) and localized to acidic endosomal/lysosomal compartments (Fig. 14.6b,c).

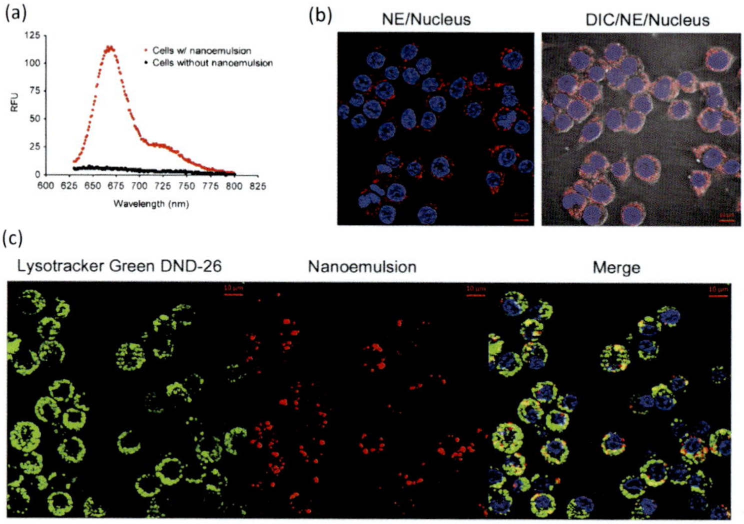

Figure 14.6 Nanoemulsion uptake in macrophages using fluorescence methods. (a) Comparison of fluorescence emission of cells treated with and without CXBNE at emission maximum 670 nm. (b) Confocal microscopy of cells incubated with nanoemulsion (30 μL/mL) for 2 h at 37°C. (c) Confocal images show co-localization of nanoemulsion (red) with acidic compartments (green) labeled with Lysotracker Green-DND. Merged image shows the co-localized pixels (yellow). Nuclei are stained with DAPI (blue). Reproduced with permission from Ref. [17].

To confirm theranostic potential in an inflammatory disease model, we employed celecoxib-loaded triphasic nanoemulsion delivered to a paw inflammation mouse model. In this study, we demonstrated feasibility of fluorescent imaging of inflammation with dye-loaded nanoemulsion in complete Freund's adjuvant (CFA) mice with follow-up ex vivo analyses from the excised site of inflammation (Fig. 14.7). Nanoemulsions were injected and mice imaged at multiple time points (Fig. 14.7a,b), showing significant differences in fluorescence signal in inflamed paws between drug-loaded and drug-free nanoemulsions. Histology confirmed that nanoemulsion (red) stayed associated with COX-2 expressing macrophages in inflamed tissues (Fig. 14.7c). The results indicate that drug delivery and macrophage imaging can be performed together. Furthermore, imaging can be utilized, not just to visualize macrophages in inflammation, but also to assess the changes in accumulation as an indicator of therapeutic effect. Essentially, features of combined therapy and diagnosis can be incorporated into these novel PFC nanoemulsions.

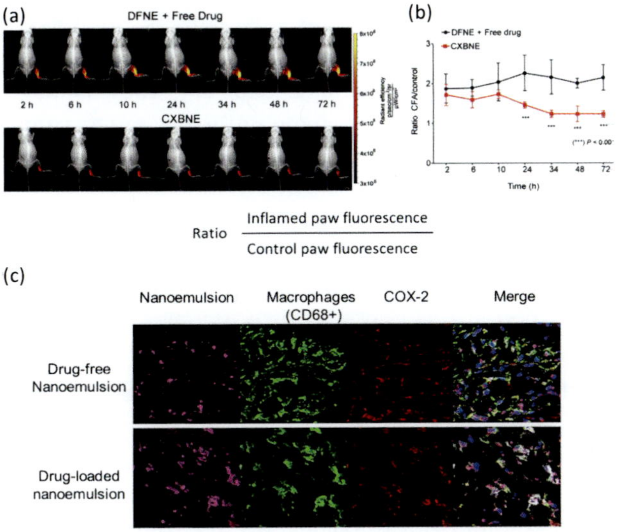

Figure 14.7 (a) Representative time-course images showing overlay of fluorescence (red) and anatomical X-ray (gray) data obtained on an IVIS Lumina XR scanner in CFA-mouse. The nanoemulsion was injected intravenously 12 h prior and right and left paw where injected with CFA and phosphate-buffered saline, respectively. (b) Fluorescence changes from CFA injected paws shows significantly less fluorescence when COX-2 inhibitor nanoemulsion was injected as compared to drug-free nanoemulsions. (c) Histological analyses indicate that nanoemulsion was selectively associated with macrophages expressing COX-2. Reproduced with permission from Ref. [17].

These results strongly suggest that macrophage-targeted PFC theranostics have potential for future personalized nanomedicines in inflammatory diseases. The nanoemulsions simultaneously deliver anti-inflammatory drugs while reporting changes in inflammation due to drug delivery. Based on these results, we hypothesize that new approaches for the treatment of inflammation can be developed utilizing multiple mechanisms to treat acute and chronic inflammation.

14.5 Pharmaceutical Perspective on Perfluorocarbon Theranostics

PFC nanoemulsion integrity directly affects in vivo imaging quality and reproducibility. For theranostic PFC nanoemulsions, stability of the droplet size is critical for success as a future pharmaceutical product. The nanoemulsion must remain in a narrow droplet size and PDI range upon storage and use. Storage stability or shelf life is typically demonstrated by repeated measurements of size and zeta potential over time [1, 13, 72]. Small changes in droplet size over time are acceptable for imaging-only applications as long as size remains within a particular range (100–200 nm). However, any size and polydispersity change over time is highly problematic for drug delivery applications and consequently for theranostics. For example, biodistribution of theranostics could be dramatically affected by changes in droplet size. Furthermore, the drug loading and release may show a differential release profile dependent on droplet size. Droplet stability is also important to ensure product homogeneity because changes in the droplet size of PFC nanoemulsions could lead to changes in droplet distribution differences, which may impact dosing. Because of the large specific weight of PFCs (1.5–1.8 g/mL), droplet enlargement resulting from Ostwald ripening during storage leads to faster sedimentation affecting the homogeneity. PFCE and PFOB nanoemulsions are known to degrade by Ostwald ripening and display a gradual droplet size increase over time driven by chemical potential differences among nanoemulsion droplets [73, 74]. Ostwald ripening is directly influenced by PFC molecular size [59]. A combination of surfactants was found to be effective in suppressing droplet growth in PFC nanoemulsions, as does adding additional water immiscible oil [10, 13, 39, 44, 59,

75, 76]. Representative data on storage stability, size distribution, and zeta potential of our earlier reported PFC nanoemulsions are shown in Fig. 14.8. Once injected in vivo, nanoemulsion is exposed to a complex biological environment in the circulatory system, which could further accelerate the nanodroplet degradation. Therefore, we developed stress tests that can evaluate PFC nanoemulsion stability under biologically relevant conditions in vitro [77]. Results from example stress tests are shown in Fig. 14.8 [39].

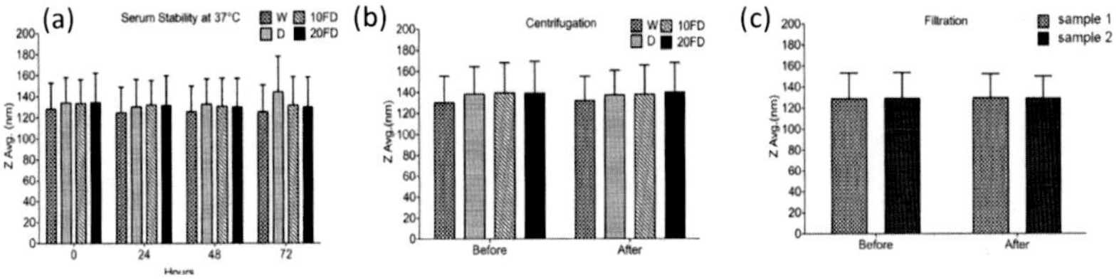

Figure 14.8 Representative stress test results for PFC nanoemulsions. (a) Bio-media stability at 37°C where W = deionized water, D = serum free media, 10FD = 10% fetal bovine serum (FBS) media; 20FD = 20% FBS media; (b) Centrifugation at 2000 rpm, 10 min at r.t.; (c) Filtration through 0.2 μm filter at r.t. In all plots standard deviation represents half width of polydispersity (PDIw/2) indicating the size distribution for each nanoemulsion. Data were obtained on a ZetaSizer Nano (Malvern, UK). Reproduced with permission from Ref. [1].

In addition to quality control, successful translation of PFC nanoemulsions into the clinic as MRI agents and theranostics, pharmaceutical process development is of critical importance. Microfluidization is fast, reliable and scalable emulsification process for kinetically stable nanoemulsions [78–81]. When compared with other high-energy emulsification methods such as sonication and high-pressure homogenization, microfluidization produces more stable nanoemulsions with smaller sizes and narrower size distribution [80, 81]. It has been shown that the droplet size of oil-in-water nanoemulsions produced by microfluidization is influenced by both formulation variables (oil volume fraction, phase viscosity, surfactant concentration) and processing conditions (pressure) [79, 82–84].

We recently reported [85] the first theranostic nanoemulsion produced on three scale levels: small (54 mL), medium (270 mL) and large (1 L) using microfluidization. No significant changes were observed in size, size distribution, and zeta potential among the nanoemulsions (Fig. 14.9). Further follow-up demonstrated high stability over 90 days at varied temperatures, exposure to filtration and centrifugation, and incubation with biological media.

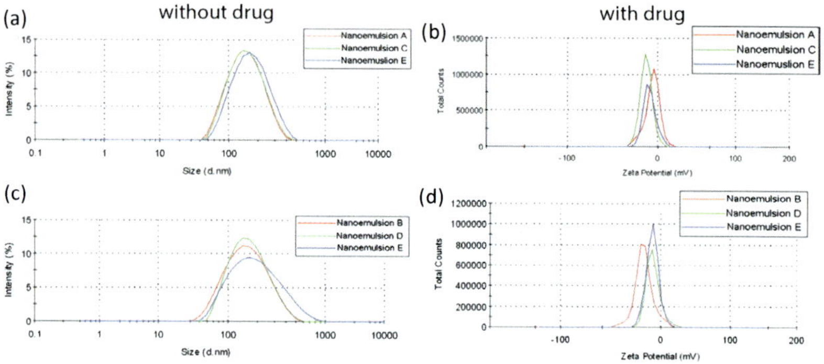

Figure 14.9 Size distribution and zeta potential comparison between small (54 mL, nanoemulsions a and b), medium (270 mL, nanoemulsions c and d) and large scale (1000 mL, nanoemulsions e and f). Reproduced from Ref. [85].

14.6 Conclusions

Successful clinical realization of theranostic systems is possible with careful selection of its building blocks and applying principles of drug design which have proven successful for designing new medicines. PFC nanoemulsions provide a great advantage for biomedical imaging due to their non-toxic nature, background-free imaging potential, and ability to be internalized by monocytes/macrophages for detecting inflamed sites. The clinical success of PFC nanoemulsions will require selection of processes amenable to large-scale production, rigorous characterization of colloidal stability under storage and exposure to biological media, toxicity profiling, and assaying potential effects on cell function. Overall, we hope to demonstrate that PFCs are highly promising materials for theranostic nanomedicines and ^{19}F MRI to take to the clinic.

References

1. Janjic, JM, et al., Perfluorocarbon nanoemulsions with fluorescent, colloidal and magnetic properties. *Biomaterials*, 2014. **35**(18): 4958–4968.
2. Lammers, T, et al., Theranostic nanomedicine. *Acc Chem Res*, 2011. **44**(10): 1029–1038.
3. Ahrens, ET, and JW Bulte, Tracking immune cells in vivo using magnetic resonance imaging. *Nat Rev Immunol*, 2013. **13**(10): 755–763.
4. Ahrens, ET, and J Zhong, In vivo MRI cell tracking using perfluorocarbon probes and fluorine-19 detection. *NMR Biomed*, 2013. **26**(7): 860–871.
5. Hitchens, TK, et al., ^{19F} MRI detection of acute allograft rejection with in vivo perfluorocarbon labeling of immune cells. *Magn Reson Med*, 2011. **65**(4): 1144–1153.
6. Janib, SM, AS Moses, and JA MacKay, Imaging and drug delivery using theranostic nanoparticles. *Adv Drug Deliv Rev*, 2010. **62**(11): 1052–1063.
7. Hajduk, PJ, and J Greer, A decade of fragment-based drug design: Strategic advances and lessons learned. *Nat Rev Drug Discov*, 2007. **6**(3): 211–219.
8. Jacoby, C, et al., Probing different perfluorocarbons for in vivo inflammation imaging by ^{19}F MRI: Image reconstruction, biological half-lives and sensitivity. *NMR Biomed*, 2014. **27**(3): 261–271.
9. Kadayakkara, DK, et al., (19)F spin-lattice relaxation of perfluoropolyethers: Dependence on temperature and magnetic field strength (7.0–14.1T). *J Magn Reson*, 2014. **242**: 18–22.
10. O'Hanlon, CE, et al., NIR-labeled perfluoropolyether nanoemulsions for drug delivery and imaging. *J Fluor Chem*, 2012. **137**: 27–33.
11. Barnett, BP, et al., Use of perfluorocarbon nanoparticles for non-invasive multimodal cell tracking of human pancreatic islets. *Contrast Media Mol Imaging*, 2011. **6**(4): 251–259.
12. Srinivas, M, et al., In vivo cytometry of antigen-specific t cells using ^{19}F MRI. *Magn Reson Med*, 2009. **62**(3): 747–753.
13. Janjic, JM, et al., Self-delivering nanoemulsions for dual fluorine-19 MRI and fluorescence detection. *J Am Chem Soc*, 2008. **130**(9): 2832–2841.
14. Decato, S, et al., Synthesis and characterization of perfluoro–butyl semifluorinated amphiphilic polymers and their potential application in hydrophobic drug delivery. *Polym Chem*, 2014. **5**(22): 6461–6471.

15. Du, W, et al., Amphiphilic hyperbranched fluoropolymers as nanoscopic ^{19}F magnetic resonance imaging agent assemblies. *Biomacromolecules*, 2008. **9**(10): 2826–2833.
16. Powell, KT, C Cheng, and KL Wooley, Complex amphiphilic hyperbranched fluoropolymers by atom transfer radical self-condensing vinyl (co)polymerization. *Macromolecules*, 2007. **40**(13): 4509–4515.
17. Patel, SK, W Beaino, CJ Anderson, JM Janjic, Theranostic nanoemulsions for macrophage COX-2 inhibition in a murine inflammation model. *Clin Immunol*, 2015 **160**(1): 59–70. doi: 10.1016/j.clim.2015.04.019. Epub 2015 May 8.
18. Willmann, JK, et al., Targeted microbubbles for imaging tumor angiogenesis: Assessment of whole-body biodistribution with dynamic micro-PET in mice. *Radiology*, 2008. **249**(1): 212–219.
19. Kadayakkara, DK, et al., In vivo observation of intracellular oximetry in perfluorocarbon-labeled glioma cells and chemotherapeutic response in the CNS using fluorine-19 MRI. *Magn Reson Med*, 2010. **64**(5): 1252–1259.
20. Patrick, MJ, et al., Intracellular pH measurements using perfluorocarbon nanoemulsions. *J Am Chem Soc*, 2013. **135**(49): 18445–18457.
21. McClements, DJ, Nanoemulsions versus microemulsions: Terminology, differences, and similarities. *Soft Matter*, 2012. **8**(6): 1719–1729.
22. Shakeel, F, et al., Nanoemulsions as potential vehicles for transdermal and dermal delivery of hydrophobic compounds: An overview. *Expert Opin Drug Deliv*, 2012. **9**(8): 953–974.
23. Kotta, S, et al., Exploring oral nanoemulsions for bioavailability enhancement of poorly water-soluble drugs. *Expert Opin Drug Deliv*, 2012. **9**(5): 585–598.
24. McClements, DJ, and J Rao, Food-grade nanoemulsions: Formulation, fabrication, properties, performance, biological fate, and potential toxicity. *Crit Rev Food Sci Nutr*, 2011. **51**(4): 285–330.
25. Rajpoot, P, K Pathak, and V Bali, Therapeutic applications of nanoemulsion based drug delivery systems: A review of patents in last two decades. *Recent Pat Drug Deliv Formul*, 2011. **5**(2): 163–172.
26. Sarker, DK, Engineering of nanoemulsions for drug delivery. *Curr Drug Deliv*, 2005. **2**(4): 297–310.
27. Mitri, K, et al., Scale-up of nanoemulsion produced by emulsification and solvent diffusion. *J Pharm Sci*, 2012. **101**(11): 4240–4247.
28. Muller, RH, D Harden, and CM Keck, Development of industrially feasible concentrated 30% and 40% nanoemulsions for intravenous drug delivery. *Drug Dev Ind Pharm*, 2012. **38**(4): 420–430.

29. Hessien, M, et al., Stability and tunability of O/W nanoemulsions prepared by phase inversion composition. *Langmuir*, 2011. **27**(6): 2299–2307.

30. Sole, I, et al., Nano-emulsions prepared by the phase inversion composition method: Preparation variables and scale up. *J Colloid Interface Sci*, 2010. **344**(2): 417–423.

31. Salager, JL, et al., Using emulsion inversion in industrial processes. *Adv Colloid Interface Sci*, 2004. **108–109**: 259–272.

32. Scholz, P, and CM Keck, Nanoemulsions produced by rotor-stator high speed stirring. *Int J Pharm*, 2015, **482**(1–2): 110–117.

33. Sivakumar, M, SY Tang, and KW Tan, Cavitation technology—a greener processing technique for the generation of pharmaceutical nanoemulsions. *Ultrason Sonochem*, 2014. **21**(6): 2069–2083.

34. Windhab, EJ, et al., Emulsion processing—from single-drop deformation to design of complex processes and products. *Chem Eng Sci*, 2005. **60**(8–9): 2101–2113.

35. Tadros, T, et al., Formation and stability of nano-emulsions. *Adv Colloid Interface Sci*, 2004. **108–109**: 303–318.

36. Salager, JL, et al., The emergence of formulation engineering in emulsion making?transferring know-how from research laboratory to plant. *Polym Int*, 2003. **52**(4): 471–478.

37. Mountain, GA, et al., Design and formulation of nanoemulsions using 2-(poly(hexafluoropropylene oxide)) perfluoropropyl benzene in combination with linear perfluoro(polyethylene glycol dimethyl ether). *J Fluor Chem*, 2014. **162**: 38–44.

38. Patel, SK, et al., Two-color fluorescent (near-infrared and visible) triphasic perfluorocarbon nanoemuslions. *J Biomed Opt*, 2013. **18**(10): 101312.

39. Patel, SK, et al., Cyclooxgenase-2 inhibiting perfluoropoly (ethylene glycol) ether theranostic nanoemulsions-in vitro study. *PLoS One*, 2013. **8**(2): e55802.

40. Janjic, JM, et al., Suppressing inflammation from inside out with novel NIR visible perfluorocarbon nanotheranostics. Proc. SPIE 8596, *Reporters, Markers, Dyes, Nanoparticles, and Molecular Probes for Biomedical Applications V*, 2013. 85960L doi: 10.1117/12.2004625.

41. Balducci, A, et al., A novel probe for the non-invasive detection of tumor-associated inflammation. *Oncoimmunology*, 2013. **2**(2): e23034.

42. Akers, WJ, et al., Noninvasive photoacoustic and fluorescence sentinel lymph node identification using dye-loaded perfluorocarbon nanoparticles. *ACS Nano*, 2011. **5**(1): 173–182.
43. Rapoport, N, et al., Ultrasound-mediated tumor imaging and nanotherapy using drug loaded, block copolymer stabilized perfluorocarbon nanoemulsions. *J Control Release*, 2011. **153**(1): 4–15.
44. Rapoport, N, Phase-shift, stimuli-responsive perfluorocarbon nanodroplets for drug delivery to cancer. *Wiley Interdiscip Rev Nanomed Nanobiotechnol*, 2012. **4**(5): 492–510.
45. Kabalnov, AS, and ED Shchukin, Ostwald ripening theory: Applications to fluorocarbon emulsion stability. *Adv Colloid Interface Sci*, 1992. **38**: 69–97.
46. Rapoport, N, et al., Focused ultrasound-mediated drug delivery to pancreatic cancer in a mouse model. *J Ther Ultrasound*, 2013. **1**: 11.
47. Bartusik, D, and B Tomanek, Application of ^{19}F magnetic resonance to study the efficacy of fluorine labeled drugs in the three-dimensional cultured breast cancer cells. *Arch Biochem Biophys*, 2010. **493**(2): 234–241.
48. Fang, JY, et al., Acoustically active perfluorocarbon nanoemulsions as drug delivery carriers for camptothecin: Drug release and cytotoxicity against cancer cells. *Ultrasonics*, 2009. **49**(1): 39–46.
49. Pan, D, et al., Anti-angiogenesis therapy in the Vx2 rabbit cancer model with a lipase-cleavable Sn 2 taxane phospholipid prodrug using alpha(v)beta(3)-targeted theranostic nanoparticles. *Theranostics*, 2014. **4**(6): 565–578.
50. Soman, NR, et al., Synthesis and characterization of stable fluorocarbon nanostructures as drug delivery vehicles for cytolytic peptides. *Nano Lett*, 2008. **8**(4): 1131–1136.
51. Soman, NR, et al., Molecularly targeted nanocarriers deliver the cytolytic peptide melittin specifically to tumor cells in mice, reducing tumor growth. *J Clin Invest*, 2009. **119**(9): 2830–2842.
52. Lanza, GM, et al., Targeted antiproliferative drug delivery to vascular smooth muscle cells with a magnetic resonance imaging nanoparticle contrast agent: Implications for rational therapy of restenosis. *Circulation*, 2002. **106**(22): 2842–2847.
53. Dewitte, H, et al., Design and evaluation of theranostic perfluorocarbon particles for simultaneous antigen-loading and (19)F-MRI tracking of dendritic cells. *J Control Release*, 2013. **169**(1–2): 141–149.

54. Waiczies, H, et al., Perfluorocarbon particle size influences magnetic resonance signal and immunological properties of dendritic cells. *PLoS One*, 2011. **6**(7): e21981.

55. Ruiz-Cabello, J, et al., In vivo "hot spot" MR imaging of neural stem cells using fluorinated nanoparticles. *Magn Reson Med*, 2008. **60**(6): 1506–1511.

56. Partlow, KC, et al., ^{19}F magnetic resonance imaging for stem/progenitor cell tracking with multiple unique perfluorocarbon nanobeacons. *FASEB J*, 2007. **21**(8): 1647–1654.

57. Zhang, H, Perfluoro-15-crown-5 ether-labeled dendritic cells. In *Molecular Imaging and Contrast Agent Database (MICAD)*. 2004: Bethesda (MD).

58. Thakkar, D, et al., Overcoming biological barriers with ultrasound. *AIP Conf Proc*, 2012. **1481**: 381–387.

59. Janjic, JM, and ET Ahrens, Fluorine-containing nanoemulsions for MRI cell tracking. *Wiley Interdiscip Rev Nanomed Nanobiotechnol*, 2009. **1**(5): 492–501.

60. Patel, SK, et al., Two-color fluorescent (NIR and visible) triphasic perfluorocarbon nanoemuslions. *J Biomed Opt*, 2013. **18**(10): 101312.

61. Stoll, G, et al., Visualization of inflammation using (19) F-magnetic resonance imaging and perfluorocarbons. *Wiley Interdiscip Rev Nanomed Nanobiotechnol*, 2012. **4**(4): 438–447.

62. Kadayakkara, DK, et al., Inflammation driven by overexpression of the hypoglycosylated abnormal mucin 1 (MUC1) links inflammatory bowel disease and pancreatitis. *Pancreas*, 2010. **39**(4): 510–515.

63. Weise, G, et al., In vivo imaging of inflammation in the peripheral nervous system by (19)F MRI. *Exp Neurol*, 2011. **229**(2): 494–501.

64. Hertlein, T, et al., Visualization of abscess formation in a murine thigh infection model of Staphylococcus aureus by ^{19}F-magnetic resonance imaging (MRI). *PLoS One*, 2011. **6**(3): e18246.

65. Kadayakkara, DK, et al., Assaying macrophage activity in a murine model of inflammatory bowel disease using fluorine-19 MRI. *Lab Invest*, 2012. **92**(4): 636–645.

66. Weibel, S, et al., Imaging of intratumoral inflammation during oncolytic virotherapy of tumors by ^{19}F-magnetic resonance imaging (MRI). *PloS One*, 2013. **8**(2): e56317.

67. Balducci, A, et al., Visualizing arthritic inflammation and therapeutic response by fluorine-19 magnetic resonance imaging (^{19}F MRI). *J. Inflammation (London, UK)*, 2012. **9**(1): 24–34.

68. Ebner, B, et al., Early assessment of pulmonary inflammation by ^{19}F MRI in vivo. *Circ: Cardiov Imaging*, 2010. **3**(2): 202–210.
69. Hitchens, TK, et al., ^{19}F MRI detection of acute allograft rejection with in vivo perfluorocarbon labeling of immune cells. *Magn Reson Med*, 2011. **65**(4): 1144–1153.
70. Balducci, A, et al., A novel probe for the non-invasive detection of tumor-associated inflammation. *Oncoimmunology*, 2013. **2**(2): e23034.
71. Patel, SK, and JM Janjic, Macrophage targeted theranostics as personalized nanomedicine strategies for inflammatory diseases. *Theranostics*, 2015. **5**(2): 150.
72. Patel, SK, J Williams, and JM Janjic, Cell labeling for ^{19}F MRI: New and improved approach to perfluorocarbon nanoemulsion design. *Biosensors (Basel)*, 2013. **3**(3): 341–359.
73. Nazarzadeh, E, T Anthonypillai, and S Sajjadi, On the growth mechanisms of nanoemulsions. *J Colloid Interface Sci*, 2013. **397**: 154–162.
74. Wooster, TJ, M Golding, and P Sanguansri, Impact of oil type on nanoemulsion formation and Ostwald ripening stability. *Langmuir*, 2008. **24**(22): 12758–12765.
75. Lee, SJ, et al., Interaction of melittin peptides with perfluorocarbon nanoemulsion particles. *J Phys Chem B*, 2011. **115**(51): 15271–15279.
76. Lim, YT, et al., Multifunctional perfluorocarbon nanoemulsions for (19)F-based magnetic resonance and near-infrared optical imaging of dendritic cells. *Chem Commun (Camb)*, 2009(45): 6952–6954.
77. Patel, SK, J Williams, and JM Janjic, Cell labeling for ^{19}F MRI: New and improved approach to perfluorocarbon nanoemulsion design. *Biosensors*, 2013. **3**: 341–359.
78. Tang, SY, P Shridharan, and M Sivakumar, Impact of process parameters in the generation of novel aspirin nanoemulsions–Comparative studies between ultrasound cavitation and microfluidizer. *Ultrason Sonochem*, 2013. **20**(1): 485–497.
79. Qian, C, and DJ McClements, Formation of nanoemulsions stabilized by model food-grade emulsifiers using high-pressure homogenization: Factors affecting particle size. *Food Hydrocolloids*, 2011. **25**(5): 1000–1008.
80. Pinnamaneni, S, NG Das, and SK Das, Comparison of oil-in-water emulsions manufactured by microfluidization and homogenization. *Pharmazie*, 2003. **58**(8): 554–558.

81. Maa, YF and CC Hsu, Performance of sonication and microfluidization for liquid-liquid emulsification. *Pharm Dev Technol*, 1999. **4**(2): 233–240.

82. Lee, L, and IT Norton, Comparing droplet breakup for a high-pressure valve homogeniser and a Microfluidizer for the potential production of food-grade nanoemulsions. *J Food Eng*, 2013. **114**(2): 158–163.

83. Tang, SY, P Shridharan, and M Sivakumar, Impact of process parameters in the generation of novel aspirin nanoemulsions–comparative studies between ultrasound cavitation and microfluidizer. *Ultrason Sonochem*, 2013. **20**(1): 485–497.

84. Mao, L, et al., Effects of homogenization models and emulsifiers on the physicochemical properties of β-carotene nanoemulsions. *J Dispersion Sci Technol*, 2010. **31**(7): 986–993.

85. Liu, L, C Bagia, and JM Janjic, The first scale up production of theranostic nanoemulsions. *Bioresearch Open Access*, 2015, 4(1): 218–228.

Index